Foundations of Biophysics

Foundations of Biophysics

A. L. STANFORD, JR.

Georgia Institute of Technology
Atlanta, Georgia

ACADEMIC PRESS

New York San Francisco London

A Subsidiary of Harcourt Brace Jovanovich, Publishers

ACADEMIC PRESS, INC.
111 Fifth Avenue, New York, New York 10003

United Kingdom Edition published by
ACADEMIC PRESS, INC. (LONDON) LTD.
24/28 Oval Road, London NW1

Library of Congress Cataloging in Publication Data

Stanford, Augustus L. (date)
Foundations of biophysics.

Bibliography: p.
Includes index.
1. Biological physics. I. Title.
QH505.S674 574.1'91 74-27792
ISBN 0-12-663350-9

PRINTED IN THE UNITED STATES OF AMERICA

Contents

Chapter IV Molecular Transport in Living Systems

UNIT II The Biophysical Approach to Special Organ Systems

Chapter V The Nervous System

Chapter VI Sound and Hearing

Chapter VII Light and Vision

Chapter VIII The Circulatory System

Chapter IX Muscle and Bone

Chapter X Viruses

UNIT III Biophysical Techniques

Preface

This is a textbook for college students at the sophomore or junior level who have completed a calculus-based sequence of courses in general physics. The object of this book is to introduce the student of physics, biology, chemistry, or engineering to the possibilities that exist for applying his skills in the physical sciences to the understanding of the principles and problems of the life sciences *at the earliest possible stage in his education*. The text evolved from lecture notes used for a number of junior-level courses that have been given for several years at Georgia Tech.

Many students of physical sciences and engineering have had no formal training in biology at the college level. This book offers these students an opportunity to learn the basic vocabulary of the life sciences. A review of the fundamentals of physics and chemistry that are especially applicable to a wide range of biological problems is also presented. Occasionally the review material is extended slightly past the level of general physics; on such occasions, the treatment is self-explanatory.

The book is divided into three units. The first one (Chapters I–IV) introduces the student to vocabulary and biological concepts necessary to initiate communication with life scientists. It directs the student's attention to some biochemical areas that require little more background in chemistry than he has acquired in freshman chemistry, capsulizes classic and molecular genetics, and presents some physical treatments appropriate to molecular transport within cells and across biological membranes.

The second unit (Chapters V–X) consists of specific studies of biological systems. In each chapter the anatomy and physiology pertinent to the system, with the necessary terminology, are accompanied by a brief review of the applicable physics. In each case, I have tried to point out the limits of the present understanding of the biological functions at the molecular level of each system discussed. Such indications will make the student aware of our awesome ignorance of living systems, assure him of the unlimited vistas awaiting his own investigations, and encourage him to make his own contributions to the understanding of biophysics.

The final unit (Chapters XI–XIII) treats a number of techniques that are fundamental to biophysical experimental studies. These techniques have been divided arbitrarily into groups dealing with wave phenomena, nuclear phenomena, and mechanical and electrical phenomena. These chapters are an attempt to acquaint the student with some of the more commonly used tools of experimental biophysics and to give him a hint of their capabilities and limitations.

A glossary at the end of the book will aid the student in his initial encounter with the language of biophysics.

The manuscript was read in its initial form by my teacher, friend, and colleague, Dr. Charles H. Braden, Regents' Professor of Physics at Georgia Tech. Both the reader and I have profited from his wise and careful consideration. His contributions have been numerous and significant, reflecting his good taste as well as his competence. As has been the case since I first met him, I am considerably in his debt.

The illustrations appearing throughout the book were patiently and expertly executed by Ms. Alyce Land, who was always cheerfully responsive to suggestions. I am pleased to thank her here. Thanks are also due to Ms. Judy Alt, who skillfully typed the manuscript. Finally, I should thank the numerous students (and often faculty members) who have patiently endured the development of the material in this text and helped in improving it.

A. L. STANFORD, JR.

Foundations of Biophysics

UNIT I Becoming Acquainted with Living Systems

CHAPTER I The Biologist's View of Life

Biophysics is the application of the techniques, approaches, and knowledge of the physical sciences to the problems of the life sciences. In a sense, it is not new. Many contributions to life science have come from physicists and chemists over a century ago. In another sense it is particularly new. The science of *biology*, the study of structure and function of animals and plants, has emerged in recent years into the realm of a molecular science. The problems now confronting biologists require the techniques of many disciplines—physics, engineering, chemistry, psychology, and mathematics. Biophysics is but one name for an effort to consolidate all our scientific and technological forces to attack the numerous new problems that have arisen in the study of living things.

The interdisciplinary nature of the current problems of the biological sciences places additional demands on the individual scientist, engineer, and student. The classic compartmentalizations of education no longer seem adequate. Every area of formal training has developed its own techniques, formalisms, and (worst of all) its own jargon. To every student falls the burden of learning entire new languages, and this is crucial. The physical scientist or engineer neither comprehends the problems of life science nor contributes to their solution until he can communicate effectively with the life scientists. It seems then that the first task of biophysical training should be that of establishing communication—learning the language and viewpoint of the life sciences.

This book seeks to familiarize the science and engineering student with the language and viewpoint of biologists—and most importantly, to disclose to him an indication of the variety of areas within the life sciences that are particularly appropriate to his interests and skills.

It may seem to the physical science student or engineering student that biophysics is inordinately biological in content. This is necessarily true at the introductory level. The same student might find some comfort in the fact that biological science students who pursue biophysics find the necessary physical and mathematical requisites a nearly impossible barrier.

As an introduction to the living world, this unit is devoted to the basic characteristics of living systems. Emphasis will fall naturally on those topics that will later be of special interest to students of the physical sciences and engineering.

Characteristics of Life

It is disconcerting that the first word encountered in biology should be so difficult to define. "Life" is an abstract word and has to be defined in abstract terms—primarily through concrete examples. It is usually accepted that one can distinguish living from nonliving matter. Biologists have established a set of characteristics whose presence or absence determines the distinction. These characteristics—*reproduction*, *growth*, *metabolism*, and *responsiveness*—are technical terms with specific meanings to the biologist.

Reproduction

That a living system can reproduce itself is an absolute requirement for its continued existence, since the thesis of *biogenesis* (that all things living today come only from living things) is universally accepted today. This is consistent with present knowledge, in that living organisms have not yet been produced in the laboratory. Sexual reproduction is accomplished when an *ovum*, or egg cell, is fertilized by a *spermatozoon*, the male reproductive cell. The resulting fertilized cell is called a *zygote*. *Asexual* reproduction occurs in lower life forms when the union of *gametes* (either ova or spermatozoa) is not involved. This form of reproduction is accomplished by processes like *fission* (the splitting of a mature cell into two or more parts) or by *budding* (in which the new individual arises from an outgrowth or bud from the parent). Figure 1-1 illustrates budding in one of the lower forms of life.

Growth

However produced, a daughter organism grows and develops into a mature individual with forms and functions similar to those of its parents. Biological growth differs from that of a crystal or a sand castle. Crystal growth or sand castle growth occurs by *accretion*, in which ready-made material outside the structure is assimilated into it. Biological growth, on the other hand, occurs from within the system. Growth of living organisms is based on *synthesis*, the formation of complex materials from simpler materials. Each living system, plant or animal, synthesizes those materials that are unique for its specific *species* (a group of individuals that resemble each other structurally and physiologically and that, in nature, interbreed, producing fertile offspring).

Metabolism

The energy changes and chemical reactions that occur within a living organism are referred to collectively as the *metabolism* of that organism. These changes provide for the growth, maintenance, and repair of the organism. The constant expenditure of energy is one of the essential attributes of living things. Metabolic processes are either *anabolic*, referring to those processes requiring an expenditure of energy with which

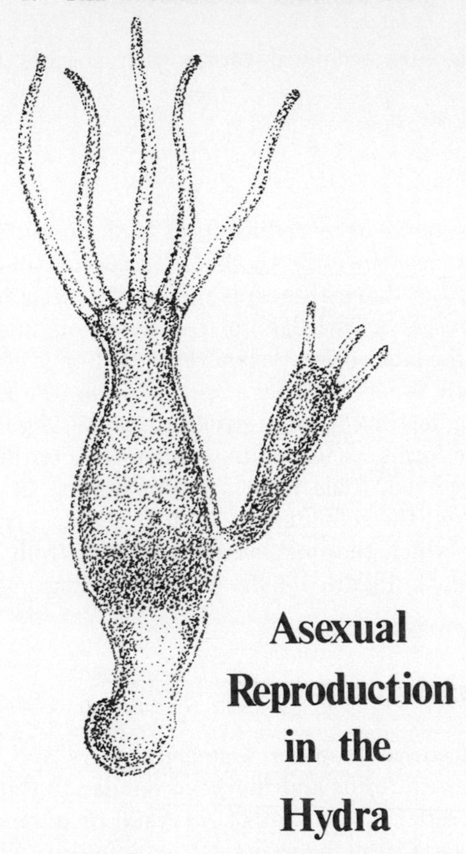

FIGURE 1-1. An example of asexual reproduction by budding. The hydra shown here is a rather simple animal which reproduces by growing a small version of itself along its side. The young hydra detaches from its parent and grows as an independent individual.

complex materials are synthesized from simple ones, or *catabolic*, in which complex chemical systems are broken down and energy is released. Plants are in general better chemical synthesizers than animals, since they can manufacture their own organic compounds from inorganic material in the soil and air. Animals must depend on plants for much of their food.

Movement

An important characteristic of living things is their ability to move themselves or cause movement of their surroundings. In most animals, movement is obvious: they run, fly, crawl, or swim using muscles that can contract and relax. Many animals spend their entire life in one place;

sponges, oysters, and corals are fixed. Yet they possess organs that move the water and its digestible foods through their bodies. Very small animals often possess microscopic hairlike projections called *cilia* and *flagella,* whose motion propels them. At an even lower level, the motion of *protoplasm,* a generic term for the living substance inside cells, causes motion of the organism by a slow oozing process called *amoeboid motion.*

Plants too are often in motion. Some leaves and flowers fold and unflold periodically. At the cellular level, the streaming motion of the protoplasm in the leaves of plants is called *cyclosis.*

Responsiveness

The characteristic responsiveness of plants and animals to their immediate and long-range environment is incorporated into traits called *irritability* (the ability to react to stimuli) and *adaptation.* The response of living things differs from that of a rock, which, when released, falls in response to gravity, or from an electron that accelerates in response to an electric field. The irritability of living organisms is not passive. The response to a stimulus by the simplest animals is the result of trial and error adjustments. Higher forms of animal life respond to stimuli rapidly, using complex organs to adjust, as when a boy catches an object unexpectedly thrown toward him. In contrast, plants respond negatively to gravity by growing upward and positively to light by growing toward light sources.

Adaptation refers to the ability of plants and animals to adjust to the demands of the changing world. Each species may become adapted by seeking out a favorable environment or by changing itself to better fit itself to its present surroundings. Adaption may be an immediate response of an organism that depends on its irritability or a long-term process involving *mutation* (change) and selection. The geographic distribution of species is dependent on the limitations of adaption within those species. The factors affecting such distribution are almost endless; they include temperature, water, predators, competitors, food availability, etc.

A Few Biological Generalizations

The biosciences share with the physical sciences a history of piecing together large bodies of facts into hypotheses, theories, and laws. The hope is that in both cases the resulting accepted body of knowledge has been inferred from careful evaluation of careful observations.

Most life scientists today agree that the phenomena associated with life can be explained in terms of the same principles that govern the physical sciences. In particular, no vital force distinguishes between living and nonliving systems. Living systems are composed of the same atoms as the nonliving ones, and they are subject to the same laws. It should be borne in mind, however, that our understanding of the laws of the physical sciences evolve with time, and the improvement of our understanding of those laws by discoveries in the biosciences is not by any means prohibited.

One of the broadest generalizations in biology is the *cell theory*, which evolved over many years as a product of many researchers. In 1665, Robert Hooke first observed evidence of cellular structure in cork oak bark. (Curiously, the discovery of the basic unit of life was first observed in the nonliving bark of a tree where only nonliving products of the living cells remained.) Dutrochet, in 1824, recognized that all biological growth is the result of increases in the volumes of cells and of the addition of new cells. Yet the cell theory is usually attributed to the botanist Schleiden and the zoologist Schwann, who in about 1839, stated that all living plants and animals are composed of cells and cell products. The cell theory at present includes the assumptions that new cells are formed by division of preexisting cells, that there exist basic similarities in the constituents and metabolic activities of all cells, and that the activity of the organism as a whole is just the sum of the activities and interactions of its individual cells.

A more controversial generalization in modern biology is the *theory of organic evolution*, whose stormy history was begun centuries before Charles Darwin published " The Origin of Species" in 1859. The theory of organic evolution is the notion that plants and animals existing today were not the product of special creation, but have descended from previously existing, simpler organisms by gradual modifications that have accumulated in successive generations. Darwin's book presented massive evidence that organic evolution had occurred. He further suggested the *theory of natural selection* to explain the means by which evolution might take place. Darwin's theory is based on the following arguments: (a) Minor variations occur in a group of plants or animals whose members are too numerous for each individual to obtain the necessities of life and survive. (b) Those individuals possessing characteristics that give them an advantage in the struggle for life are more likely to survive. (c) The survivors will transmit the advantageous characteristics to their offspring, ensuring the transmission of that characteristic to future generations. Inherent in Darwin's theory is the struggle for existence, the cliché "survival of the fittest," and the inheritance of advantageous characteristics by progeny of the surviving individuals. The central role of the theory of organic evolution in present-day biological sciences is indeed impressive. Modern courses in zoology and botany use the theory as the basis of organization, classification, and study. It is usually discussed as though it were an immutable fact, but it might be noted that there are still scientists and layman who seriously contest the theory.

A related generalization arose from observations by *embryologists*, who study the development of organisms from the fertilized egg. This concept is called either the *biogenetic law* or the *law of recapitulation*, which expresses the observation that in the course of embryonic development, organisms tend to repeat some of the corresponding stages of their evolutionary ancestors. Thus, a human embryo, during the various stages of its development, resembles a fish, an amphibian, a reptile, and so on. The implication is that the history of evolution of the species is recorded in the individual's embryonic development. This is occasionally enunciated by biological buffs in their own jargon as "ontogeny recapitulates phylogeny."

Other generalizations have become part of the biosciences and have exerted various influences on teaching and research in these areas. Some, like the recent "one gene–one enzyme–one reaction" theory of Beadle and Tatum, are widely accepted and will be discussed subsequently. All are tested repeatedly and are subject to constant revision, as are all concepts in the sciences.

Cellular Organization

A series of discoveries and observations spanning two centuries established the *cell* as the fundamental unit of living material. There are, however, several levels of organization above and below the cell level. Groups of cells—perhaps of many different varieties—make up a *tissue* which serves a specialized function. Muscle, skin, or nerve tissues are examples in animals, while plants have bark or water conduction tissues. An even higher level of organization is realized in an *organ*, a group of tissues that performs a relatively complex function, such as the brain or heart in an animal or the flower or root in a plant. An *organism* usually refers to an entire individual of a particular species.

At a lower level of organization than the cell are the *organelles*, which are identifiable structures inside a cell that perform a particular function. The organelles and their component parts comprise the *ultrastructure* of cells.

Cells are usually described in terms of their *morphology*, or structure and form, and their *physiology*, or function.

Observation of Cellular Structures

A model of a typical animal cell is depicted in Figure 1-2. Most of the components of the cell that we shall be discussing are represented in this illustration.

Cells and tissues are usually transparent when observed by a light microscope, and most organelles are also transparent. To visualize the cellular components, techniques of *fixing* and *differential staining* have been developed. A cell or tissue is fixed by killing it in a rapid and careful manner so that the position of the components are minimally disturbed. Thin slices of the tissue are prepared by a knife called a *microtome.* The slices are prepared by applying stains or dyes that color only certain chemicals in the cell; thus, particular components are discernible. *Vital stains* are taken up by living tissue and are useful in studying cell behavior in live systems.

Phase contrast microscopy and electron microscopy are modern adjuncts to light microscopy that permit visualization of normally transparent cellular components and the resolution of much smaller structural detail. Their role in the observation of cellular ultrastructure has revolutionized

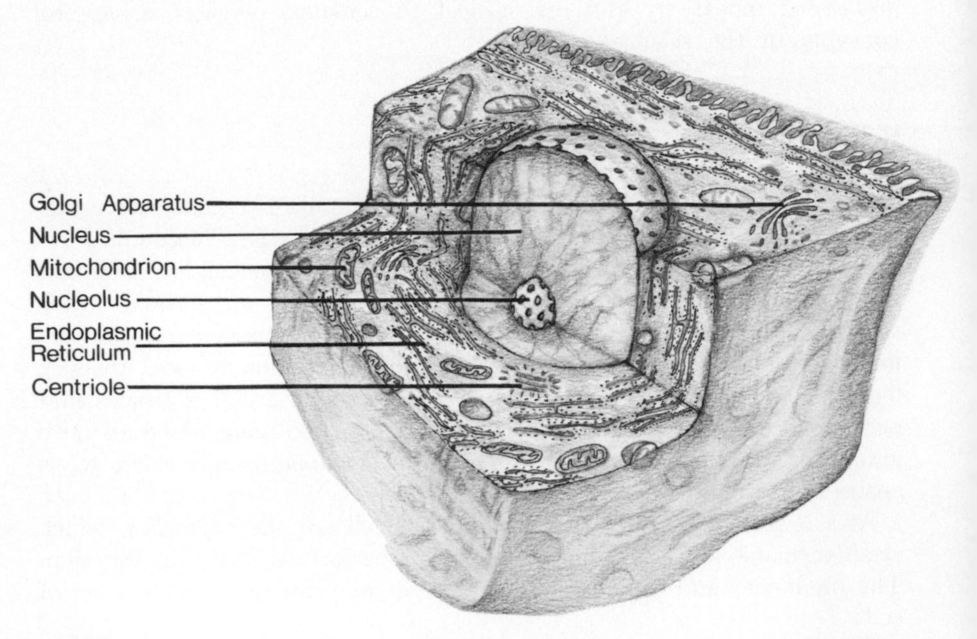

FIGURE 1-2. A model of a typical animal cell showing the prominent organelles of the cell.

cytology, the study of cells, in modern years. These instruments will be considered further in Chapter XI.

Of particular aid in observing cellular growth and function has been the technique of *tissue culture*, in which living tissues are provided all the necessary nutrients to grow and reproduce outside the organism. These techniques have been particularly useful in studies of cell movement, cell division, nutritional requirements of cells, and the effects of drugs and radiation on cells.

Cellular Structure and Components

Almost every cell has a *plasma membrane*, a semipermeable covering (one that permits passage of some materials but not others) that encloses the cellular contents and functions in the regulation of incoming food and water and outgoing wastes and secretions. Typically, the plasma membrane is about 10 nm thick (nm is the abbreviation for nanometer, 10^{-9} meter) and is formed like a sandwich of lipids between layers of protein molecules. (Lipids are molecules that are insoluble in water and are often used in membranes as the waterproofing material. Both lipids and proteins are discussed in Chapter II). In electron photomicrographs, the membranes appear as two dark parallel lines in cross section when the cell has been fixed in osmium tetroxide.

THE NUCLEUS

The most obvious organelle visible in cells is the *nucleus*, a roughly spherical or oval body that is the chemical control center of the cell. It is surrounded by a porous *nuclear membrane*, which controls the chemical messengers that proceed in and out of the nucleus. The fluid inside the nucleus is called *karyoplasm* to distinguish it from the *cytoplasm*, the medium outside the nucleus including the plasma membrane. Inside the nucleus are one or more smaller bodies, each of which is called a *nucleolus*. The number present in a nucleus is constant in the cells of a particular species. The nucleoli disappear when a cell begins to divide and later reappear. The function of the nucleolus has been somewhat mysterious, but recent evidence seems to indicate that the nucleolus is the site at which *ribosomes* are formed. Ribosomes are small bodies, rich in nucleic acids, that leave the nucleus and become attached to structures in the cytoplasm. There they play an important role in the synthesis of proteins.

When a cell is not dividing, the nucleus contains a network of filamentous material called *chromatin*, whose most important component is deoxyribonucleic acid (DNA). When the cell begins to divide, the chromatin

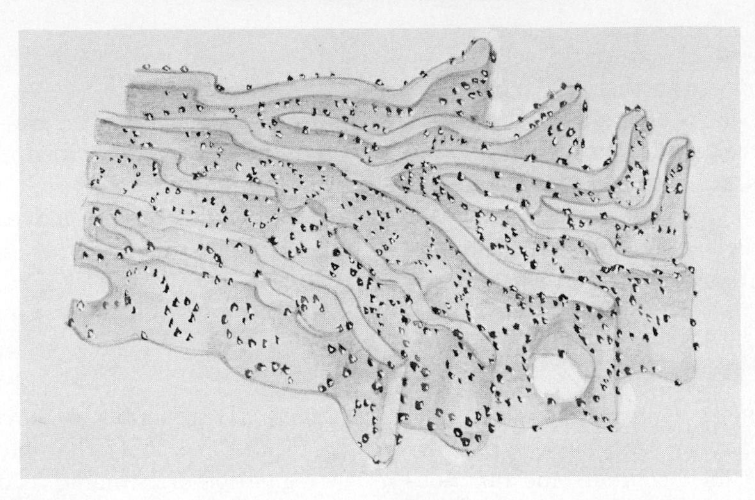

FIGURE 1-3. Ribosomes attached to some fragments of the endoplasmic reticulum in the cytoplasm of a cell. The ribosomes are the sites of protein synthesis within the cell.

forms into thick, short bodies, which are then called *chromosomes*. Each chromosome has a constriction somewhere along its length called a *centromere*, which is functionally related to the movement and alignment of chromosomes as they are distributed during cell division. The chromosomes carry and transmit the hereditary information of a species, i.e., they contain the *genes*, the submicroscopic hereditary factors. The role of these hereditary materials will be discussed further in Chapter III.

THE CYTOPLASM

The unstructured portion of the cytoplasm, the *hyaloplasm*, is interspersed with a variety of organelles. Throughout the cytoplasm is a tortuous system of tubules that have a ribbonlike appearance. The tubules are membranous material that interconnect among themselves and the plasma membrane. They are collectively called the *endoplasmic reticulum*, which literally means "network within the fluid." A portion of endoplasmic reticulum is illustrated in Figure 1-3. The presence of these tubules greatly increases the membranous surface area—and membrane surfaces are where most chemical reactions occur in a cell. Thus the endoplasmic reticulum enhances the cell's efficiency in carrying on its metabolism. The tubules also transport certain cellular fluids to the plasma membrane where they are secreted.

On the surfaces of some of the endoplasmic reticulum are found small granules called *ribosomes*. They are composed of ribosomal ribonucleic acid (rRNA) and are the sites of protein synthesis (see Chapter III).

OTHER ORGANELLES

Some of the prominent organelles in the cytoplasm are *mitochondria* (singular, mitochondrion). They are usually ovoid in shape and are composed of a double-walled membrane. The structure of mitochondria is shown in Figure 1-4. The inner membrane folds inward, forming interlocking walls called *cristae* on the interior of the mitochondrion. This is presumably to increase the surface area of the interior. Again, the purpose of a large surface area is probably to enhance the efficiency of chemical reactions. Very important reactions take place in the mitochondria, for they are the site of the reactions that release energy for the cell's various needs. Because of their primary function, the mitochondria are thought of as the "power plants" of cells. The process of releasing energy within a cell and its utilization by the cell is called *respiration.*

Among the cytoplasmic organelles that are surrounded by some mystery are the *Golgi complexes* and the *centrioles.* The Golgi complex usually appears in photomicrographs as thin flat sacs piled upon one another. They are prominent in cells whose primary function is secretory, and it has been assumed that they are primarily associated with secretions. They conduct their secretions to the endoplasmic reticulum, which conducts them to the plasma membrane. Only recently, Golgi complexes have been observed in plant cells as well as animal cells. They are also found in nonsecreting cells, and it is presumed that they have other functions that

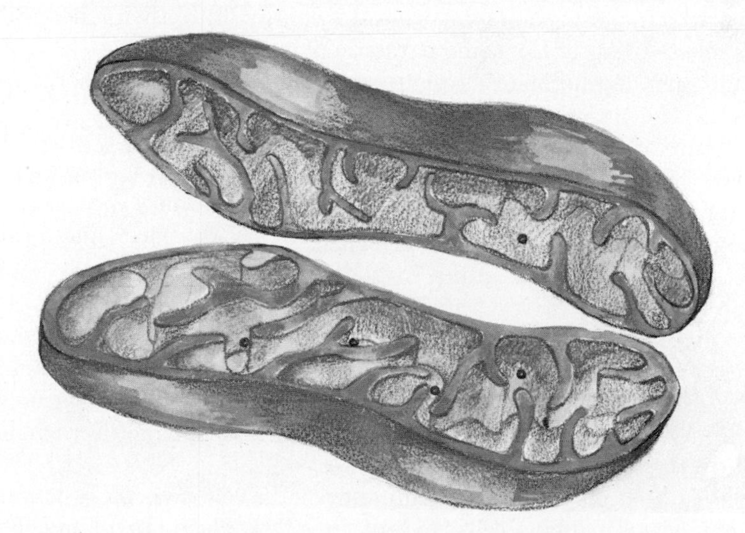

FIGURE 1-4. A mitochondrion. The interlacing walls (cristae) inside this structure provide a large surface area. Mitochondria are the reservoirs of energy molecules used by the cell in its metabolic processes.

have not yet been identified. Situated near the nucleus are two small cylindrical bodies, the centrioles. These play a prominent role at the time of cell division (*mitosis*), separating, migrating to opposite poles of the cell, and serving as centers of organization for the division process. Each *centriole* divides during mitosis, so that each daughter cell receives a pair, but the means of division is not understood.

PLANT CELLS

In higher animals, the support of cells, tissues, and organs is accomplished by a skeleton or an exoskeleton (as in the case of many insects and crustaceans). In lower animals, the cells are often covered by external skeletons of hard materials (like SiO_2) on the surface of cells or by a tough but flexible *pellicle* that maintains the shape of the animal but permits freedom of movement. Figure 1-5 illustrates a typical plant cell. In plant cells, support is accomplished by a *cell wall* of cellulose, which is secreted by the cell inside it. The intact walls are impermeable to water and the chemicals in the surrounding medium, but small connecting tubes called *desmosomes* between cell walls allow chemical communication between cells while maintaining structural strength.

Plant cells contain free organelles called *plastids*, which are generally ovoid in shape. Most plastids contain pigment and are then referred to as *chromoplasts*. The most common chromoplasts contain chlorophyll and are called *chloroplasts*, which play the essential role in *photosynthesis*, the process by which most plants incorporate energy into their chemicals from the sun (see Chapter II). Some parasitic plants, such as fungi, have no plastids and must depend on other plants to assimilate the sun's energy for them.

The structure of chloroplasts is known in considerable detail. An orderly array within the plastid is composed of dense bodies called *grana* that have the appearance of stacks of coins. The grana are supported by and between membranous sheets, each of which is called a *stroma*. The chlorophyll molecules occur within these grana in sheets along with layers of lipids and fats.

Another distinguishing characteristic of plant cells is the occurrence of *vacuoles*, cavities within the cell body filled with a watery fluid and surrounded by a vacuolar membrane. A few lower animals have food vacuoles that contain food in the process of being digested and contractile vacuoles that regulate the water content of cells.

Centrioles, which play such a prominent role in cell division of animal cells, do not occur in plant cells, except in a few algae. In plants, cell division occurs not in regions where the rigid cell walls are present but in special *meristematic* regions where rapid growth takes place.

Nuclear Control of Cellular Processes

Biologists often define a cell as a mass of cytoplasm under the control of a nucleus. A number of classical experiments have demonstrated the crucial role of the nucleus in the functioning of cells. In every case, the nucleus acts as the coordinator and director of cellular activities; in return the nucleus is nourished by chemicals from the cytoplasm.

In experiments on *protozoa* (one-celled animals, e.g., the amoeba and paramecium), the cell has been cut into halves—one half with the nucleus, the other without. The plasma membrane rejoins and seals each half. The

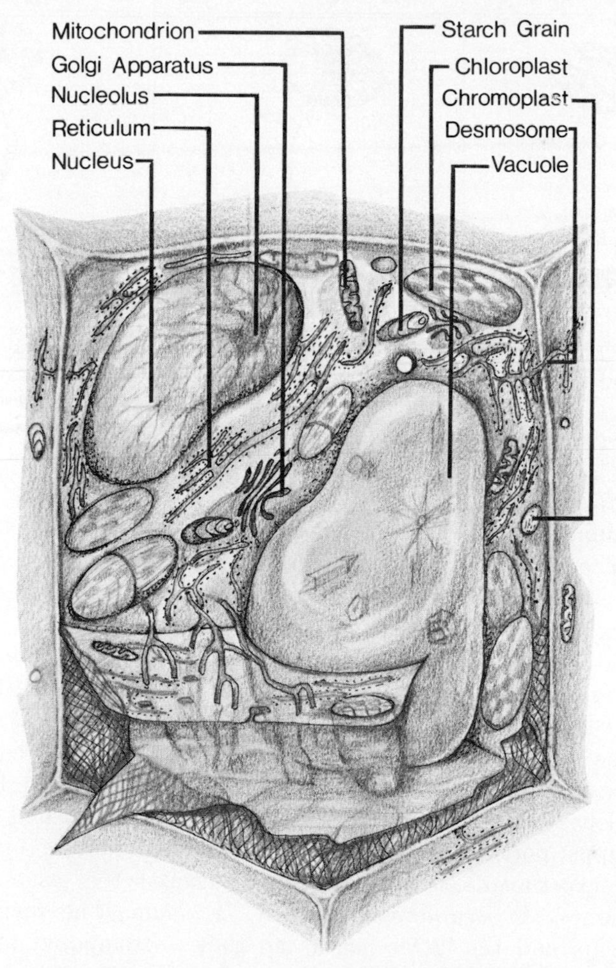

FIGURE 1-5. A model of a typical plant cell. The presence of the cellulose cell wall is the most distinguishing feature of a plant cell.

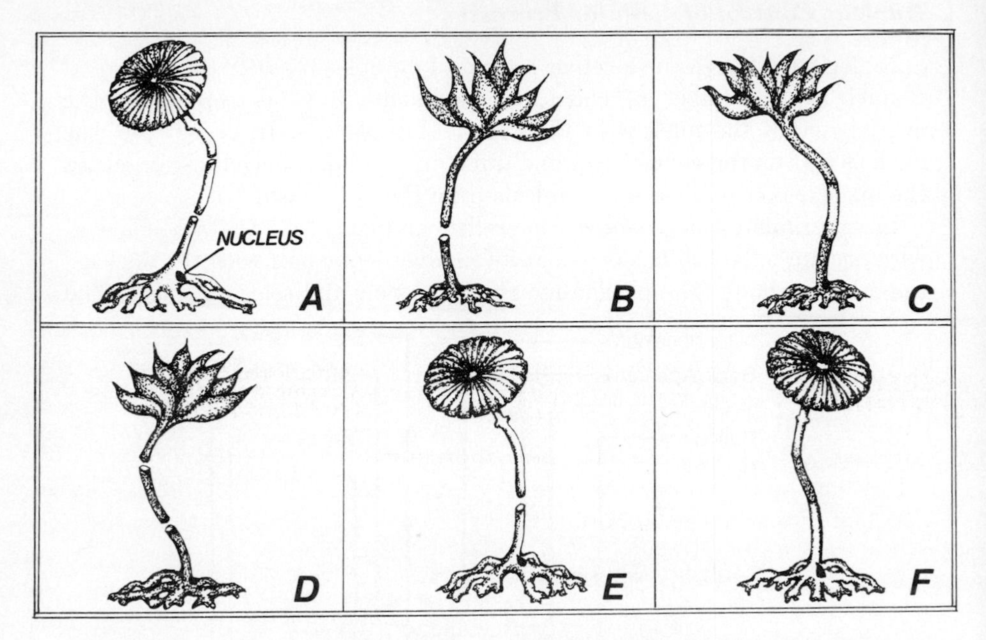

FIGURE 1-6. The grafting experiments with the single-celled alga *Acetabularia*. (A), (B), and (C) illustrate the grafting of a segment of the *med* stalk onto the base (containing the nucleus) of a *cren* plant. In (C) the *cren* plant has regrown a characteristic *cren* umbrella. (D), (E), and (F) illustrate the counter-experiment in which a section of the *cren* plant is grafted to a *med* base, which produces a *med* umbrella.

half containing the nucleus continues its normal functions, feeding and growing. The enucleate (having no nucleus) half survives for a while but soon dies. Many similar experiments imply that chemical messages from the nucleus direct the functions of the cytoplasmic systems.

A group of experiments by Hämmerling, using the single-celled alga *Acetabularia*, demonstrates the control that the nucleus exerts over the growth of cells. This plant, though only a single cell, is several centimeters tall and has an "umbrella" atop a stalk. The nucleus of this cell is located near the base of the stalk. When the stalk was severed above the location of the nucleus, the cell would regenerate another umbrella and continue to live. The upper portion was not able to regenerate a base, and it soon died.

Further experiments with *Acetabularia* utilized the existence of two similar species, *A. mediterranea* and *A. crenulata*. The former has an umbrella cap, and the latter has a cap that is crenulated (scalloped in appearance similar to the arrangement of petals on a daisy). Sections from these species can be grafted onto each other without significant cell damage. Figure 1-6 schematically illustrates the experimental procedures. In the upper figures, the stalk of a *med* cell that had been dissected away from its base and umbrella (A) was grafted onto the base (containing the nucleus) of a *cren* cell (B). The cell regenerated a new umbrella (C) of the *cren* form, indicating nuclear dominance over the nearer cytoplasm of the *med* cell. A counterexperiment, shown in (D), (E), and (F), using a *med* base and a *cren* stalk, resulted in the generation of a *med* umbrella.

Still another set of experiments with *Acetabularia* demonstrated the influence of the cytoplasm on the nucleus. In a mature plant, the nucleus undergoes division in preparation for cell division. If the umbrella of a mature plant is removed, however, the nuclear division is suppressed until a new umbrella is regenerated. Similarly, if an immature plant with a nondividing nucleus has a mature umbrella grafted onto it, the nucleus will begin to divide. These results indicate that the nucleus receives information from the cytoplasm that affects the function of the nucleus.

It seems that a balance of control exists between the nucleus and the cytoplasm. It will be seen later how the nucleus effects some of its direction over cytoplasmic processes.

Cellular Division

The process of cell division is called *mitosis*. It is the means by which tissues grow and repair themselves. Mitosis is a remarkably orderly and precise sequence of events in which a parent cell divides into two cells and transmits to each daughter cell all of the characteristics necessary to continue the life process.

What Causes Cells to Divide?

Considerable study has been devoted to those factors that influence cells to divide. It is difficult to explain that one cell in a tissue divides and another does not. In humans, the cells producing red corpuscles divide at a rapid rate, while nerve cells are never replaced if destroyed. Cancer is the uncontrolled growth and division of cells, and the continuing menace of cancer is testimony that the factors influencing cell division are not well understood. Still, a number of the factors affecting the initiation and rate of division are recognized.

The surface to volume ratio declines as cells enlarge, and this may serve as a stimulus for cells to divide. The surface area may become insufficient to handle the supply of food and oxygen necessary to the cell's needs. Similarly the surface area of the nucleus may become inadequate to supply the needs of the cytoplasm. Cell division may be the means by which a more favorable ratio is restored.

Other factors—some chemical, some physical—affect the initiation and rate of mitosis. Temperature is an observable factor. In general, the mitotic rate is directly proportional to the temperature, until an optimum temperature is reached; above this optimum, higher temperatures decrease the rate of division and become destructive. Mechanical stimuli can increase growth rate. Calluses are evidence of accelerated cell division where repeated irritation of the cells of the skin has occurred. Certain growth hormones are recognized that produce pronounced effects in plants and animals. *Auxin* is a hormone that promotes growth in plants. The pituitary gland in animals secretes a growth hormone. Other effects are probably influential in prompting cells to divide—gravity, pressure, etc.—but very little detail is understood about how such mechanisms cause mitosis.

The Stages of Mitosis

The primary purpose of mitosis is the production from a parent cell of two daughter cells, each of which has a full complement of chromosomes. This process is so fundamental to the continuation of life, it is not surprising that it has been intently studied. Mitosis is conveniently described in terms of stages, which give the impression that it is a segmented and discontinuous process. Actually it is a highly coordinated, smoothly flowing process. There are five recognized stages or phases in the mitotic process: *interphase*, *prophase*, *metaphase*, *anaphase*, and *telophase*. In this description, processes and characteristic figures refer to a generalized, or "typical" animal cell. Figure 1-7 illustrates the distinctive features in the stages of mitosis.

INTERPHASE

The period during which the cell is not dividing is referred to as interphase. Some important functions necessary to the division process occur at this stage. In addition to conducting its normal functions, the cell duplicates the genetic materials in its chromatin during interphase. The genetic material is deoxyribonucleic acid (DNA); it has the capability of self-duplication, using materials from the karyoplasm. In Chapter III, the details of the self-duplication of genetic materials are discussed in further detail.

PROPHASE

The first stage of cell division is marked by a number of simultaneous processes. The chromosomes appear as shortened, thickened bodies with

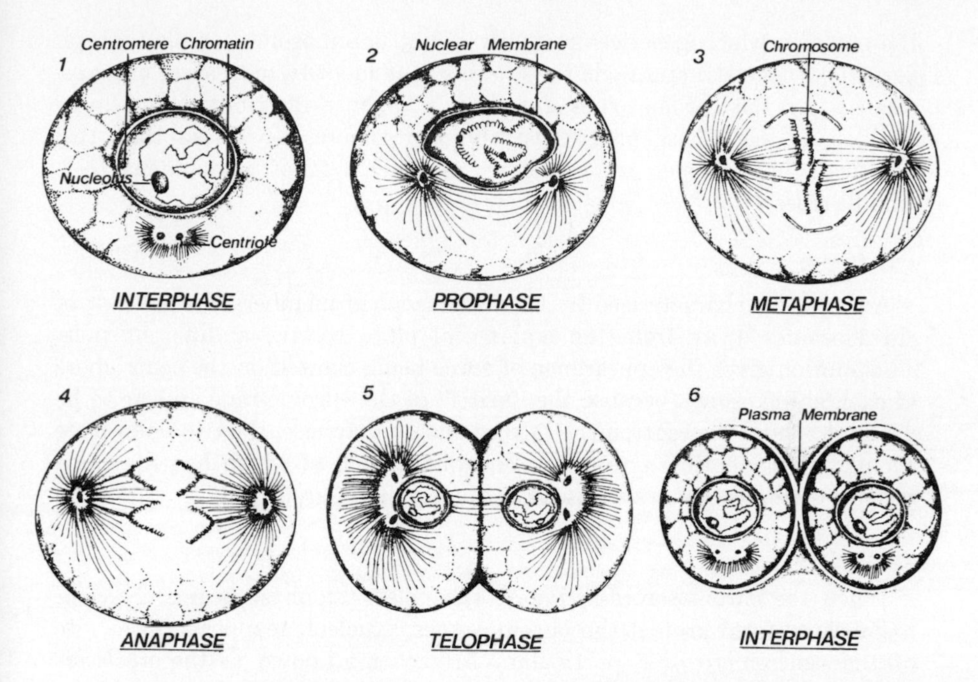

FIGURE 1-7. The stages of mitosis. A description of the characterizing features of each stage is given in the text.

two pairs of *chromatids*. During prophase, the nuclear membrane breaks down and the nucleolus disappears. As the nucleus begins to lose its form, the two small centrioles migrate to opposite sides of the nucleus. A clear area, called the *centrosome* is visible around the centriole, and radial projections, called *aster rays*, project outward all around the centrosome. Then the *spindle* forms between the centrosomes. The spindle is composed of microtubules called *spindle fibers* that arrange themselves in a pattern that reminds one of the lines of electric field intensity between two point charges—one positive, the other negative. Near the end of prophase, the chromosomes are scattered in what appears to be a random pattern along the spindle fibers. Further, there seems to be no relation of the chromosomal positions to their location or orientation in the nucleus before the spindle fibers had formed.

METAPHASE

When the chromosomes become arrayed in a planar configuration near the center of the spindle fibers, the cell is said to be in metaphase. Because of the position of the array between the centrioles (sometimes casually referred to as poles), it is called the *equatorial plate*. Early metaphase is the stage in which geneticists usually count chromosomes or photograph *karyotypes* (the structural characteristics that identify individual chromosomes), for the chromosomes are then uniformly distributed and most visible. As metaphase progresses, the centromeres divide so that two identical chromosomes are connected at each site of attachment to a spindle fiber.

ANAPHASE

Anaphase is characterized by the movement of members of each pair of chromosomes away from the equatorial plate toward a different pole. This motion gives the appearance of force being exerted on the centromere of each chromosome, because the "arms" of the chromosome appear to be dragged behind the centromere. At the end of anaphase, the two poles have accumulated identical sets of chromosomes, each of which has the same genetic material that was present in the original parent cell.

TELOPHASE

When the chromosomes arrive at the poles, telophase begins. Nuclear membranes form around the chromosomes. Nucleoli reappear at the site of the centromere of a particular chromosome known as the *nucleolar organizer*. At the periphery of the cell, along an imaginary extension of the equatorial plate, a constriction, or indention, begins to form on the plasma membrane. As the constriction proceeds inward, the cell gives the appearance of being pinched in two. During the constriction, the spindle disappears. Then the constriction reaches the center of the cell, the plasma membranes join on each side of the constriction, completing the division process.

Each daughter cell is now complete, having not only genetic material necessary to grow and continue the life process, but also a portion of the cytoplasmic constituents necessary to its functions. It is not yet clear how the cytoplasm is apportioned between daughter cells; this mechanism may be influential in determining when some daughter cells become different from parents, or specialized, during certain stages of growth in an organism.

Cell Differentiation

Closely related to growth and mitosis is the phenomenon known as cell *differentiation*, in which cells become modified and specialized to perform a specific function. Since an organism grows from a single zygote, and all of the information necessary to form the variety of cells in the adult organism must be present in that single cell, it becomes of interest to learn what factors cause differentiation among the cells and control the timing of the process. This continues to be one of the more fascinating problems in biology.

The process of mitosis ensures that all the cells of a developing embryo have the full DNA complement of the species. The process of differentiation implies that sister cells often receive different cytoplasmic complements, and it further implies that the unequal distribution of non-DNA elements of the parent is an orderly and controlled process.

An example may serve to show the relationship between the classical embryological and the molecular approach to the problems associated with differentiation. When a section of the outer tissue of a frog embryo is transplanted to the future mouth region of a salamander embryo, the frog tissue forms labial palps (embryonic lip and mouth formations). The palps have the characteristic shapes and features of frog embryos, yet the tissue would not form the mouth parts unless surrounded by the salamander tissue appropriate to the mouth region. It seems that the salamander cells have "induced" the frog tissue to form labial palps. The fact that the palps are froglike indicates that their actual formation is dependent on species-specific molecules that are synthesized in cells transplanted from frogs under the control of the frog DNA. The *induction* (the development of a specific structure as a result of a chemical transmitted from one part of an organism to another) by the salamander tissue suggests that the DNA, or genetic, activity is initiated by an extranuclear *"regulator" substance.* Experiments today are directed at learning the nature of the regulator substances, on the one hand, and of the species-specific molecules, on the other hand.

Biological Classification

In the past, the science of biology has been criticized and often belittled by the practitioners of more quantitative sciences for being a "descriptive science." The thought of learning vast lists of Latin names for wildflowers has cooled the ardor of many a young person for whom biosciences might have been an exciting area of study. Many modern biologists speak condescendingly of the classical descriptive aspects of biology and tend to think of *taxonomy,* the science of classification, as an outmoded sport of ancient scholars. The fact remains that taxonomy, which places plants and animals into groups based on natural relationships, provides organization and order to an otherwise bewildering diversity of living things. Equally importantly, biological systematics, as taxonomy is sometimes called, stresses both the unifying similarities as well as the diversity of living things. The student is encouraged to recognize that any efforts toward acquainting himself with the systematics of plants and animals will be surprisingly rewarding.

All life is categorized into two *kingdoms*—plant and animal. In general, the distinction is made on the basis of nutrition, mobility, and responsiveness. Plants synthesize their food from sunlight by photosynthesis and animals ultimately must depend on plants for their food. Plants are relatively stationary, while animals move about rapidly. Plants respond slowly to stimuli, while animals are quick to respond. There are exceptions to all these statements, and some organisms are difficult to place, though usually this is not a problem. A few modern systems include two additional kingdoms to emphasize certain aspects of organisms. *Protista* has been suggested as a third kingdom to distinguish those organisms having no tissues, e.g., one-celled animals (protozoa) and algae. The fourth proposed kingdom is called *Monera*, which includes organisms such as bacteria that have no nuclear membrane, i.e., no barrier between the genetic material and the cytoplasm. The older system of two kingdoms, plant and animal, is used here.

Classification is based on the system devised by Carl Linnaeus in the eighteenth century. It has been modified to emphasize the evolutionary relationships that seem to exist between organisms. The system uses Latin nomenclature to standardize names throughout the world. It also utilizes a system of *paired characteristics* to distinguish species. Paired characteristics refer to the presence or absence of a structure in an organism. Thus, an animal either has a vertebral column or it does not, which places it among the chordates or it does not; the same animal either has feathers or it does not, which further specifies it as being a bird or otherwise, and so on. In other words, an organism can be classified by answering a set of "yes or no" questions.

A further characteristic of the system is its use of *binomial nomenclature* to specify a *species*. Two words are associated with each species name; the first is the *genus* (plural, genera) to which the organism belongs, and the second is the species. The genus represents the next higher unit of classification that might encompass a large number of species. However, the species name is not used alone, but always in conjunction with the genus name. As an example, *Felis leo* is the lion, a species of the genus *Felis*, to which all cats belong. *Felis tigris* is the tiger, also a cat. Note that the genera are capitalized, while the species are not. Occasionally, the species is followed by a third name, in which a variation within the species is specified, e.g., *Homo sapiens africanus* identifies a race among living men, the African Negro.

A superstructure of categories extends from species up to kingdom. A *family* is a group of genera; an *order* is a group of families, etc. As examples the classification of a honeybee and the sugar maple are given below.

Classification of the Common Honeybee (*Apis mellifera*)

Kingdom:	Animal
Phylum:	Arthropoda—jointed legs, segmented bodies, exoskeleton
Class:	Insecta—insects
Order:	Hymenoptera—four membranous wings
Family:	Apidae—all honeybees
Genus:	*Apis*—hive bees
Species:	*mellifera*—the common worker honeybee

Classification of the Sugar Maple (*Acer saccharum*)

Kingdom:	Plant
Subkingdom:	Tracheophyta—vascular plants
Division:	Anthophyta—ovules enclosed in ovaries
Class:	Angiospermae—flowering plants
Subclass:	Dicotyledonae—embryo with 2 leaf-covers
Order:	Sapindales—woody plants
Family:	Aceraceae—simple, opposite leaves, doubly winged fruit
Genus:	*Acer*—all the maples
Species:	*saccharum*—the sugar maple

References and Suggested Reading

Brachet, J. (1961). The living cell. *Sci. Amer.* **205,** 51. (Offprint 90).

Brachet, J., and Mirsky, A. E., eds. (1961). "The Cell," Vol. 2. Academic Press, New York.

DeRobertis, E. D. P., Nowinski, W. W., and Saez, F. A. (1970). "Cell Biology," 5th ed. Saunders, Philadelphia, Pennsylvania.

Guttman, B. S. (1971). "Biological Principles." Benjamin, New York.

Hämmerling, J. (1963). Nuclear-cytoplasmic interactions in *Acetabularia* and other cells. *Annu. Rev. Plant Physiol.* **14,** 65.

Jaegar, E. C. (1955). "A Source-Book of Biological Names and Terms," 3rd ed. Thomas, Springfield, Illinois.

Keeton, W. T. (1972). "Biological Science," 2nd ed. Norton, New York.

Lehninger, A. L. (1964). "The Mitochondrion: Molecular Basis of Structure and Function." Benjamin, New York.

Leowy, A. G., and Siekevitz, P. (1969). "Cell Structure and Function," 2nd ed. Holt, New York.

Nelson, G. E., Robinson, G. G., and Boolootian, R. A. (1967). "Fundamental Concepts of Biology." Wiley, New York.
Oparin, A. I. (1961). "Life: Its Nature, Origin and Development." Academic Press, New York.
Steen, E. B. (1971). "Dictionary of Biology." Barnes & Noble, New York.
Stern, H., and Nanney, D. L. (1965). "The Biology of Cells." Wiley, New York.
Villee, C. A. (1967). "Biology," 5th ed. Saunders, Philadelphia, Pennsylvania.
Weisz, P. B. (1967). "The Science of Biology," 3rd ed. McGraw-Hill, New York.

CHAPTER II The Chemistry of Living Systems

The vast diversity among living organisms and the intricacy of the workings of the cellular components within the cells of these organisms are awesome. When one considers the molecular levels of these systems—the millions of chemical compounds involved, many of which have molecular weights exceeding a million—it is difficult to conceive of ever understanding in detail all the biological macromolecules, much less their relative locations within cells. The organic chemists have brought remarkable order to this immense puzzle by recognizing two important, simplifying concepts of organization. The first is that all biological macromolecules are polymeric, i.e., all biological macromolecules are made up of a relatively few smaller units called monomers. The other is that there exist well-defined sequences of reactions, called *metabolic pathways*, that occur repeatedly in cells.

In this chapter, a number of macromolecules will be introduced in much the same order as they are synthesized in cells. In cells the small subunits are formed, then systematically linked together. Here the major subunits will be considered; then the linking processes will be described. A brief description of some of the most important metabolic pathways is also included. And by way of introduction, a number of general aspects of the bonding and reacting properties of chemicals are included.

Chemical Bonds

Types of Bonds

When atoms are joined together as the result of electronic interaction, the force that holds them together is called a *chemical bond*. Several characteristics are associated with bonds, viz., the bond strength, the maximum number of bonds an atom can make, the bond angle, and the freedom of rotation permitted by the bond.

The number of *covalent bonds* an atom can make is called its valence. A single covalent bond is formed when two atoms share a pair of electrons. In a covalent bond the number of electrons that are available to be shared is definite; oxygen can form only two covalent bonds, carbon only four. The angle between two bonds originating from a single atom is called a bond angle. Covalent bond angles are usually definite; bond angles formed by weaker bonds are more variable.

Ionic bonds are formed when electrons are first transferred from one atom to another and the ions thus formed are bound together by electrostatic attraction. The ionic bond is considerably weaker than the covalent bond. Other weak bonds may be formed that are of considerable importance in biological systems. *Van der Waals forces* are nonspecific attractive forces (i.e., the forces may occur between any kinds of atoms) that are present when atoms come close to each other. Van der Waals bonds are created by the fluctuating charge distributions caused by the nearness of molecules. When atoms get very close, they experience a repulsive force due to the interpenetration of electronic shells. When this repulsive force exactly balances the van der Waals attraction, the distance (specific for each type of atom) is called the van der Waals radius for that atom.

A very important bond in biological macromolecules is the *hydrogen bond,* also a weak bond. Hydrogen bonds arise when a covalently bound hydrogen atom that has some positive charge encounters a covalently bound acceptor atom (one with a negative charge). The most important hydrogen bonds in biological systems are those in which the bond is between H and O or between H and N. Usually the hydrogen itself is also covalently bonded to either O or N. Unlike van der Waals bonds, hydrogen bonds are highly directional; they have maximum strength when pointing directly at the acceptor atom. The water molecule, which comprises nearly three-fourths the mass of a living cell, is a polar H—O—H group that forms hydrogen bonds with other water molecules as well as with acceptors in other biological molecules.

Bond Strengths

MAKING AND BREAKING BONDS

In a chemical reaction, the bond formation involves a change in energy. Energy is released from the form of internal energy in the unbound atoms to another form of energy. The stronger the bond, the greater the energy released when the bond forms. The bonding of atom A to atom B is thus described by

$$A + B \rightarrow AB + \text{energy} \tag{2-1}$$

in which AB is the bonded aggregate. Chemical bonds may also be broken, often by heat energy or by collisions. The first law of thermodynamics (which says that energy can neither be created nor destroyed, except as it is interconvertible with mass) assures us that if a bond is broken, the energy required is equal to the energy released upon formation of that bond. This breaking of a bond is indicated by

$$AB + \text{energy} \rightarrow A + B \tag{2-2}$$

In a closed system, i.e., a system with a constant number of atoms, when equilibrium is reached, the number of bonds breaking and the number being formed are equal. Regardless of the initial concentrations of A, B, and AB, at equilibrium the concentrations are expressed by K_{eq}, the equilibrium constant, in the formula

$$K_{eq} = \frac{[AB]}{[A][B]} \tag{2-3}$$

where [A], [B], and [AB] are the concentrations of A, B, and AB in moles/liter. Equation (2-3) is called the mass action formula.

FREE ENERGY

A thermodynamic quantity of great usefulness in biochemistry is called free energy. In general, the change in free energy ΔG is the energy that becomes available to do work, if there are appropriate means, as a system proceeds toward equilibrium. The spontaneous approach of a system to equilibrium is in the direction of *negative* free energy. Once at equilibrium, the free energy available to a system is zero.

Returning to chemical bonds, it is now seen that the stronger the bond, the greater the free energy ΔG accompanying its formation, and the greater the fraction of atoms in a system that are bound. These facts are expressd analytically by

$$K_{eq} = e^{-\Delta G/RT} \tag{2-4}$$

where R is the universal gas constant, approximately 2 cal/deg mole. The free energy associated with bond formation is negative and is usually expressed in kilocalories per mole. Recall from Equation (2-3) that a large value of K_{eq} means that practically all the atoms will be bound when equilibrium is reached, assuming the reactants are present at molar concentrations. In fact, if ΔG values are as small as -2 kcal/mole, K_{eq} is 28 at 27°C ($T = 273° + 27° = 300°$K), and a bond-forming reaction will be driven to virtual completion.

The free energies accompanying the formation of the strong covalent bonds between free atoms like H and O are typically large negative values, like -80 kcal/mole. Applying Equation (2-4), one finds the bonding reaction to be correspondingly large ($K_{eq} = e^{133}$), so that the concentrations of H and O atoms that are not bound will be vanishingly small.

On the other hand, van der Waals bonds are very weak, having ΔG values of -1 to -2 kcal/mole. Hydrogen bonds and ionic bonds typically range from -3 to -7 kcal/mole. These bonding energies may be compared to the average kinetic energy of molecules due to thermal motion at physiological temperatures: $\frac{3}{2}RT = \frac{3}{2}(2 \text{ cal/mole °K})(300 \text{ °K}) = 900$ cal/mole $= 0.9$ kcal/mole. Since there is considerable spread in the molecular kinetic energy about the average, some molecules will have enough energy to break the weak bonds occasionally. This is an important factor in living systems, for these secondary bonds often bind certain molecules to selective groups of molecules. The weak bonds are of an appropriate strength—large enough to permit preferential binding to appropriate molecules, but not so large as to permit the development of rigid lattice arrangements in cells. Larger bond energies would permit the bonds to break only seldom, a condition that we will find to be incompatible with cellular existence because of the high rates of diffusion required of cellular materials.

BONDS AND MOLECULAR SHAPES

It will be seen subsequently that the weak bonds are important in determining the shapes of molecules—and shapes are crucial among biological macromolecules.

The shape of many molecules is automatically determined by the distribution of covalent bonds. Double and triple bonds, in which two and three electron pairs are shared, respectively, are rigid and do not permit rotation about the bond. Accordingly, double-bonded ring compounds are flat with the atoms in a plane, e.g., the purine and pyrimidine rings seen in Figure 2-3. On the other hand, molecules having single covalent bonds permit parts of the molecule to rotate about those bonds, suggesting that covalently bound molecules exist in a wide variety of shapes. This possibility seldom is the case, however, because among the various possible configurations, different numbers of weak bonds can be formed, and usually there will be one configuration with the least free energy. Thus, the weak bonds are very important in determining the structure of biological macromolecules.

Subunits of Macromolecules

Most organic compounds in living systems are derivatives of *hydrocarbons* (compounds containing only carbon and hydrogen). Usually oxygen and nitrogen have been substituted into the hydrocarbons, but often phosphorus, sulfur, and certain metals are present. The added atoms are usually primarily responsible for the reactivity of the organic compounds. For this reason the additives are called functional groups. Despite the vast number of compounds in cells, most of the reactions that occur in cells are of relatively few types, depending on the functional groups involved. Some of the common functional groups are briefly discussed here.

Organic Functional Groups

The —OH group is called the *hydroxyl,* or sometimes the *alcohol* group. When this group takes the place of a hydrogen in a hydrocarbon, the resulting molecule becomes an alcohol.

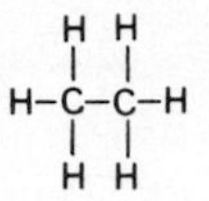

```
  H H            H H
  | |            | |
H-C-C-H        H-C-C-OH
  | |            | |
  H H            H H
```

Ethane

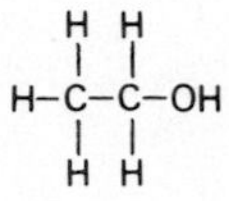

Ethanol (Ethyl Alcohol)

The *carbonyl* group is the double-bonded carbon-oxygen group, which replaces a carbon and two hydrogens in a hydrocarbon. If the double bonded O is at an end of a hydrocarbon chain, the molecule is called an *aldehyde;* if it occurs other than at an end, the molecule is called a *ketone.*

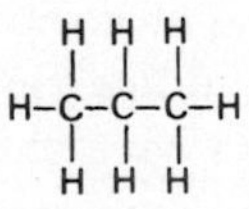

```
  H H H           H H              H O H
  | | |           | |    O         | ‖ |
H-C-C-C-H       H-C-C-C⟨         H-C-C-C-H
  | | |           | |    H         |   |
  H H H           H H              H   H
```

Hydrocarbon (*n*-Propane)

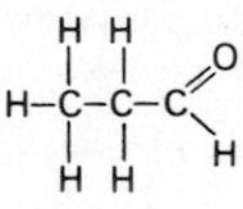

Aldehyde (Propionaldehyde)

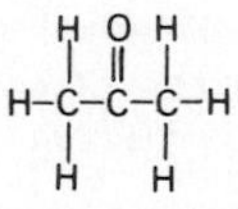

Ketone (Acetone)

The organic acids are compounds containing the *carboxyl group* —COOH. The acid group produces hydrogen ions (—COOH $\rightleftharpoons$ COO^-/ + H^+), but the acids are weak, ionizing to a relatively small extent in water.

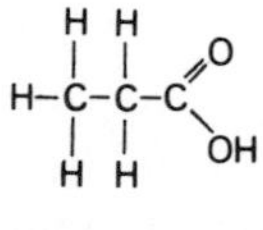

```
  H H
  | |    O
H-C-C-C⟨
  | |    OH
  H H
```

Propionic Acid - C_2H_5COOH

Organic acids that have a carbon chain exceeding four carbons in length are called *fatty acids,* which make up part of the important biological molecules called lipids. Fatty acids are characterized as *unsaturated* if the molecule has one or more double or triple covalent bonds between any of its carbon atoms, or *saturated* if it has none. Examples of both are shown:

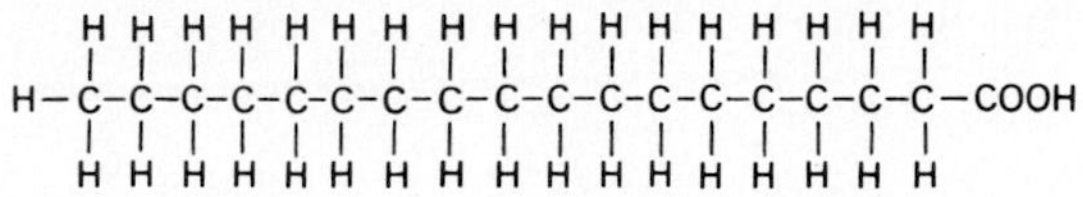

```
   H H H H H H H H H H H H H H H H H
   | | | | | | | | | | | | | | | | |
H—C-C-C-C-C-C-C-C-C-C-C-C-C-C-C-C-C—COOH
   | | | | | | | | | | | | | | | | |
   H H H H H H H H H H H H H H H H H
```

Stearic Acid (Saturated)

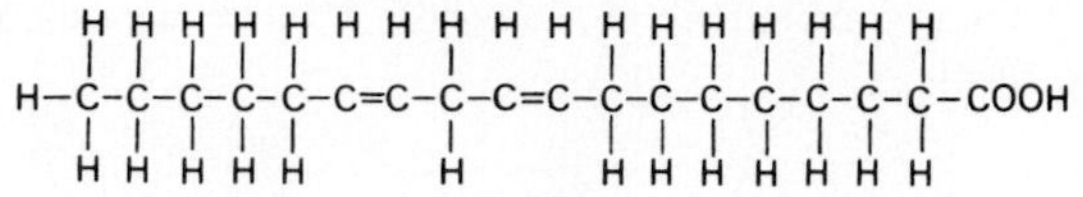

```
   H H H H H H H H H H H H H H H H H
   | | | | |     |     | | | | | | |
H—C-C-C-C-C-C=C-C-C=C-C-C-C-C-C-C-C—COOH
   | | | | |     |     | | | | | | |
   H H H H H     H     H H H H H H H
```

Linoleic Acid (Unsaturated)

The *amino group* —NH_2 is a derivative of ammonia, i.e., it can be obtained starting from ammonia, and replaces a single H in hydrocarbons. *n*-Propylamine is an example.

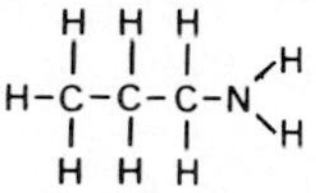

```
   H H H
   | | |  /H
H-C-C-C-N
   | | |  \H
   H H H
```

Amino Acids

Organic compounds that contain both the carboxylic acid group (often called the carboxyl group) and an amino group are called *amino acids.* There are 20 naturally occurring amino acids. In most of these, the carboxyl group and the amino group are attached to the same carbon. A hydrogen is also usually attached to this same carbon. The remaining group attached to the carbon is a group of varying structure, which is the characteristic of each of the different amino acids:

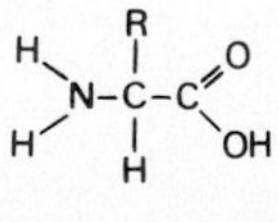

```
       R
H      |   O
 \     |  //
  N-C-C
 /     |  \
H      |   OH
       H
```

The letter R represents the side group that determines a particular amino acid. When R = H, the amino acid is called glycine (Gly), the simplest of the twenty. When R = CH_3, the amino acid is alanine (Ala), the next simplest. Some of the side groups become quite complex.

When two amino acids combine, the carboxyl group at the end of one joins the amino group at the end of the other, forming a water molecule

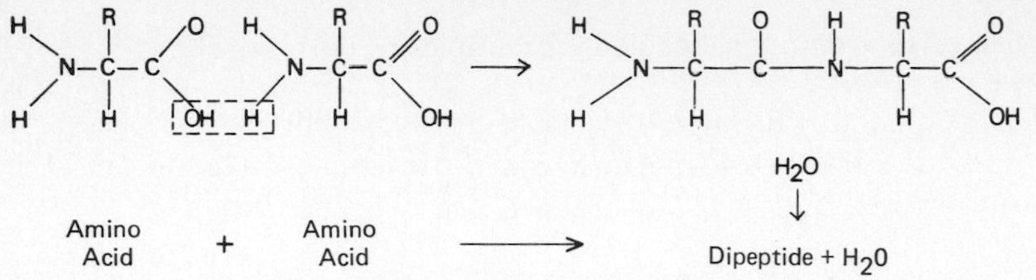

Amino Acid + Amino Acid ⟶ Dipeptide + H_2O

FIGURE 2-1. A peptide bond is formed when amino acids combine. The peptide bond is the covalent bond between the nitrogen in the amino group of one amino acid and the carbon in the carboxyl group of the other. A water molecule is freed when the peptide bond is formed.

that is separated from the two reacting molecules. The covalent bond formed between the amino group N and the carboxyl group C is called a *peptide* bond. (See Figure 2-1.) A pair of amino acids joined by a single peptide bond is called a *dipeptide.* A combination of more amino acids joined in this manner is called a *polypeptide,* until about fifty amino acids are in the chain, at which point chains of longer length are called *proteins.*

Nucleotides

Nucleotides are composed of a five-carbon sugar, an inorganic phosphate group, and a base substance. There are ten different nucleotides that occur in living cells. They were discovered in the nucleus of cells, which accounts for their name, but they are now known to be present in the cytoplasm as well.

Two varieties of sugar form part of nucleotides—ribose and deoxyribose—both of which are pentose (containing five carbon atoms) rings and differ only by the absence of a single oxygen atom in the deoxyribose variety. Figure 2-2 shows both the ribose and deoxyribose sugar structures.

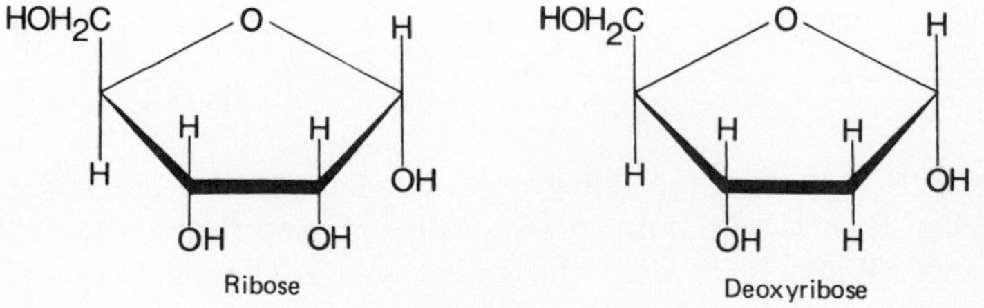

Ribose

Deoxyribose

FIGURE 2-2. Ribose sugar and deoxyribose sugar are the two five-carbon sugars that are part of naturally occurring nucleotides. Notice that they differ only in the absence of a single oxygen atom in the deoxyribose sugar.

Five bases (so called because they give an alkaline or basic reaction in aqueous solution) occur in nucleotides. These bases fall into two classes called *purines* and *pyrimidines*. Purines are double ring structures, and pyrimidines are single rings. Both classes contain nitrogen atoms within the ring structure.

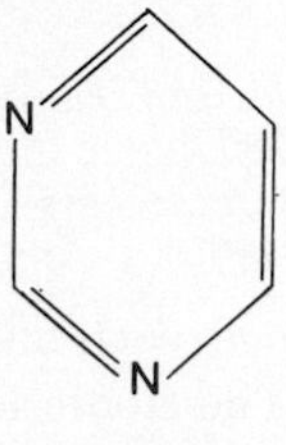

Pyrimidine

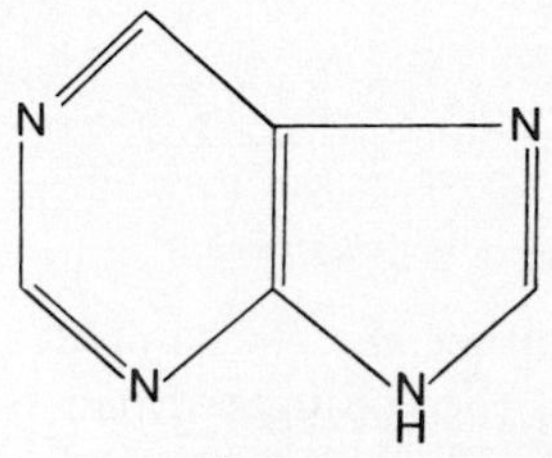

Purine

The above illustrations use some conventions that organic chemists often employ to simplify structural diagrams. Each intersection of the bonds is assumed to contain a carbon atom unless otherwise specified, and the carbon atoms are assumed to have hydrogen atoms attached in sufficient number to make the total number of bonds at each carbon equal to four. Sometimes the bonds are left out when they are obvious (like the N-H bond or the C-H bond). The purine structure, for example, can be depicted in either of the following ways:

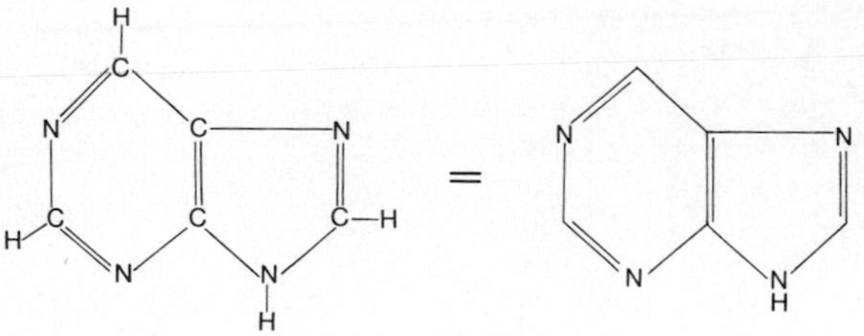

Two purines, *adenine* and *guanine*, occur in nucleotides:

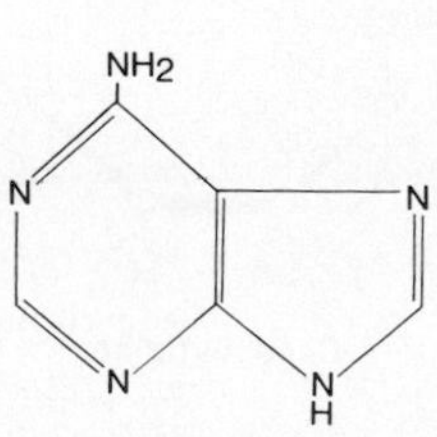

Adenine

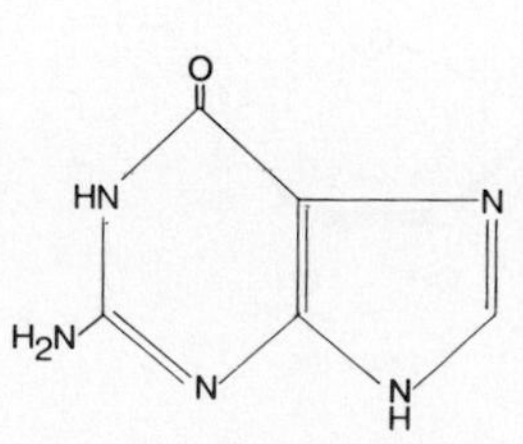

Guanine

The three pyrimidines are *cytosine, thymine,* and *uracil:*

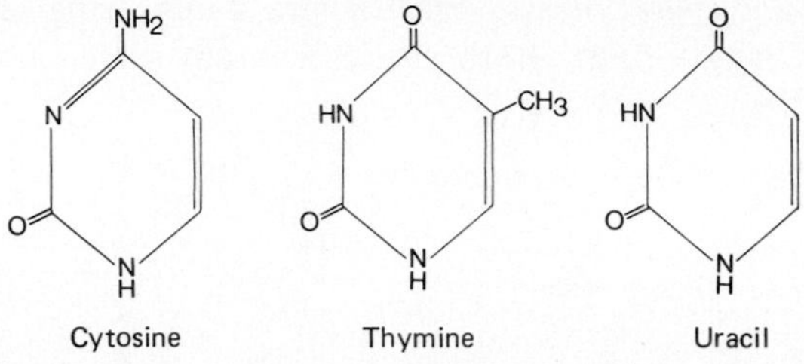

Cytosine Thymine Uracil

When either the ribose or deoxyribose sugar is joined with one of the five bases, the conjugated pair is called a *nucleoside.* The ribose linked to adenine is called *adenosine* and when linked to guanine is called *guanosine.* Similarly, the ribose pyrimidine nucleosides are *cytidine* and *uridine.* The analogous nucleosides formed with deoxyribose are called deoxyribonucleosides: deoxyadenosine, deoxycytidine, etc. The exception to this naming procedure is the deoxynucleoside of thymine, which is called simply thymidine—not deoxythymidine—since this pyrimidine is almost never found except in DNA. Figure 2-3 shows examples of two nucleosides.

Nucleotides are formed when a phosphate group, PO_4^{3-}, is attached to the sugar of a nucleoside. The result is a strongly acidic compound, whose name is taken from its corresponding nucleoside. The ribose nucleotides

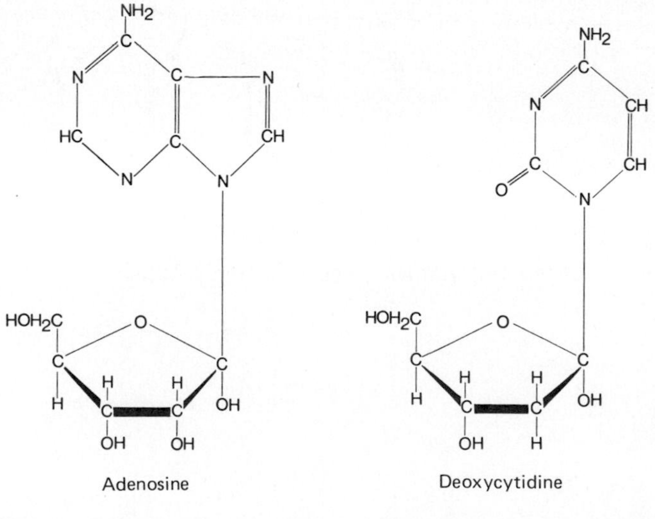

Adenosine Deoxycytidine

FIGURE 2-3. Examples of nucleosides, conjugated pairs of either ribose or deoxyribose sugar and one of the purines or pyrimidines. Adenosine is the combination of ribose sugar and adenine; deoxycytidine combines deoxyribose sugar and cytosine.

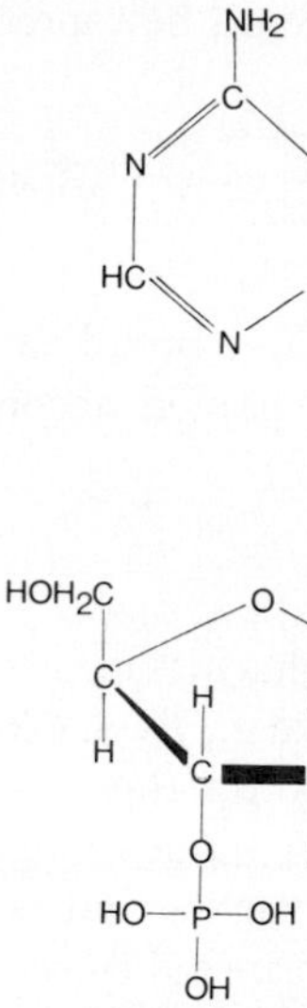

FIGURE 2-4. A complete nucleotide, deoxyadenylic acid. A nucleotide consists of a pentose sugar, a purine or pyrimidine (the base group), and a phosphate group. In this case, the sugar is deoxyribose and the base group is adenine.

are adenylic acid, guanylic acid, cytidylic acid, and uridylic acid. The deoxyribose nucleotides are deoxyadenylic acid, deoxycytidylic acid, deoxyguanylic acid, and thymidylic acid. Figure 2-4 shows a complete nucleotide.

Biological Macromolecules

Of the many classes of chainlike molecules that provide the endless diversity found in living systems, five will be considered briefly here: saccharides, lipids, proteins, the nucleic acid chains (DNA and RNA), and porphyrins. All of these play crucial roles in the chemistry of the living world.

Saccharides

Carbohydrates are organic compounds that contain a carbonyl group in addition to two or more alcohol groups, or that yield such compounds upon hydrolysis. Hydrolysis is the "breaking up by water" of a compound. For example, the hydrolysis of table sugar ($C_{12}H_{22}O_{11}$) into simple sugars by the addition of a water molecule would be represented by

$$C_{12}H_{22}O_{11} \underset{-H_2O}{\overset{+H_2O}{\rightleftharpoons}} 2\ C_6H_{12}O_6$$

Carbohydrates occur in nature in three broad categories—*monosaccharides*, *disaccharides*, and *polysaccharides*—named according to the complexity of their structures.

MONOSACCHARIDES

The simple sugars, or monosaccharides, are classified according to the number of carbons in their chains, e.g., five-carbon sugars (pentoses) and six-carbon sugars (hexoses), which are the most common types in living organisms. Two of the important monosaccharides have already been introduced: ribose and deoxyribose sugars that occur in nucleic acids. Most natural sugars occur in the ring form. Glucose and fructose are the most important monosaccharides from a biological point of view. They have the same atomic composition ($C_6H_{12}O_6$) and are both hexose sugars, though glucose has a hexagonal ring and fructose a pentagonal ring:

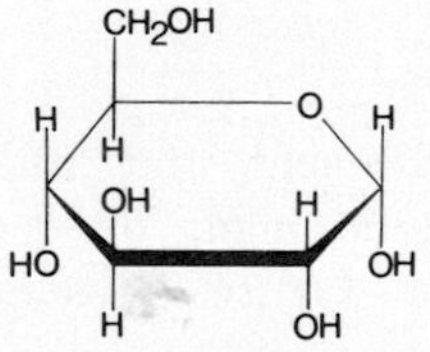

Glucose

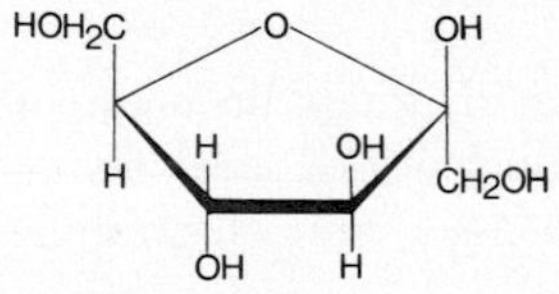

Fructose

DISACCHARIDES AND POLYSACCHARIDES

Two simple sugars are combined by removing a hydrogen atom from one monosaccharide and a hydroxyl group from the other (effectively removing a molecule of water) and joining the two groups by way of an oxygen atom. The resulting compound is a disaccharide. The bond linking the two simple sugars together is called a *glycosidic bond. Sucrose* (table sugar), for example, is a disaccharide composed of a glucose and a fructose group joined by a glycosidic bond (see Figure 2-5).

The vast majority of carbohydrates existing in nature are in the form of polysaccharides, in which long chains of monosaccharides are linked by glycosidic bonds. Glucose is the most common, and often the only, monosaccharide unit in polysaccharides. The many natural polysaccharides differ among themselves in length, degree of branching, and type of glycosidic bond, as well as in their constituent monosaccharide composition.

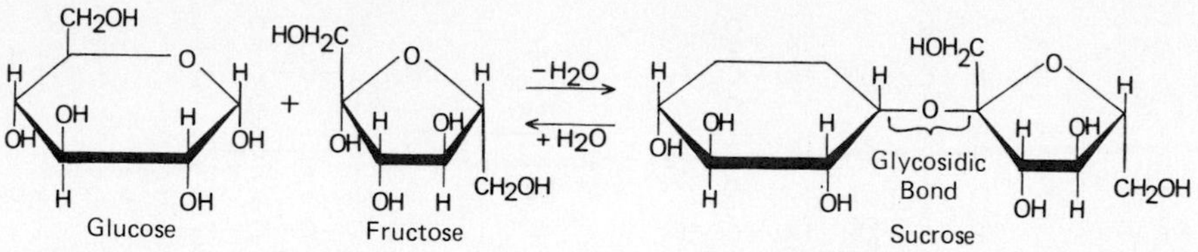

FIGURE 2-5. Formation of the glycosidic bond. As sugar groups are linked together, a water molecule is released and an oxygen bridge connects the two sugar rings together. The diagram also illustrates that the addition of a water molecule is required when a glycosidic bond is broken.

Structural polysaccharides are those that, in nature, are employed as a mechanical support or protection. Cellulose is the most important of these, since it is the material of cell walls in plants. In fact, cellulose is the most abundant organic compound found on earth. The cellulose molecule is an unbranched chain of 300 to 3000 glucose units with molecular weights ranging from about 5×10^4 to 5×10^5. The exoskeleton of insects and crustaceans is made of *chitin*, an amino derivative of the glucose polysaccharide chain. More complex derivatives, called *mucopolysaccharides*, are found in a wide range of animal tissues, like skin, tendons, heart valves, bones, etc.

Metabolic polysaccharides (those having a significant role in metabolism or nutrition) include the *starches*, a storage form of glucose units in plants, and *glycogen*, the storage form of glucose units in animals. Starches occur in both branched and unbranched chains of glucose, and the glycosidic bonds in starches are different from those found in cellulose. Glycogen is always branched. The metabolic polysaccharides sometimes have molecular weights exceeding 10^8.

Lipids

A heterogeneous group of organic compounds that are soluble in common organic solvents and in each other, but are essentially insoluble in water, are called lipids. Two important classes of lipids are considered here: *triglycerides*, the neutral fats in living systems, and *steroids*, complex molecules that serve diverse functions in metabolism and development.

TRIGLYCERIDES

Trigylcerides, or neutral fats, are the most abundant lipids in nature. They are made up of a three-carbon alcohol, *glycerol*, combined with three fatty acids (see Figure 2-6). The saturated nature of some fatty acids, most commonly *stearic acid* and *palmitic acid* in animals, causes most animal fats to be solids at room temperature. The unsaturated fatty acids (like

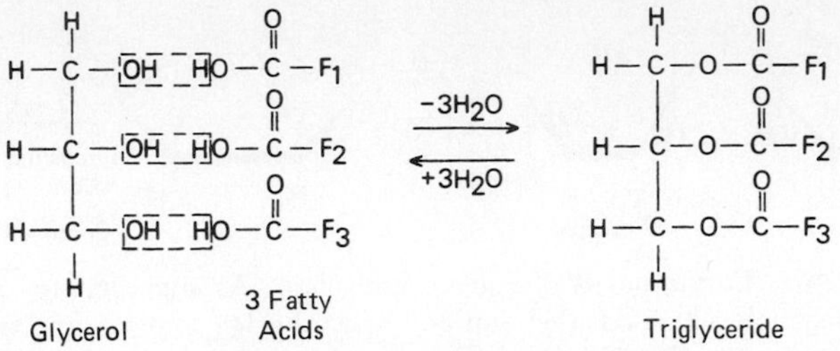

$-3H_2O$ / $+3H_2O$

Glycerol | 3 Fatty Acids | Triglyceride

FIGURE 2-6. The hydrolysis of a triglyceride. Glycerol is a three-carbon compound that may be joined to three fatty acids to form a triglyceride by releasing three water molecules. The groups labeled F_1, F_2, and F_3 indicate that the fatty acids that combine with a glycerol molecule are not necessarily the same one.

oleic acid) that occur in vegetable triglycerides are liquid at room temperature, and they are called *oils*.

STEROIDS

Another class of lipids, which has no particular structural relationship to triglycerides, is called the steroids. They are rather complex, having four fused carbon rings to which is usually attached a long carbon chain. A typical steroid is *cholesterol* (see Figure 2-7), which has become a household word in recent years because of its suspected implication in heart and circulatory diseases. Steroids have profound effects on living organisms, even when they occur in small quantities. Many hormones are steroids or derivatives of steroids.

THE ROLE OF LIPIDS IN CELL STRUCTURES

The fact that lipids contain relatively long nonpolar chains of hydrocarbons makes them insoluble in water. *Polar molecules* are those in which the electronic distribution over the molecule is arranged such that one part of the molecule has a local net positive charge. In water molecules, for example, the hydrogen atoms are positive with respect to the oxygen atom. As a result, when other polar molecules are placed in water, the positive regions are attracted to the negative regions of the water molecule (and similarly the negative regions to the positive regions of water), and the molecules are separated, i.e., dissolved, by the water. The weak bonds holding the similar solute molecules togther are overcome by their electrical attraction to water. Nonpolar molecules do not possess such charged regions and are therefore relatively insoluble in water.

Their nonpolar characteristic makes lipids an ideal structural material for membranes in cells. The chemical structure of lipids permits another useful structural function—self-assembly. When long nonpolar chains are present, they tend to join to each other by weak interactions, such as van

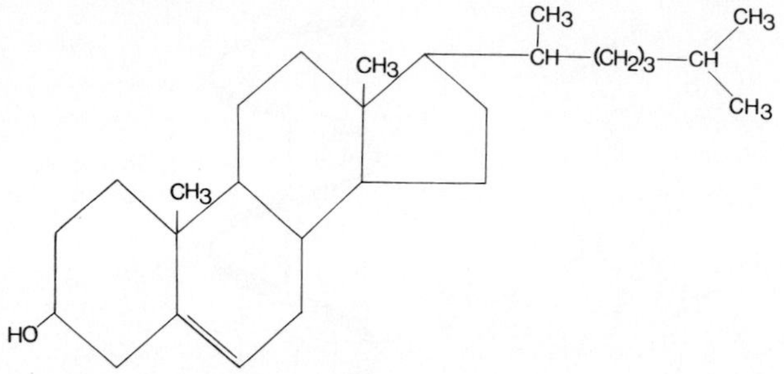

FIGURE 2-7. The cholesterol molecule, a typical steroid. Steroids are one of the classes of lipids.

der Waals interactions. Because of this characteristic, lipids tend to drift away from the aqueous (water is a polar molecule) regions of the cell's interior to the cell walls where they are incorporated into the wall structure by weak bonding.

Proteins

Proteins occupy a central position in the architecture and function of living systems, because they are prominent in practically every phase of chemical and physical activity in cellular life. Even quantitatively, proteins are the primary material of tissues, constituting nearly 75% of the dry weight in animal tissue.

STRUCTURE

As mentioned earlier in this chapter, proteins are long chains of amino acids connected by peptide bonds. The sequence of amino acids (with twenty to choose from) determines the specific protein. Since proteins often have up to several thousand amino acids in the chain, the number of possible different proteins is staggering. Some proteins are even made of more than a single amino acid chain.

The protein chain of amino acids is called the *primary structure* of protein because it is an essentially linear chain. However, the chains do not occur in nature in linear form. The most common configuration is a twisted form of the linear chain called the *α-helix,* which is the *secondary structure* of proteins. The chain is held in the α-helix form by hydrogen bonds between the atoms of the backbone (the side groups are on the outside), which have identical orientations when in the helical configuration. An amino acid is 1.5 Å in length (between peptide bonds), and the distance along the axis of the helix is 5.4 Å per turn, so there are 3.6 amino acids per turn. The

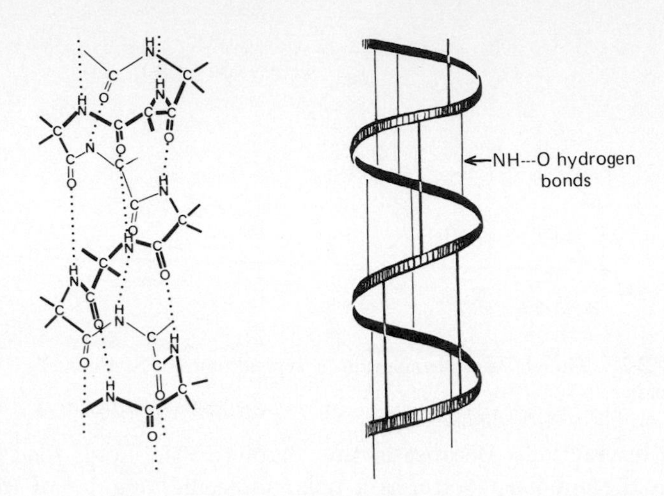

FIGURE 2-8. Peptide chain of a protein coiled to form an alpha helix. The configuration of the helix is maintained by hydrogen bonds, shown as approximately vertical lines. The helix on the left shows the detailed atomic structure of the peptide chain (although the side groups are not shown on each amino acid). The helix on the right is a schematic representation without structural detail. From Roberts and Caserio (1967). Copyright 1967 by W. A. Benjamin, Inc. Reproduced by permission.

hydrogen bonds occur between the carboxyl group in an amino acid and the amino group of the fourth amino acid along the chain. (See Figure 2-8.)

The three-dimensional structure of a protein is called its *tertiary* structure. Practically no proteins exist in an extended form of the α-helix. The helix is usually folded about in three dimensions and pinned into a fixed shape by *disulfide bonds.* The amino acid *cysteine* (Cys) contains a SH group (sulfhydryl group) that allows S—S (disulfide) bridges to form between the Cys units in the chain. Experiments have shown that after proteins are denatured (uncoiled from their natural three-dimensional form), the disulfide bonds nearly always reattach to the same places that they occurred in the original "native" protein. So although the S—S bonds pin the structure into its three-dimensional spatial configuration, the three-dimensional structure of a protein is still determined solely by the amino acid sequence, which locates the Cys positions and thus locates the S—S bridges.

ENZYMES

Nearly all chemical reactions in living systems occur through the mediation of a class of proteins called *enzymes,* which are biological catalysts (substances that accelerate a chemical reaction without being consumed in the overall process). Enzymes enable the chemical reactions necessary to living systems to proceed efficiently at life temperatures far below the temperatures at which they would otherwise take place. There seems to be an endless number of enzymes, judging from the biochemical literature; apparently almost no reactions are left to chance.

The substance upon which an enzyme acts is called the *substrate,* and most enzymes are named by adding "ase" to the substrate name. For instance, sucrase is the enzyme that catalyzes the hydrolysis of the disaccharide sucrose into glucose and fructose. Others that perform a more general function may have "ase" added to the function name, e.g., a *reductase* is an enzyme that reduces, or transfers, a hydrogen atom (or electron) from one substrate to another.

An important characteristic of every enzyme is its *specificity* (referring to the fact that the enzyme catalyzes only one specific reaction or at most a class of reactions). Nonbiological catalysts do not have such specificity. What accounts for the highly specific nature of biological catalysts? The tertiary structure of the amino acid chain that makes up the enzyme forms a particular region, called the *active site,* on the enzyme that has a very special spatial configuration. This configuration exactly fits the appropriate region of the substrate like a "lock-and-key" arrangement or pieces of a jigsaw puzzle. (See Figure 2-9.) When the enzyme and substrate are combined, the resulting group is called the *enzyme–substrate complex.* The position of the reactants is then such that the reaction proceeds rapidly.

Coenzymes are nonprotein components that are often required to be associated with an enzyme in order to effect enzymatic activity on a

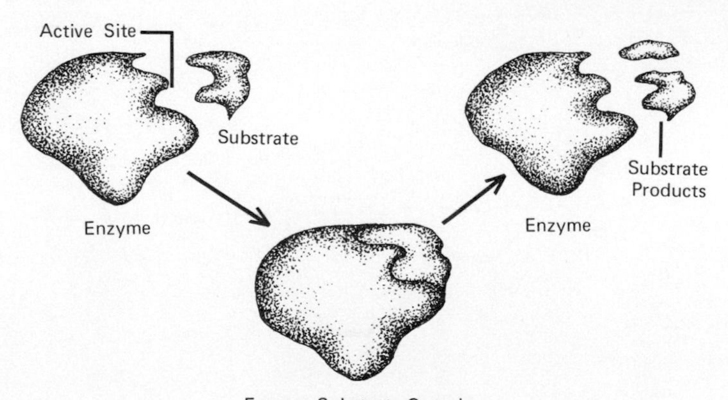

FIGURE 2-9. A schematic illustration of the interaction between an enzyme and its substrate. The active site of the enzyme specifically fits the configuration of the appropriate region of its substrate. The relative sizes of enzyme and substrate shown here are not significant.

substrate. The protein portion with which a coenzyme joins is called an *apoenzyme*. *Vitamins* are organic substances that are usually required in minute quantities in normal cellular functions, and many vitamins serve as coenzymes. *Thiamine* is a vitamin used in brain tissue as a coenzyme with the enzyme *carboxylase* in the process that breaks down glucose to produce energy for the brain cell functions. Without thiamine the reaction cannot take place.

DNA and RNA Chains

The molecules that contain the inherited traits, transmit genetic information from parent to progeny, and govern the chemical activity of the cells of living organisms have been of special interest to biologists, chemists, and physicists for two decades. The nucleic acids, DNA and RNA, play a fundamental role in what has come to be considered the "seat of life."

DNA

Deoxyribonucleic acid (DNA) is a long-chain polymer composed of nucleotides linked together. The links between the nucleotides occur between the phosphate group and the deoxyribose sugar (see Figure 2-10). Only the two purines, adenine (A) and guanine (G), and the two pyrimidines, cytosine (C) and thymine (T), are normally found in DNA. In a schematic representation of a DNA chain, the squence of the bases, A, T,

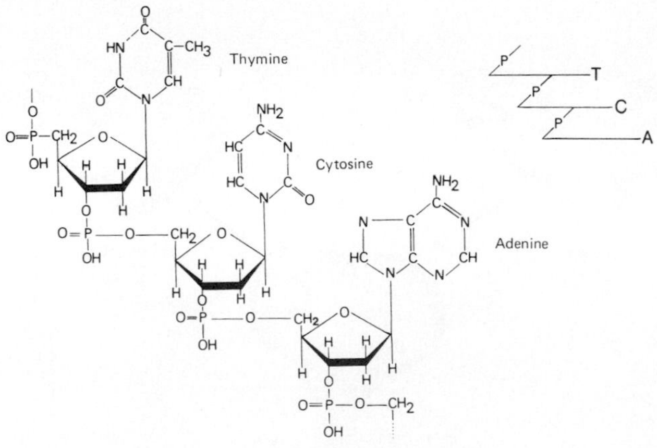

FIGURE 2-10. A portion of a strand of DNA. On the left, the detailed structure of each nucleotide is shown; on the right, a schematic representation indicates the base in each nucleotide and emphasizes that the nucleotides are linked through the phosphate groups.

G, and C, is all that is required; the backbone of the sugar and phosphate is assumed.

The molecular weight of native DNA (i.e., DNA naturally occurring in living systems) ranges from 6 to 12 million. Since the molecular weight of a nucleotide is of the order of 300, it is evident that a DNA molecule contains 20,000 to 40,000 nucleotides.

The analysis of naturally occurring DNA from different species of bacteria and from specific animal tissues shows a striking regularity. The ratio of (A + T)/(G + C) is characteristic of the species or tissue, but even more important, the ratios A/T and G/C are always equal to unity.

The details of the structure of DNA were elucidated by Watson and Crick (1953) using X-ray diffraction data obtained by Maurice Wilkins. Watson and Crick suggested that DNA is a double helix composed of two intertwining chains of polynucleotides. In the double helix, the bases are on the interior and are joined (purine to pyrimidine) by hydrogen bonds. The three hydrogen bonds occurring between G and C and the two occurring between A and T are shown in Figure 2-11. The selectivity of these

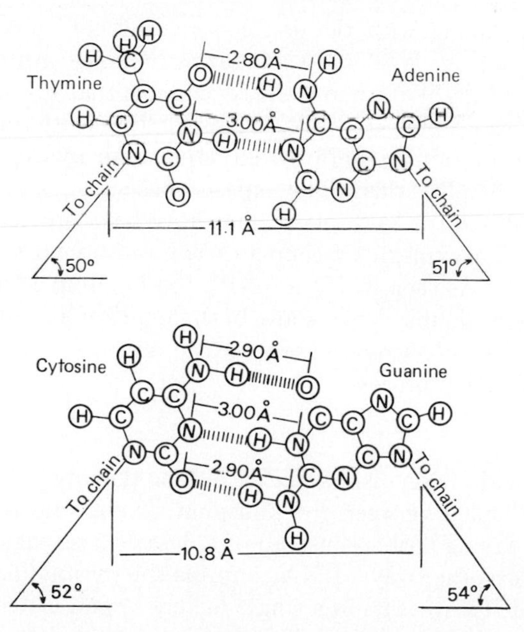

FIGURE 2-11. The hydrogen-bonded base pairs that occur in DNA. Thymine and adenine are always joined by two hydrogen bonds, while cytosine and guanine are always joined by three hydrogen bonds. Each hydrogen atom in all the hydrogen bonds point directly at its acceptor atom (nitrogen or oxygen), so all the hydrogen bonds in DNA are strong. From Watson (1970). Copyright 1970 by J. D. Watson, W. A. Benjamin, Inc. Reproduced by permission.

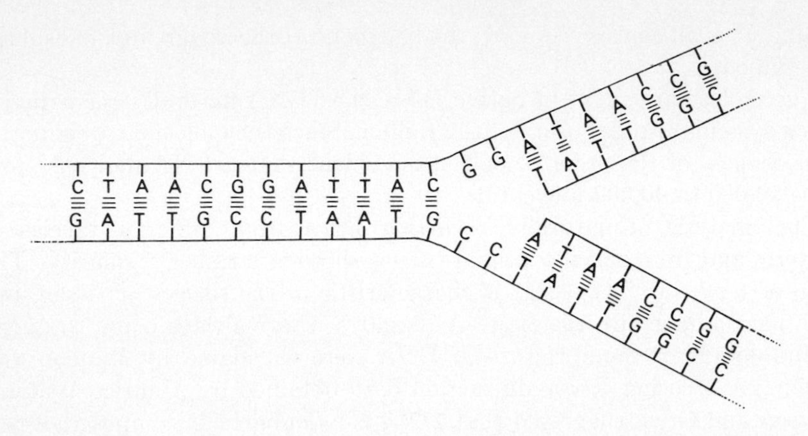

FIGURE 2-12. Schematic representation of double-stranded DNA replicating itself. Because of the obligatory bonding of A to T and G to C, the two daughter molecules are identical to the original, or parent, molecule. The hydrogen bonding between bases is indicated by series of bars.

hydrogen bonds, along with the double-stranded configuration of native DNA, accounts for the A/T and G/C ratios of unity. An even more impressive feature of the Watson and Crick model is that it provides a means by which DNA can dulplicate itself. When the duplication process begins, the double helical strands separate (only the weak hydrogen bonds must be broken to accomplish this). The exposed base groups are then free to attract nucleotides in the karyoplasm that have been previously assembled (with the help of enzymes). Of course, A can only bond to T, G to C, T to A, and C to G. As seen in Figure 2-12, the bonding selectivity assures that the two new double helices are both identical to the original one. An enzyme, *DNA polymerase,* catalyzes the assembly of the nucleotides during the duplication process.

RNA

Ribonucleic acid (RNA) is, like DNA, a long polynucleotide. The ribose nucleotides are linked between the phosphate group and the ribose sugar group. Besides having a ribose instead of a deoxyribose sugar, RNA differs from DNA in two other ways: RNA contains the pyrimidine uracil instead of thymine, and RNA occurs in a single helical strand instead of in double strands with complementary base pairs.

Three distinct types of RNA are known, and all of them play roles in the synthesis of protein (see Chapter III). All three kinds are *transcribed* from DNA through the mediation of the enzyme *RNA polymerase,* which means that a single strand of DNA was used as a template in ordering the base sequence of the RNA by complementary base pair bonding (C to G, A to U, etc.).

Messenger RNA (mRNA) is a linear form of RNA that is used to transmit information from the nuclear DNA to the site of protein synthesis in the cytoplasm. *Transfer RNA* (tRNA), sometimes called soluble RNA (sRNA), is a form that contains about 70 nucleotides in a fixed, rather complex configuration. The tRNA molecules are used to guide the appropriate amino acid into the proper position for incorporation into proteins that are being synthesized. The third variety of RNA is called *ribosomal RNA* (rRNA), because it comprises about half of the composition of ribosomes, which are the sites of protein synthesis. The structure and function of rRNA are not very well understood. Further details of these macromolecules will be discussed in Chapter III.

Porphyrins

Almost all species, animal and plant, synthesize complex ring systems called *porphyrins*. These molecules are usually found in association with a protein, and most natural porphyrins contain a metal. Porphyrins contain small four-carbon rings called *pyrroles* as a subunit:

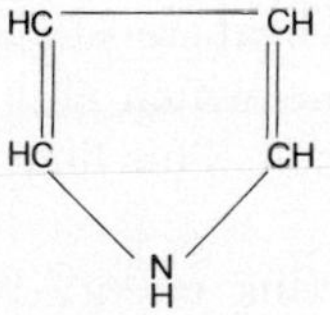

Pyrrole

Four pyrroles are arranged symmetrically about a metal ion (usually divalent) to form a porphyrin, as shown in Figure 2-13. The two most important metal porphyrins are the *hemes* and *chlorophylls*, in which the metals are Fe and Mg, respectively.

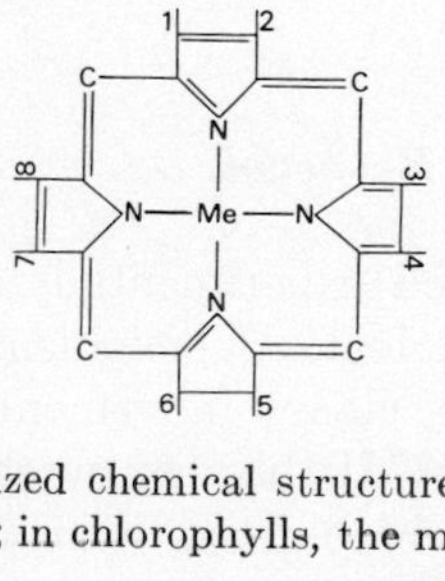

FIGURE 2-13. The generalized chemical structure of a porphyrin. In hemes, the metal at the center (Me) is iron; in chlorophylls, the metal is magnesium.

Hemes and their protein combinations include *hemoglobin*, the pigment in the blood that transports oxygen, *cytochromes*, compounds important in certain metabolic processes, and a number of enzymes.

The chlorophylls are necessary to the process of photosynthesis. Four types, identified as chlorophyll a, b, c, and d, occur in plants. All the chlorophylls are similar, usually differing in the side chains connected to the pyrrole rings.

Cellular Energy Conversion

The functions of living cells depend upon the availability of a usable form of energy. The energy that serves all living systems ultimately comes from the sun in the form of electromagnetic waves. One method of conversion of solar energy occurs in the cells of plants by a process called photosynthesis, in which the energy of light quanta is converted into "food molecules," i.e., into chemical energy stored in molecules of the cells. At this stage, the energy is essentially stored in the covalent bonds of carbon atoms. The energy is finally released for utilization by the cell in a series of reactions that oxidize these carbon compounds. (Recall that in a chemical reaction, the atoms losing or donating electrons are said to be oxidized; those accepting electrons are said to be reduced). The released energy is not all in the form of heat; most of its is converted into new chemical bonds in a compound from which all cells can quickly and efficiently utilize the energy in these bonds. This final storage compound is *adenosine triphosphate* (ATP).

Another method of converting energy to forms that are useful to an organism is respiration, a rather long series of reactions that start with glucose and result in ATP molecules.

In this section, a brief account of these cellular energy conversion processes is presented. In order that the ultimate products of the processes are recognized whenever they occur (they appear in the early stages of conversion as well as at the end), the presentation will begin with ATP, the end product.

ATP—Biological Energy Reservoir

It has been demonstrated experimentally that the immediate energy source in many essential biological phenomena, such as protein synthesis, mitosis, muscle contraction, ciliary movement, and many others, is ATP. *Adenosine monophosphate* (AMP) is very much like the nucleotide adenylic acid, except that the phosphate group is attached at a different point on

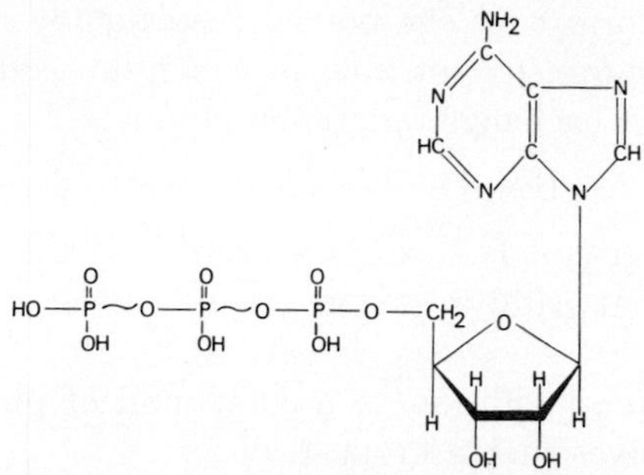

FIGURE 2-14. The adenosine triphosphate (ATP) molecule.

the nucleoside adenosine. *Adenosine diphosphate* (ADP) and ATP are made by the serial phosphorylation (addition of phosphate groups to a compound) of AMP. Figure 2-14 shows a complete ATP molecule.

ATP is nearly always described as an "energy-rich" compound, and "high energy phosphate bonds" is an often used term that implies that the available energy resides in the P—O bond and becomes available when the bond is broken. The idea of releasing energy by breaking a bond normally makes no sense—energy is required to break a bond. When ATP is converted to ADP by hydrolysis, the reaction

$$\text{ATP} + \text{H}_2\text{O} \rightleftharpoons \text{ADP} + \text{P} \tag{2-5}$$

where P is the symbol used for an inorganic phosphate or phosphoric acid, has a ΔG value of -7 kcal/mole. This means that the hydrolysis has freed energy for useful work by transferring the phosphoryl group:

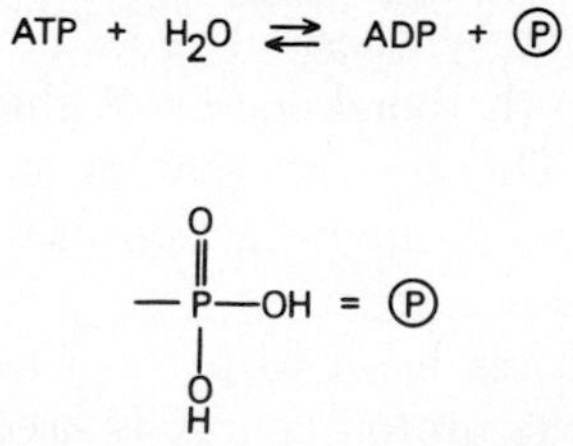

$$\text{ATP} + \text{H}_2\text{O} \rightleftharpoons \text{ADP} + \text{Ⓟ}$$

$$-\text{P}(=\text{O})(\text{OH})-\text{OH} = \text{Ⓟ}$$

The breakdown of ATP to ADP has a negative ΔG, and the formation of polymeric chains from monomeric building blocks requires a positive ΔG. By coupling the breakdown of ATP with the formation of a peptide bond ($\Delta G = +0.5$ kcal/mole), for example, there is a net ΔG of -7.5 kcal/mole. This is more than sufficient to ensure that the approach of the system (breakdown plus bond formation) to equilibrium is in the direction of protein synthesis.

The coupled reaction described above cannot be accomplished by allowing the ATP breakdown to free its energy as heat, because heat energy cannot be used to form covalent bonds. Instead, the coupled reaction is achieved by a *group-transfer reaction,* in which no oxidation or reduction occurs, but molecules exchange functional groups:

$$A—B + C—D \rightarrow A—C + B—D \qquad (2\text{-}6)$$

In Equation (2-6), groups B and D are exchanged with the components A and C. Enzymes that catalyze these group-transfer reactions are called *transferases.*

The *phosphorylation* [addition to a compound of phosphate group(s)] of a compound involves such a group-transfer:

$$ATP + X \rightarrow ADP + X \sim P + \Delta G \qquad (2\text{-}7)$$

Compounds such as X in Equation (2-7) are said to be "activated" by ATP. The symbol $\sim$ is used to represent the "high energy" bond.

Photosynthesis

The utilization of solar energy in the living world begins with the process of photosynthesis, which depends on the presence of the porphyrin chlorophyll in chloroplasts of plants. The overall process of photosynthesis can be summarized by the expression

$$6\ H_2O + 6\ CO_2 \xrightarrow{h\nu} C_6H_{12}O_6 + 6\ O_2 \qquad (2\text{-}8)$$

in which $h\nu$ is the energy of a photon that supplies the energy for the entire process. Equation (2-8) expresses the fact that sunlight converts water and carbon dioxide into glucose and free oxygen. This equation accounts for the materials required and the products formed, but it fails to express the complex nature of photosynthesis. It does not tell us how the photon energy is transformed into the bond energy of glucose. It does not suggest the role of chlorophyll. The process that is summarized by Equation (2-8) is quite involved, and its investigation has been an exciting area of study for over three decades.

The photosynthesis process has two parts. The first is the *photophase,* or light phase, in which the proton energy is used to split H_2O molecules into O_2 that is released as a by-product and hydrogen. The hydrogen is trapped by an acceptor molecule, which holds the hydrogen until it is used in the second phase. The photon energy is also used to convert ADP to ATP, which is used in the second phase. The second part of the process is usually referred to as the "*dark reactions,*" which implies that the reactions in this phase can proceed without additional inputs of energy from photons. The dark phase fixes the CO_2 into a carbohydrate, finally forming glucose and releasing water as a by-product. Figure 2-15 is a schematic representation of the photosynthesis process.

In the photophase, the electrons of chlorophyll are raised to excited levels by absorption of photons in the red and blue regions of the visible spectrum (which explains the green color of chloroplasts since the green wavelengths are predominantly reflected). The deexcitation of the chlorophyll electrons provides the energy for splitting water molecules and for phosphorylizing ADP into ATP. The oxygen is freed, but the hydrogen is trapped by a coenzyme called *NADP* (nicotinamide adenine dinucleotide phosphate). NADP is called a hydrogen acceptor because it readily combines with hydrogen to form $NADPH_2$, which passes the hydrogen on to another compound in the dark phase. The ATP formed in the light phase is also used in the dark reactions. After complete deexcitation, the chlorophyll molecule can absorb other photons, restarting the cycle.

In the dark phase, CO_2 is fixed by a pentose sugar phosphate that is present in chloroplasts. This carbon dioxide acceptor is called *RDP* (ribulose diphosphate), and it incorporates the CO_2 into a hexose sugar that is very unstable. Very quickly the sugar splits to form two molecules of a three-carbon compound *PGA* (phosphoglyceric acid), which are combined with the two hydrogens supplied by $NADPH_2$. The products of this reaction are *PGAL* (phosphoglyceraldehyde) molecules. The formation of PGAL from PGA requires energy input, and this energy is supplied from the ATP made in the photophase. PGAL can be used directly by the plants as a nutrient (plants artifically nourished with PGAL can survive without photosynthesis) or converted to RDP, which is recycled to fix CO_2 in the chloroplasts. The third possible fate of PGAL molecules is their conversion into glucose by substitution of a hydrogen for each phosphate group:

$$\underset{\text{PGAL}}{2\,C_3H_5O_3 \sim P} + 2\,H \rightarrow \underset{\text{glucose}}{C_6H_{12}O_6} + 2\textcircled{P} \qquad (2\text{-}9)$$

The photosynthesis events described above are only an outline of a very large series of complex reactions that are mediated at every step by en-

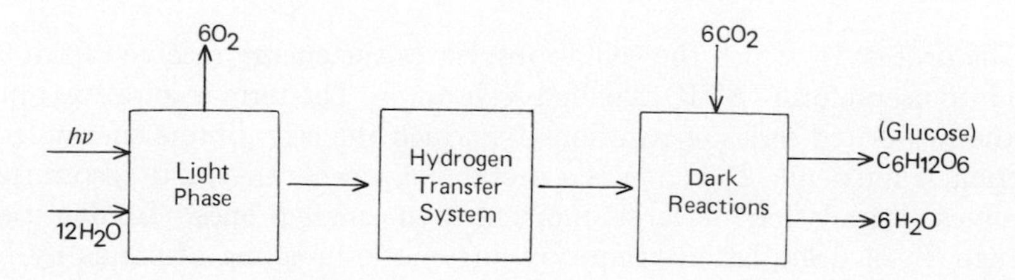

FIGURE 2-15. Schematic representation summarizing the process of photosynthesis. Photons ($h\nu$) and CO_2 are absorbed by a plant undergoing photosynthesis, and glucose and oxygen are produced.

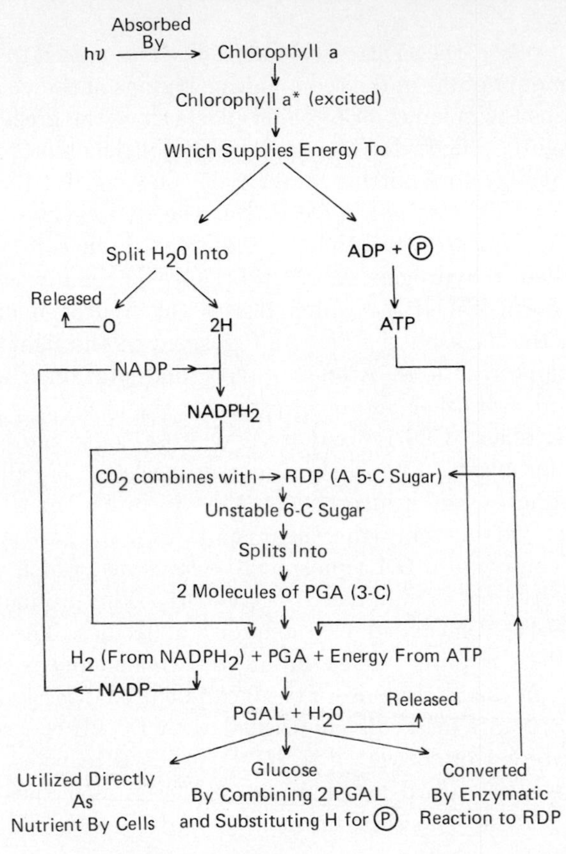

FIGURE 2-16. Abbreviated flow chart of the chemical process in photosynthesis.

zymes. A schematic representation of this abbreviated description is shown in Figure 2-16.

Respiration

The process by which the cell appropriates the energy received from the sun into useful form (ATP) is called *respiration*. The term respiration refers to the integrated series of reactions by which the cell obtains energy from particular nutrients. *Digestion* is a metabolic process that, like respiration, involves degradation of large molecules to smaller ones. In digestion, though, most degradation occurs by enzymatic breaking of bonds by hydrolysis, resulting in energy dissipated as heat. The salient feature of respiration is that an appreciable portion of the energy it releases is stored in ATP to be used for the many activities of the cell.

One of the common cellular nutrients is glucose, the product of photosynthesis. Glucose may be metabolized by several pathways, but one is a well-established route that is apparently the most widely utilized throughout the animal and plant worlds. This pathway consists of an *anaerobic* phase and an *aerobic* phase. The former is independent of oxygen, while the latter requires molecular oxygen. The anaerobic phase always precedes the aerobic phase. The combination of both phases results in the degradation of the carbohydrate glucose into CO_2 and water, with the released energy stored in ATP. A very abbreviated version of this metabolic pathway is presented here.

ANAEROBIC RESPIRATION

The anaerobic respiratory pathway is called *glycolysis* when it takes place in animals and the higher plants and *fermentation* when it occurs in microorganisms. The essential features of this pathway may be summarized as follows:

The six-carbon glucose molecule is split into two three-carbon groups that form *pyruvic acid,* which is converted to *lactic acid* and ethyl alcohol. During this process, a net gain of two ATP molecules is produced.

Lactic acid is produced in large quantities when a human exerts himself very strenuously and suddenly. About 4 gm per second are produced (about 40 gm in a 100-yard dash). It diffuses into the bloodstream, and the increase of lactic acid in the bloodstream can cause severe pain. An *oxygen debt* is established during heavy exercise, because the body cannot acquire oxygen at a sufficient rate to supply its needs; it therefore acquires energy via the anaerobic cycle. After labored breathing, the incoming oxygen acts as a hydrogen acceptor and converts the lactic acid to pyruvic acid, which can be resynthesized into glucose.

AEROBIC RESPIRATION

The primary metabolic pathway is aerobic respiration, a very complex but efficient scheme for converting energy stored in glucose into ATP with the minimum loss of energy in the form of heat. The major steps, shown in Figure 2-17, may be summarized as follows:

1. Glucose, after acquiring a phosphate group from ATP, splits into two three-carbon compounds that form pyruvic acid. This is the same as glycolysis up to this point.
2. A coenzyme, called *coenzyme A,* is formed from pyruvic acid and is converted into a two-carbon compound called *acetyl-CoA.* CO_2 is given off during this process.
3. Acetyl-CoA enters a portion of the respiration process known as the *citric acid cycle,* sometimes called the *Krebs cycle.* The two-carbon acetyl-

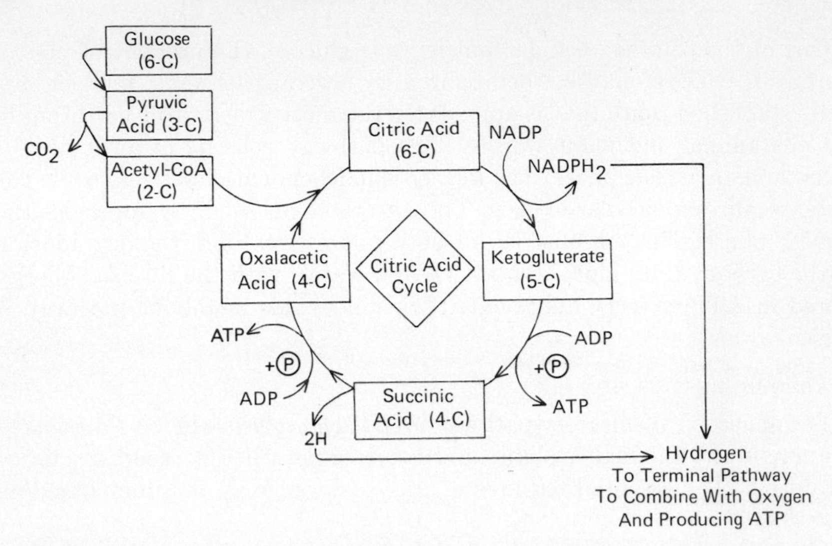

FIGURE 2-17. The aerobic respiration cycle, including the citric acid cycle.

CoA joins a four-carbon *oxalacetic acid* to form citric acid (6-C). CO_2 and H_2O are released, and ADP is phosphorylized to ATP a number of times throughout the cycle as the citric acid is degraded to a five-carbon (5-C) and then a four-carbon (4-C) compound, completing the cycle when oxalacetic acid is formed.

4. The *terminal respiratory pathway* is the final group of processes of aerobic respiration, which consists of enzymatically controlled stepwise transfer to O_2 of the hydrogens produced in the previous stages. The union of H and O_2 to form water is an energy liberating process. While this union is being effected step-by-step, several ATP molecules are formed.

When the two systems—aerobic and anaerobic—are compared, it is found that a net yield of 38 ATP molecules per mole of glucose are formed in the aerobic process compared to 2 in the anaerobic process. When a mole of glucose is completely burned in a calorimeter, it produces 686 kcal. Each conversion of ATP to ADP yields about 7 kcal, so that the energy stored in 38 ATP molecules is about 266 kcal. The aerobic respiration pathway conserves $266/686 = 39\%$ of the energy of glucose for useful purposes; the remainder is dissipated as heat.

References and Suggested Reading

Arnon, D. I. (1960). The role of light in photosynthesis. *Sci. Amer.* **203,** 104 (Offprint 75).

Bassham, J. A. (1962). The path of carbon in photosynthesis. *Sci. Amer.* **206,** 89 (Offprint 122).

Bresler, S. E. (1970). "Introduction to Molecular Biology." Academic Press, New York.

Calvin, M., and Bassham, J. A. (1962). "The Photosynthesis of Carbon Compounds." Benjamin, New York.

Crick, H. F. C., and Watson, J. D. (1954). The complementary structure of deoxyribonucleic acid. *Proc. Roy. Soc., Ser. A* **223,** 80.

Dawkins, M. J. R., and Hull, D. (1965). The production of heat by fat. *Sci. Amer.* **213,** 62 (Offprint 1018).

Pauling, L., Corey, R. B., and Branson, H. R. (1951). The structure of proteins: Two hydrogen-bonded helical configurations of the polypeptide chain. *Proc. Nat. Acad. Sci. U.S.* **37,** 205.

Rabinowitch, E. I. (1948). Photosynthesis. *Sci. Amer.* **179,** 24 (Offprint 34).

Racker, E. (1965). "Mechanisms in Bioenergetics." Academic Press, New York.

Roberts, J. D., and Caserio, M. C. (1967). "Modern Organic Chemistry." Benjamin, New York.

Stumpf, P. K. (1953). ATP. *Sci. Amer.* **180,** 85 (Offprint 41).

Sumner, J. B., and Myrbäck, M., eds. (1951). "The Enzymes," Vol. 1, Part 2. Academic Press, New York.

Watson, J. D., and Crick, H. F. C. (1953). *Nature (London)* **171,** 737.

Vol'kenshtein, M. V. (1970). "Molecules and Life." Plenum, New York.

CHAPTER III Genetics

The science of the transmission and distribution of hereditary traits from parent organisms to their progeny is called *genetics*. Recently the field of genetics has been broadened to include the study of the hereditary determinants (now known to be DNA) and the molecular basis for their duplication, variation, and coding of information. The foundations of classical genetics are due to the monumental works of Gregor Mendel

(1822–1884), and much of the terminology of genetics is based on the Mendelian view of inheritance. It is therefore desirable that the student of modern genetics understand the principles of Mendelian genetics. In this chapter, both classical and molecular genetics will be discussed.

Mendelian Genetics

The description of genetic experiments involves a number of terms that might well be defined in advance. The concept of a *gene* originated with Mendel, who referred to an unknown "factor" that controls a particular inherited trait. The idea that a number of genes, like "beads on a string," govern all inherited characteristics has been replaced by a biochemical formulation, but the gene concept is still useful, even in the Mendelian form.

A particular gene is said to occur at a particular *locus* (position) on a chromosome. A cell that contains a set of chromosomes from the male parent and a set from the female parent is said to be a *diploid* cell. Sperm cells or ova contain only one set and are said to be *haploid*. The number of chromosomes in a set is constant for a species (humans have 23 in a haploid cell, 46 in a diploid cell). In diploid cells, each chromosome of a set has a *homologous* chromosome in the other set, i.e., one in which the genes for the same characteristic are found at equivalent loci. The genes found at the equivalent locus on each of two homologous chromosomes are called *alleles*.

Certain *distinctive characteristics* of an organism may be recognized, e.g., Mendel selected height, seed color, etc., in plants that were either tall or short and whose seeds were either green or yellow, with no intermediate sizes or colors to confuse the classification. *Crosses*, or matings, are made among organisms that have been *bred true* with respect to a particular characteristic (i.e., an individual organism shares with its parent a particular characteristic that has shown no deviation through several generations). The *parental generation* includes the organisms of an initial cross in a genetic experiment. The first generation produced by the parents is called the *first filial*, or F_1 generation, and the second generation is called the *second filial*, or F_2 generation, and so on.

In Mendel's experiment, it was found that when plants that were bred true for tallness were crossed with those bred true for shortness, the F_1 generation was entirely made up of tall plants. It was assumed that one characteristic was *dominant* over the other, which was called *recessive*. Mendel assigned the symbol T for the dominant tall characteristic (this system is still used). A lower case letter may be thought of as a negation of the characteristic for which the letter is the initial. In terms of the genes that control the characteristic of tallness, *T* and *t* are alleles. T is *expressed* in the F_1 generation; t is not expressed. The *phenotype* of an organism is its expressed appearance (T in the case of the F_1 generation of this example), while the *genotype* is the genetic composition (TT in the parents bred true for tallness, tt in the parents bred true for shortness, and Tt in the F_1 generation). For a recessive trait to be expressed, both alleles must be recessive; otherwise the dominant trait will be expressed.

A *homozygote* is an organism with two genes of the same kind, one on each of its homologous chromosomes. Thus, TT and tt are both homozygotes and the organisms to which they refer are said to be homozygous with respect to tallness. A *heterozygote* (an organism whose genotype is Tt is *heterozygous* with respect to tallness) has different alleles on each of its homologous chromosomes. Mendel called heterozygotes "hybrids," a word retained today in various kinds of genetic experiments. A few kinds of genetic experiments will be considered in this chapter. But first the process by which the genetic composition of sex cells is acquired will be considered.

Meiosis

Sex cells of all species are produced by a process of cell division known as *meiosis*. The process differs from mitosis in that two serial cell divisions occur in meiosis, resulting in cells that have half the normal number of chromosomes. A cell destined to become a *germ cell*, or *gamete* (either an ovum or a spermatozoon), has at first a *diploid* number of chromosomes (double set of chromosomes), but the gametes themselves have a *haploid* number (a single set). Thus, in humans all the cells, except ova and spermatozoa, are diploid with 46 chromosomes; the sex cells are haploid with 23. When an ovum is fertilized by a sperm cell, the zygote is again diploid. Without the process of meiosis, the chromosome number would increase from one generation to the next.

Meiosis consists of two successive divisions of a diploid cell. The two divisions are usually designated as meiosis I and meiosis II. A schematic illustration of some of the important aspects of meiosis is shown in Figure 3-1.

Before the first meiotic division takes place, the DNA in the chromosomes duplicates itself by the unzippering and replicating process described in Chapter II. The long threadlike strands of DNA are then double. The homologous chromosomes—one set of which came from the female parent and one from the male parent—then begin to form a very precisely aligned pair. The pairing process, called *synapsis*, aligns the chromosomes end by end and "zippers" the pair of bivalents together all the way along its length. After synapsing, the chromosomes coil, shortening and thickening

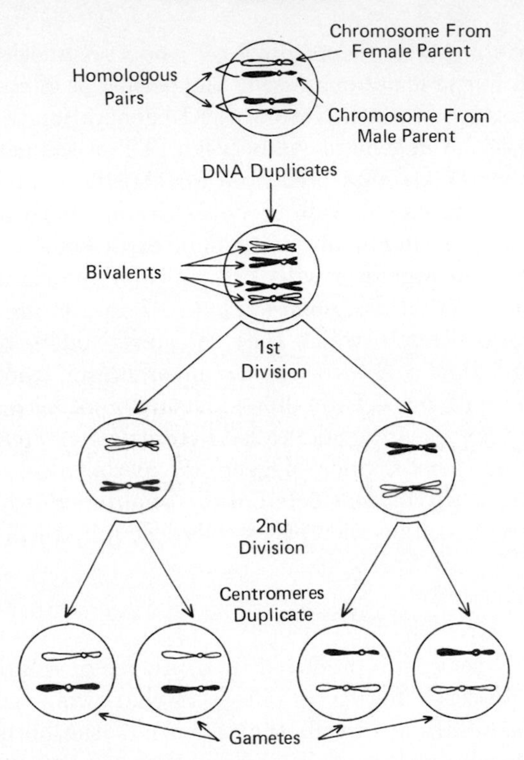

FIGURE 3-1. A schematic representation of meiosis, the process by which haploid gametes are produced.

themselves. It can then be observed that four "arms" of each doubled chromosome extend from a single centromere. Each pair of arms is called a *chromatid*, and each unit composed of one centromere and the doubled DNA is called a *bivalent*. During meiosis I, it is often said that "the DNA duplicates, but the centromeres don't."

The first division occurs at this stage. During the division, the bivalents are divided equally between the two resulting cells. Each cell gets a bivalent from each homologous pair of chromosomes, but the number of paternal and maternal bivalents that go to each of the cells cannot be predicted.

In meiosis II the centromeres of each bivalent divide, and each pair of chromatids becomes a complete chromosome with its own centromere. When division occurs in meiosis II, each daughter cell receives one chromosome from each pair of chromosomes. But since the DNA did not duplicate during meiosis II, the four resulting gametes now contain a haploid number of chromosomes.

Besides converting a diploid cell into four haploid cells, the meiotic process provides for extensive variation among gametes of a single organism. Since the distribution of bivalents in the first division with respect to parental origin of the chromosomes is random, the permutations among the numbers of paternal and maternal chromosomes in each gamete becomes very large, especially in species that have a large chromosome number. This variation is an important feature of meiosis with respect to heredity.

Another phenomenon that occurs during meiosis adds further variability to the genetic composition of gametes. When two homologous chromosomes are paired during synapsis, all four chromatids lie in the same plane. This means that only one chromatid of a homologue (one member of a pair of homologous chromosomes) has a pairing partner with the other homologue, and these are the inner chromatids. While paired in this manner, genetic material may be exchanged between the paired chromatids. This exchange is called *crossing-over,* and the place at which it occurs is called a *chiasma* (plural, chiasmata). The locations of the chiasmata are unpredictable, though usually they appear relatively far from the centromere. Figure 3-2 illustrates how a crossover changes the genetic content of the chromatids. The effect is to incorporate some genetic material from the

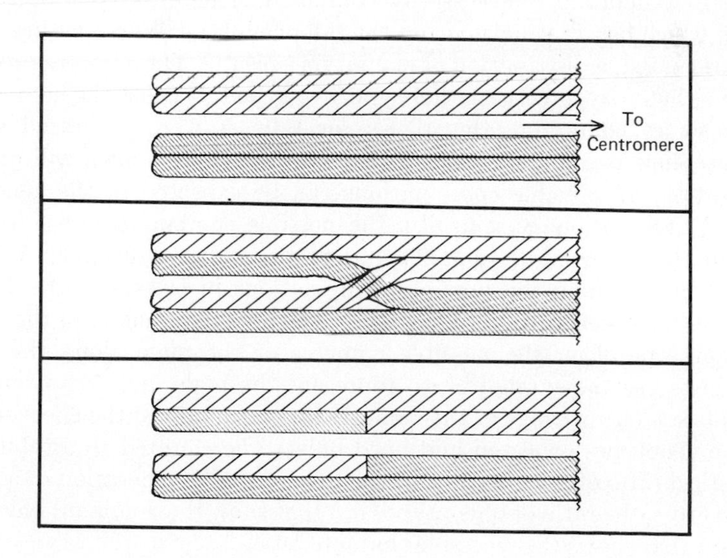

FIGURE 3-2. Crossing-over. Genetic material on chromatids of paired chromosomes is exchanged so that the genetic content of the resulting chromosomes is altered.

male parent into the chromosome that is derived predominantly from the female parent and vice versa. It is the random nature of the positions of chiasmata that contributes to the genetic variability among gametes.

Crosses

The fundamental experiments of genetics are crosses. When distinctive characteristics are observable in an organism, relatively simple breeding experiments can yield considerable information about the function and influence of genes. Some of the common experiments are discussed in this section.

THE MONOHYBRID CROSS

When two organisms each of which is homozygous for a particular trait are mated, the mating is called a monohybrid cross. Of course, the cross is nontrivial only if one organism has the dominant gene and the other the recessive gene. In the example mentioned earlier, the mating of the pure-bred tall plants with the TT genotype and the pure-bred short plants with the tt genotype will serve as an example of the monohybrid cross. The results of the cross through the F_2 generation are seen in the diagrams of Figure 3-3.

The F_1 generation is necessarily composed of heterozygous organisms, each of which has Tt genotype and the tall phenotype. When two organisms of F_1 are mated, a distribution of phenotypes results. The gametes produced by the F_1 heterozygotes are either type T or type t, because the homologous chromosomes containing the alleles for tallness were separated during meiosis. Since it is impossible to predict which two gametes will unite in fertilization, all possible combinations can be expected in the genotypes of the F_2 generation. Specifically, the possible combinations are TT, Tt, tT, and tt. The combinations Tt and tT are indistinguishable. A simple method of recording the possible combinations in crosses is the *Punnett square*, or "checkerboard" of Figure 3-3. The possible gametes of the female are along one edge; the possible gametes of the male along the other. (Biologists use the symbol ♂ to represent the male and ♀ to represent the female.) The pairings of the gametes in the squares of the checkerboard are the genotypes expected and need only to be counted to establish the theoretical ratio of phenotypes. In the case of the F_2 generation of a monohybrid cross, the ratio of phenotypes is 3 that show the dominant characteristic to 1 that shows the recessive characteristic.

It was from such an experiment as this monohybrid cross, in which he found an actual ratio of 2.84 : 1, that Mendel proposed what is now known as the *law of segregation: In the formation of gametes, the two genes of*

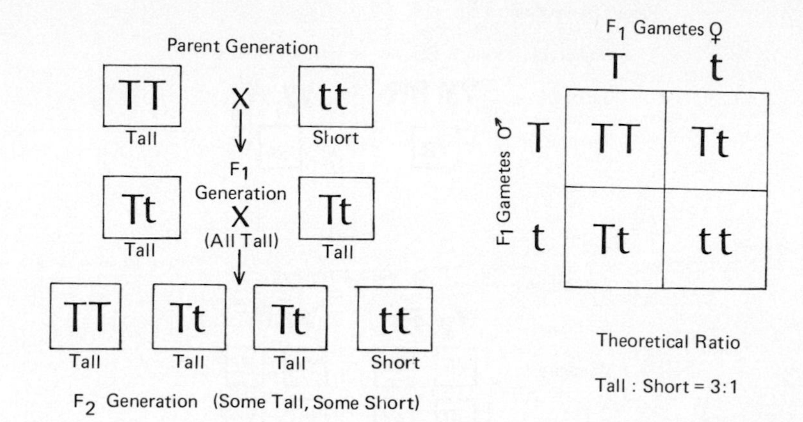

FIGURE 3-3. A monohybrid cross. Two homozygous parents, one dominant (capital letters), the other recessive for the same characteristic (lower case letters), are crossed. The F_1 generation is completely heterozygous and all individuals in F_1 express the dominant characteristic. The F_2 generation is the result of crossing two individuals of the F_1 generation. The Punnett square at the right includes all the possible assortments of the male and female gametes from the F_1 generation; the square therefore permits a direct count of the F_2 phenotypes.

each pair are segregated (or separated) *from one another with each gamete receiving only one gene of a given pair.* Mendel's law, deduced from experimental observation, is consistent with what is now known from observation of the chromosomes in meiosis.

THE DIHYBRID CROSS

Mendel studied the simultaneous inheritance of two distinctive traits in his plants. A mating of an organism with two pure-bred dominant characteristics and another organism with two pure-bred recessive characteristics is called a *dihybrid cross.* Mendel used two traits of seeds—color and shape. One parent plant had been bred true for the dominant yellow seeds (YY) and round seeds (RR); the other had been bred true for green (yy) and wrinkled seeds (rr). Figure 3-4 shows a representation of the results of a dihybrid cross. The F_1 generation are all yellow and round, but the theoretical ratio of phenotypes in the F_2 generation is 9 : 3 : 3 : 1, i.e., 9 double dominants, 3 with one dominant and one recessive, 3 with the other dominant and the other recessive, and 1 double recessive. The checkerboard can be applied to more complex crosses, such as a trihybrid cross.

From the results of dihybrid cross experiments, Mendel postulated what has become known as the *law of independent assortment: each pair of*

Parent Generation:

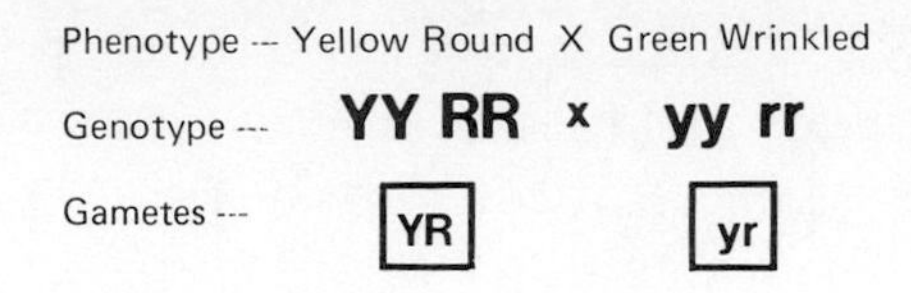

F_1 Generation:

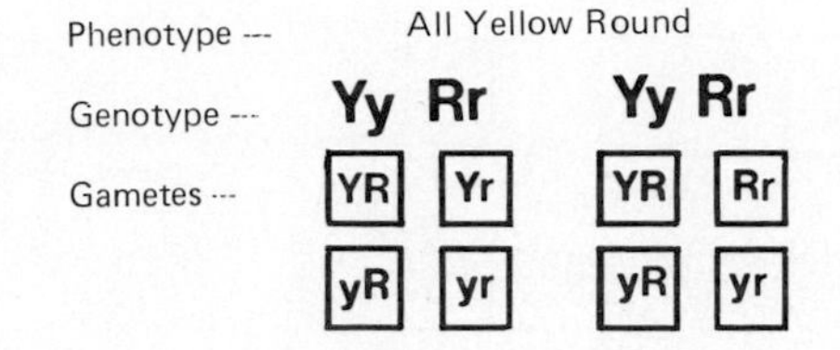

F_2 Generation:

Genotypes

	YR	Yr	yR	yr
YR	YYRR	YYRr	Yy RR	Yy Rr
Yr	YYRr	YYrr	Yy Rr	Yy rr
yR	Yy RR	Yy Rr	yy RR	yy Rr
yr	Yy Rr	Yy rr	yy Rr	yy rr

Phenotypes — 9 Yellow Round: 3 Green Round:
3 Yellow Wrinkled: 1 Green Wrinkled

FIGURE 3-4. A dihybrid cross. A cross between two parents, one of which is homozygous for two dominant traits and the other homozygous, but recessive, for the same two traits, is illustrated.

genes is assorted at random and is inherited independently of the other pairs. This law is valid only if each of the pair of genes whose traits are observed happen to be on different chromosomes. Mendel was indeed fortunate that all of the dihybrid crosses he chose to observe happened to be controlled by genes located on different chromosomes.

THE TEST CROSS

Often it is desirable to determine whether an organism exhibiting a dominant characteristic is a homozygote or a heterozygote. A *test cross* (or *backcross*) is a simple genetic technique that makes this distinction. In this experiment, an individual of unknown genotype for a particular trait is crossed with an individual that is homozygous recessive for the same trait. If the organism of unknown genotype is homozygous dominant, all of the offspring will be of the dominant phenotype; if the unknown is heterozygous for the dominant characteristic, half the offspring will be dominant and half will be recessive in the characteristic.

The test cross can be used to analyze two or more pairs of genes, but such crosses are more difficult to perform in practice. Such crosses require pure-bred lines of homozygous recessive individuals in two or more traits.

Incomplete Dominance

So far it has appeared that dominant genes are completely effective in obscuring the recessive traits. This complete dominance of one allele over another is rarely, if ever, the case. If examined minutely, heterozygotes invariably show some evidence of recessive traits. Usually, the effects of both alleles are relatively obvious. The presence of characteristics of both alleles in a heterozygote is called *incomplete dominance.* The distribution of gametes is still consistent with Mendel's findings, and Mendel's laws are still valid, except that the assumptions of absolute dominance and recessiveness no longer apply.

The traditional example of incomplete dominance is the cross between pure-bred four o'clock plants, one with red flowers and the other with white flowers. The F_1 generation is composed entirely of pink flowers. The two alleles for flower color are equally effective. A cross between two plants of the F_1 generation produces red, pink, and white flowers in the ratio of 1 : 2 : 1. These results are consistent with the law of segregation, because although the effects of the genes blend, the genes themselves do not blend but remain discrete and segregate according to Mendel's law.

Sex Linkage

In humans and most other animals, the sex of an organism is determined by a pair of *sex chromosomes,* so called to contrast with the remainder of the chromosome complement known as *autosomes.* The sex chromosomes are recognizably different and are called X and Y chromosomes. A normal ovum always carries the X chromosome as part of its haploid set. Sperm cells may carry either an X or Y in its haploid set. Fertilization of an ovum by sperm carrying the X results in a female (XX); when fertilized by sperm bearing the Y, the zygote results in a male (XY).

A range of "maleness" or "femaleness" is possible in animals (including humans). Chromosomal analyses of humans with unusual sex characteristics have revealed females with an XO combination, i.e., lacking a male chromosome. Certain males who have external male genitalia, but evidence of female secondary sex characteristics, have been found to have XXY chromosome composition. Some XYY males have been recognized recently.

They have been termed "supermales" because of their particularly aggressive nature, exceptional musculature, and size. Unfortunately, these characteristics are simultaneously associated with mental deficiencies; the genetic abnormality apparently affects the central nervous system as well as the sex characteristics.

Males and females can now be distinguished without a chromosomal analysis because of the presence of a stainable body found on the edge of the nuclear membrane of a cell called a *Barr body*. The cells of a male do not show Barr bodies, but the cells of normal females exhibit one. These bodies can be exhibited, if present, by taking scrapings of the mouth tissue and staining them to reveal nuclear structures. The sex of unborn fetuses can be determined by examining cells from the amniotic fluid (the fluid surrounding a fetus) for the presence of Barr bodies.

Genes carried on sex chromosomes control traits that are said to be *sex-linked.* Because of the longer length of the X chromosome, many more genes are found on it than on the Y chromosome. (This unequal length of homologous chromosomes is peculiar to sex chromosomes.) As a result, the genes on an X chromosome are often expressed regardless of their dominant or recessive nature because they are not matched by alleles on the Y chromosome. This is the reason that recessive anomalies often appear in males and not in females—because the female, with two X chromosomes, usually has one normal allele; the male has no back-up normal allele on his Y chromosome.

Two abnormal genetic expressions in humans that are sex-linked are *red-green color blindness* and *hemophilia* (a blood disease in which clotting time is so slowed that even a minor wound may be fatal). The genes of both abnormalities are recessive. Males exhibit these abnormalities much more frequently than females. The females who have the defective X chromosome (but do not usually express it because of the presence of the additional normal X) are called "carriers" because their male descendants can aquire the defective X chromosome but have no counteracting allele on their Y chromosome.

Sex linkage has been discussed in terms of animals rather than plants because plants are often *hermaphrodites* (having both male and female sex organs in the same organism), which complicates genetic experiments concerning sex linkage in plants.

Multiple Alleles

The term multiple alleles applies to more than two alternative alleles that may be located at the same locus on a pair of homologous chromosomes in a population. This distinguishes multiple alleles from *multiple genes*, which is a term that applies to more than one gene affecting a particular trait in which the genes for that trait are found at different loci and have cumulative effects. An interesting and practical example of multiple alleles is the typing of human blood.

Blood cannot be transfused indiscriminately among humans because sometimes, when blood from one person is incompatible with that of another, the blood will clump together, clog capillaries, and cause rapid death. It is now known that there are four different phenotypes of blood, called A, B, AB, and O. The phenotypes are controlled by three alleles, any two of which may be found at the same locus on a pair of homologous chromosomes. The different members of a series of alleles like this are designated by a single base symbol with superscripts. The three alleles controlling blood type are I^A, I^B, and I^O. Of these, I^A and I^B are both dominant to I^O, but they are codominant with each other.

A very elementary understanding of the *antigen–antibody reaction* is necessary to understand blood type reactions. *Antigens* are usually foreign disease organisms that enter the body, causing a response in the body called *antibody formation*. The antibodies are specific for the antigen that cause their formation, and the antibodies are effective in disabling the antigens. Some antigens, however, are produced naturally within the body; this is the case with blood types. I^A is responsible for producing antigen A; I^B produces antigen B. I^O produces no antigen. Of course, a person with a given natural antigen does not possess antibodies that react with that antigen but can produce antibodies that react with the other antigens. Thus, a person with antigen A can produce antibody B, and one with antigen B can produce antibody A. Type O blood has neither antigen and can produce both antibodies. Type AB blood has neither antibodies.

The genotypes associated with blood types A, B, AB, and O are I^AI^A or I^AI^O, I^BI^B or I^BI^O, I^AI^B, and I^OI^O, respectively. Since type AB blood contains no antibodies and cannot react with any of the antigens, a person with blood type AB is known as a *universal recipient* and can be transfused with any blood type. Type O can be used to transfuse persons with any type blood, and these people are called *universal donors*. The presence of antibodies in the type O blood, which is transfused into persons of another type, is harmless because they are diluted in the recipient's circulation and have negligible effect.

A person's blood type is obviously dependent on the genotypes of his mother and father. A type A child cannot have parents who are type B and type O, for example. Because some genetic crosses exclude certain types, blood typing can sometimes be used to prove innocent (but never to convict) defendants in paternity suits.

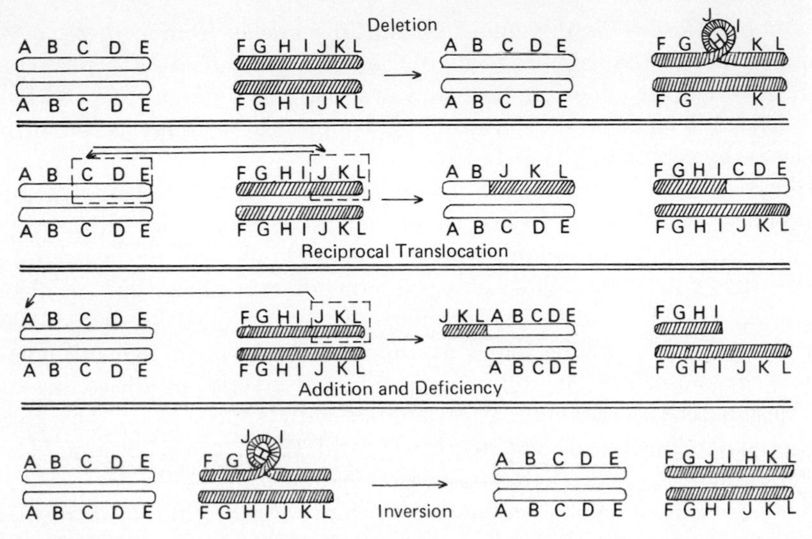

FIGURE 3-5. Four types of chromosomal variation. The letters represent identifiable regions on chromatids. In each case, identical homologous pairs of chromatids are made different by one type of alteration.

Further Chromosomal Variations

The sources of genetic variation have been stressed throughout the preceding sections of this chapter. Further means by which chromosomal variation occurs will be discussed briefly. Each of these is illustrated in Figure 3-5.

CHROMOSOMAL REARRANGEMENT

There are a number of ways that genes can be arranged on chromosomes. Genes can be removed from a chromosome completely by double breaks in the chromosome. This defect is called *deletion* and can be observed during synapsis when one of the homologous chromosomes forms a loop at the location where its corresponding allele is missing.

Reciprocal translocation can occur between two pairs of homologous chromosomes. A section of one member of a pair is exchanged with a section from a member of another pair.

A section of one chromatid may break off and become attached to the end of the chromatid of another chromosome. This type of rearrangement is referred to as *addition and deficiency.*

A chromosome may form a loop on one of its chromatids, twist, break, and rejoin so that the sequence of genes on the looped section becomes inverted. This abnormality is called *inversion.*

POINT MUTATIONS

Changes in the chemical structure of a gene that lead to alteration of the physiology or morphology of an organism are called *point mutations.* A chemical change in a gene produces new characteristics. What is more, the gene is then as stable as its original form.

Most mutations are harmful, because the genetic content of every organism has been selected over the course of time for survival characteristics. It must be assumed that the organism has acquired a favorable gene pool, so that most (by far) mutations are likely to be detrimental. Favorable mutations can occur. It is by this process that evolution of species is presumed to take place.

Mutations may be caused by a number of means, all of which probably have not been identified. *Thermal agitation* of molecules may cause a particularly energetic molecule to strike the DNA and produce a chemical change. *Mutagenic chemicals,* if they penetrate the gonads, can cause mutations. Nitrogen mustard and formaldehyde are two among a number of well-known mutagenic chemicals.

Ionizing radiation is a well-established mutagenic agent. Gamma rays and X-rays produce mutations in proportion to the dose, and the effects of ionizing radiations are additive. Several small doses produce almost the same number of mutations as an equal amount of radiation delivered in a single dose. High-energy electromagnetic waves (such as X- and gamma rays) can penetrate an organism and deliver considerable energy to the DNA molecule in germ cells and cause chemical change. It should be noted that mutations can occur in any cell and cause damage in varying degrees to that cell, but germ cell damage is transmitted to all cells of the offspring and obviously has a more pervasive effect on the organism. Mutagenic effects of ionizing radiation will be considered further in Chapter XII.

Molecular Genetics

It has been stressed repeatedly up to this point that DNA is a key molecule in living systems. It has been pointed out that DNA is the molecule that directs the chemical activities of cells, that DNA controls genetic inheritance, that changes in DNA can cause radical variations in the organism, and that DNA is capable of replicating itself. How does this molecule so completely dominate the course of life? The answer is unequivocal—enzymes!

Almost every chemical reaction that takes place in living things is controlled by enzymes. The presence or absence of enzymes governs both the appearance and function of every detail of living systems. And enzymes are produced by DNA—though not directly. DNA is not the direct template that orders the sequence of amino acids. There is an intermediate template, namely, RNA. The relationship between DNA, RNA, and the protein product is called the *central dogma*, which is stated by the formula

$$\text{DNA} \xrightarrow{\text{transcription}} \text{RNA} \xrightarrow{\text{translation}} \text{protein}$$

in which the arrows indicate the direction of transfer of genetic information. Further, the arrows are unidirectional, emphasizing that the sequence of nucleotides on RNA are never determined by protein as a template and those on DNA are never determined using RNA as a template.

The concepts of the central dogma have been corroborated by experiment and have been extended by the work of Beadle and Tatum (1941) to include the "*one gene–one enzyme–one reaction*" hypothesis. The hypothesis is almost self-explanatory. Each enzyme, which controls a specific kind of reaction, is controled by the DNA that comprises a single gene. Considerable genetic evidence has been compiled that substantiates this hypothesis.

Synthesis of RNA

The fact that both DNA and RNA are very similar molecules, having almost the same bases and always composed of four nucleotides, suggests information carried by DNA is transferred to RNA through the sequence of bases. Since this is the case, DNA must separate the two strands of its double helix in order to act as a template for the RNA. Then one or another (or both) DNA strands serves as the template by base pairing—guanine to cytosine and adenine to uracil—with the ribose nucleotides forming a complementary RNA strand. But if the DNA strands are separated, what controls whether or not each strand makes a complementary DNA strand (replication) or a complementary RNA strand (transcription)? This ambiguity was resolved by the discovery of an enzyme that links together ribonucleotides by joining their sugar-phosphate backbone. This enzyme is called *RNA polymerase*, and it can be detected in virtually all cells. Furthermore, it was found that RNA polymerase binds itself to the DNA and joins ribonucleotides only in the presence of DNA, facts that strongly suggest that DNA must align the RNA units (by base pairing) before the RNA polymerase can join them together. This was proved to be the way that RNA is transcribed from DNA when it was found that the RNA (A + U)/(G + C) ratio is about the same as the DNA (A + T)/(G + C) ratio in every enzymatic synthesis of RNA. The equality of these ratios demands that G is bound to C on both the RNA strand and the DNA strand, that the A on the DNA is bound to the U on the RNA, and that the T on DNA is bound to A on RNA. Only for such an arrangement would the relation

$$\left(\frac{A + U}{G + C}\right)_{RNA} = \left(\frac{A + T}{G + C}\right)_{DNA}$$

hold true for every reaction in which DNA serves as the template for making RNA.

Only one of the two DNA strands is used as a template for transcription of RNA. It is not clear how the particular strand that is to be copied is chosen, but a number of experiments indicate that only one is copied. The fact that only one DNA strand is copied explains why the base ratios of RNA are not complementary. A does not equal T, nor does G equal C in a given strand of DNA, and since RNA is copied from only one strand, there is no reason that A should equal U or G equal C. Analysis of RNA confirms that, in general, no such base relations exist.

The transcription of RNA from DNA is schematically shown in Figure 3-6. The RNA polymerase opens a section of the DNA double helix, moves along the appropriate strand, and peels off the growing RNA strand. After the enzyme passes a section of the DNA, the hydrogen bonding reforms the two complementary DNA strands.

RNA polymerase is a very complex enzyme, composed of six separate polypeptide chains held together by secondary (noncovalent) bonds. Five of these chains comprise the *core enzyme*, so called because it can catalyze the joining of ribonucleotides with or without the sixth strand, which is called the *sigma factor*, or simply σ. Therefore, σ does not function as a

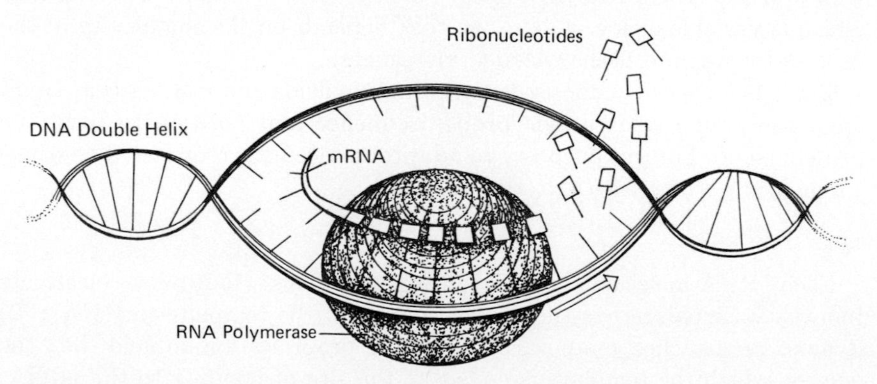

FIGURE 3-6. The transcription of mRNA from DNA. The double helix of the DNA is opened, the ribonucleotides are positioned by obligatory pairing with the bases of one strand of the DNA, the ribonucleotides are joined together to form a single strand of RNA, and the two strands of DNA are rejoined into the original double helix configuration.

catalyst; its function is to recognize "start signals" along the DNA molecule. Start signals are points along the DNA at which the transcription of RNA begins, so that only a definite section of DNA (a gene) will be transcribed by the RNA. Once σ has identified a start signal, transcription begins, and σ is released from the core enzyme and is available to attach to another core enzyme.

The signal to stop transcription along the DNA molecule is the function of a protein (but not associated with RNA polymerase) called the *rho factor*, or ρ. This protein has been identified, and experiments have shown that its absence causes transcription to proceed past the end of a gene, causing transcription of the next genetic region. The details of the stop signals and ρ function have not been determined. There may be more than one ρ and more than one type of stop signal on the DNA.

The Role of RNA in Protein Synthesis

It was pointed out in Chapter II that there are three known types of RNA—mRNA, tRNA, and rRNA. All of these are produced from DNA templates, and they all play roles in protein synthesis—but quite different roles.

MESSENGER RNA

As the name implies, *messenger RNA* (mRNA) conveys the information encoded in the DNA base sequence to the site of protein synthesis. Naturally the mRNA is itself encoded; this is accomplished at the time of its transcription. The mRNA is a linear chain of ribonucleotides, the length of which is variable over a wide range that depends on the complexity of the protein for which it is the encoded messenger.

There is no specific affinity between amino acids and mRNA that could align the amino acids in the proper sequence and position to form the protein chain. For this purpose, an adapter molecule is required. The adaptor molecule is *transfer RNA* (tRNA).

TRANSFER RNA

Each tRNA molecule must serve three functions: (1) it must selectively join (by base pairing) to the proper section of the template (mRNA); (2) it must be attached temporarily to the appropriate amino acid, i.e., the one for which the template is coded at the site of bonding to the tRNA; and (3) it must have an appropriate three-dimensional configuration in order to space properly the adjacent amino acids so that peptide bonds can form. Since there are twenty amino acids, there must be at least twenty different types of tRNA (actually, for reasons that become apparent when the code is described, there are more than twenty types of tRNA).

Although the code itself has not yet been discussed, it is convenient to mention at this point that the *codon* (the unit composed of a number of bases on the mRNA template that specifies a specific amino acid) is a triplet, i.e., it consists of three bases such as ACU. The three bases of tRNA that attach to a codon are called *anticodons*, the complementary base pairs of the codon. For example, for the codon ACU the anticodon is UGA.

The structure of each tRNA molecule consists of a cloverleaf configuration (see Figure 3-7), with three major loops and a "handle" that attaches to the appropriate amino acid. The loop opposite the handle contains three exposed bases that constitute the anticodon. The other two loops are now thought to be involved in the binding of tRNA to enzymes that catalyze the reactions necessary to form the protein chain and perhaps in the binding of tRNA to the surface of ribosomes (more will be said about ribosomes subsequently). In any event, the loops are important in the three-

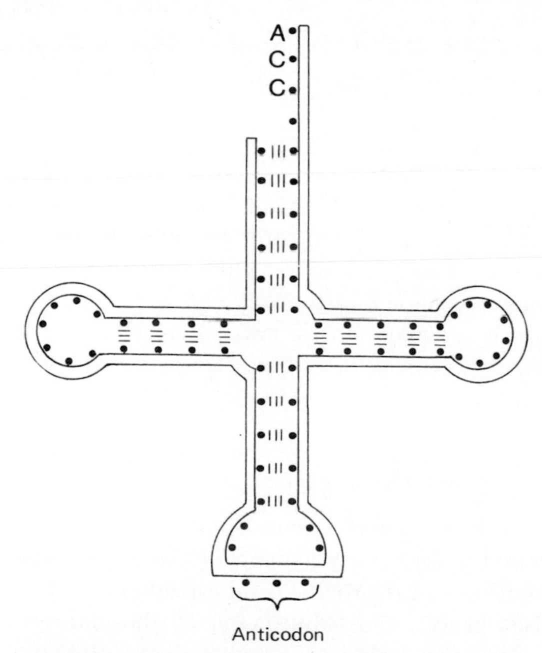

FIGURE 3-7. A schematic representation of the tRNA molecule. The solid dots represent purine or pyrimidine bases, and the hydrogen-bonded base pairs are represented by series of bars. The actual structure of tRNA is a complex three-dimensional shape rather than the planar configuration shown here. The "handle" of every tRNA molecule has the base sequence ACC, as shown here.

dimensional structure of tRNA that aligns the amino acids. The stems are actually double helices in three dimensions but are shown two-dimensionally in Figure 3-7.

The handle of the cloverleaf always terminates in the nucleotide sequence CCA, with adenosine at the end. The carboxyl group of an amino acid attaches to the end of the adenosine (with the aid of an enzyme that is specific for one and only one amino acid). Amino acid molecules attached to the tRNA are said to be *activated.* The enzyme that binds the amino acid to tRNA must be able to recognize (i.e., bind to) both the amino acid side group and the tRNA molecule specific for that amino acid. This enzyme an *aminoacyl synthetase,* therefore, has two active sites, and there must be at least twenty different enzymes of this type—one for each amino acid. It is interesting that it is an enzyme, not tRNA, that recognizes the correct amino acid. Thus, the ultimate accuracy with which protein synthesis can be carried out depends on the accuracy with which the activating enzymes can select the various amino acids.

All tRNA molecules have molecular weights of about 25,000 and consist of about 80 nucleotides. Mixed species of tRNA have been solidified into very regular crystals, which suggests that the shape of the adaptor molecules are all the same.

RIBOSOMAL RNA

Protein synthesis never occurs when the activated amino acids and the mRNA template are free in solution but only at the surface of a ribosome. Ribosomes are composed of roughly half protein and half rRNA. The rRNA carries no genetic information, and its function is still unknown. The ribosome itself is composed of two subunits that come apart after protein synthesis has occurred but reform when the ribosome attaches to a mRNA molecule.

The Protein Synthesis Mechanism

The details of the formation of a protein chain at the surface of a ribosome are relatively complex, and many aspects of the process are current topics of research. An outline of the process is described here.

When synthesis begins, the subunits of a ribosome join to a mRNA template. The activated amino acid (amino acid–tRNA, or AA–tRNA) groups can bind to two special sites on the ribosome's surface. The ribosomal surface sites can accept any AA–tRNA since the ribosomes bind to a nonspecific region of the tRNA (presumably one of the side loops). The specific AA–tRNA that is actually bound is selected by the codon on the mRNA. Figure 3-8 shows that the two bound AA–tRNA groups (labeled 2 and 3) are held in position while their amino acids are enzymatically joined by a peptide bond. At the same time the bond between tRNA 2 and its amino acid is broken. Then tRNA 2 is freed from the ribosome as the mRNA moves to the left, and tRNA 3 moves to the bound site formerly occupied by tRNA 2. In turn, tRNA 4 is selected by the next codon on the mRNA and becomes bound to the site formerly occupied by tRNA 3. Another peptide bond is formed and the process continues in this manner, producing a continuous polypeptide chain.

The energy for the protein synthesis process is supplied by *guanadine triphosphate* (GTP), which is similar to ATP (see Chapter II), except that guanadine replaces adenine. It takes about one minute to make a typical protein.

Several ribosomes may attach to the same mRNA and produce identical proteins simultaneously. This is illustrated in Figure 3-9. When several ribosomes are attached to the same mRNA, they are called *polyribosomes.*

When the protein synthesis process is complete, the ribonucleotides of mRNA are broken apart by the enzyme *ribonuclease* and are reused in the transcription process on DNA molecules to make more mRNA with a different code. The tRNA molecules are recombined with appropriate amino acids and repeat their role of transporting the amino acids to the site of protein synthesis. The energy for recombination is provided by ATP. Ribosomal subunits are separated until they rejoin on a new mRNA molecule to begin the synthesis process anew. Figure 3-10 summarizes the process of protein synthesis.

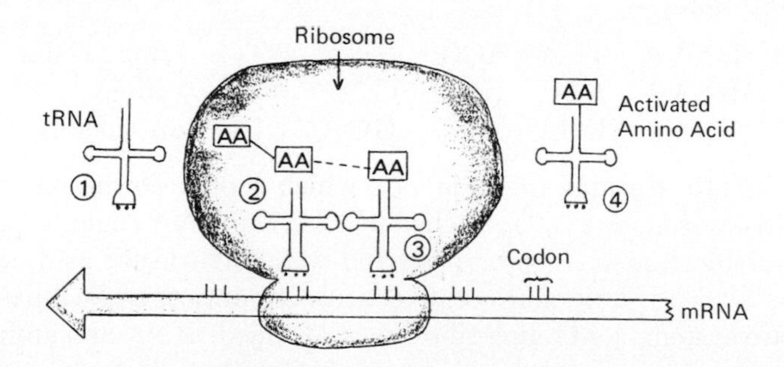

FIGURE 3-8. Protein synthesis at the ribosome. The activated amino acid (tRNA–amino acid group) is positioned by complementary base pairing of the anticodon on the tRNA to the codon on the mRNA. On the ribosome, the amino acids join together by peptide bonds and disjoin from the tRNA. The tRNA is freed to recombine with the appropriate amino acid and participate in further synthesis. As the process proceeds, the chain of amino acids for which the mRNA was coded evolves from the ribosome.

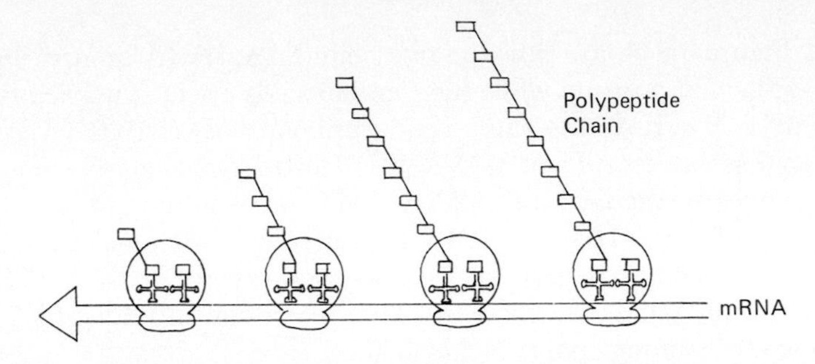

FIGURE 3-9. A polyribosome. Synthesis of polypeptide chains proceeds from several ribosomes which are simultaneously attached to the same mRNA strand.

The Genetic Code

Since DNA contains the genetic information to direct the production of proteins by ordering the sequence of amino acids, it was assumed that DNA was encoded by the nucleotide sequence along the DNA chain. The genetic code is a triplet code (consisting of three-letter words). It permits the specification of $4^3 = 64$ different words. It was thought that perhaps all the possible three-letter words would not be used (a conjecture that turned out to be wrong). The code as it occurs on DNA is called the *triplet code*, the complementary base pair triplet on mRNA corresponding to a triplet code word on DNA is the *codon*, and the triplet label on tRNA that base pairs to mRNA is the *anticodon*. Recall that thymine in DNA is replaced by uracil in RNA; then a series of coded triplets would appear as the following.

DNA	TGA	ACC	GGA	TTG	triplet codes
mRNA	ACU	UGG	CCU	AAC	codons
tRNA	UGA	ACC	GGA	UUG	anticodons

The means for learning the code, i.e., which codons correspond to which amino acids, was made available when synthetic mRNA could be produced. The first artificial mRNA to be prepared was polyuridylic acid, known as "poly(U)," which is a ribonucleotide chain with only the uracil base. In an appropriate system, containing ribosomes, purified tRNA and amino acids, and energy-supplying molecules, it was found that the only amino acid chains produced were of the single amino acid *phenylalanine*. Thus, the code word for phenylalanine was concluded to be UUU. Similar experiments with poly(C,) poly(G), and poly(A) provided other pieces to the code's solution. Synthetic mRNA was prepared with sequences such as AUAUAU..., then with more unusual arrangements, and in every case the product amino acids were analyzed. The results of this kind of detective work

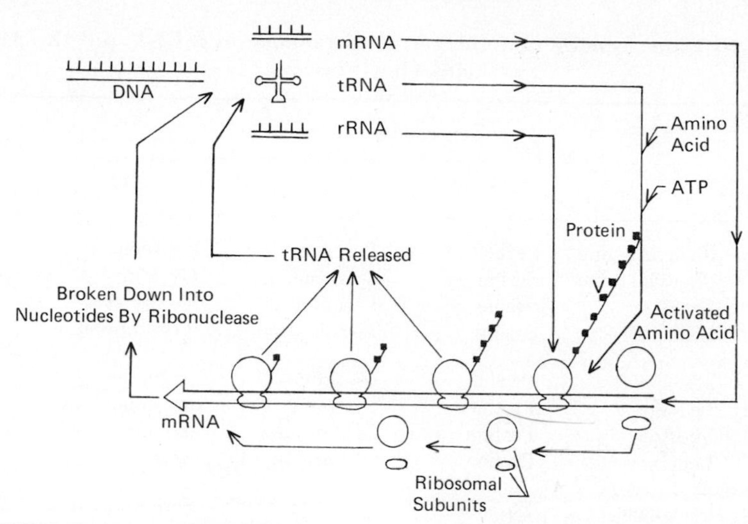

FIGURE 3-10. Summary of the protein synthesis process from DNA to the finished amino acid chain. All three types of RNA are transcripted from DNA. Adapted from Watson (1970). Copyright 1970 by J. D. Watson and W. A. Benjamin, Inc. Reproduced by permission.

provided a number of unambiguous results:

1. The code is *degenerate,* i.e., more than one codon often codes for the same amino acid. This degeneracy is not surprising, since there are 64 codons and only 20 amino acids.
2. The code is read in one direction only along the mRNA. This explains why AUG and GUA are distinct codons.
3. Three of the codons, which were at first thought to be nonsense codons, are chain terminators. These are not read by tRNA anticodons, but by specific proteins called *release factors.* Two release factors have been identified. One is specific for UAG and UAA and the other for UAA and UGA, but it is not known how the release factors recognize two codons.

The complete genetic code in terms of the mRNA codons is shown in Table 3-1.

Regulation of Protein Production

What mechanism controls when certain enzymes are to be synthesized? What triggers the DNA to produce proteins or inhibits it from continuously producing them? These problems are presently at the forefront of molecular genetic research. Much of what is presently known of these problems has been learned only from studies of microorganisms, and a great

Table 3-1 The Genetic Code in Terms of Codons on mRNA and the Amino Acids They Specify

First position	Second position U	C	A	G	Third position
U	Phenylalanine	Serine	Tyrosine	Cysteine	U
	Phenylalanine	Serine	Tyrosine	Cysteine	C
	Leucine	Serine	Terminator	Terminator	A
	Leucine	Serine	Terminator	Tryptophan	G
C	Leucine	Proline	Histidine	Arginine	U
	Leucine	Proline	Histidine	Arginine	C
	Leucine	Proline	Glutamine	Arginine	A
	Leucine	Proline	Glutamine	Arginine	G
A	Isoleucine	Threonine	Asparagine	Serine	U
	Isoleucine	Threonine	Asparagine	Serine	C
	Isoleucine	Threonine	Lysine	Arginine	A
	Methionine	Threonine	Lysine	Arginine	G
G	Valine	Alanine	Aspartic acid	Glycine	U
	Valine	Alanine	Aspartic acid	Glycine	C
	Valine	Alanine	Glutamic acid	Glycine	A
	Valine	Alanine	Glutamic acid	Glycine	G

deal must still be considered speculation. Not surprisingly, a wealth of new terminology accompanies the new area of research, and this section is meant to acquaint the student with the concepts and terminology associated with the area.

A *structural gene*, which is a section of DNA responsible for the production of a particular enzyme, or a group of structural genes are considered to be under the influence of an *operator gene*, or simply an *operator*. The operator may stimulate one or more structural genes into activity under proper conditions, and the operator is situated immediately adjacent (on the same chromosome) to the structural genes under its influence. The operator and its associated structural genes are collectively called an *operon*. The entire operon is controlled by a substance called a repressor, which is a protein whose production is directed by still another section of DNA (on the same chromosome) called the *regulator gene*.

The repressor has two functional states (active or inactive), and it assumes one or the other of these states by combining with either a *corepressor*, which activates the repressor, or an *inducer*, which inactivates the repressor. Corepressors and inducers are external molecules that are present or absent depending on the nutritional or metabolic state of the cell. When the repressor is activated, it can bind to the operator, which in turn prevents the binding of RNA polymerase and thus specifically prohibits the initiation of mRNA synthesis. If the repressor is combined with an inducer, it cannot bind to the operator and so cannot inhibit mRNA synthesis.

The control mechanism described here is basically a negative system, one in which antagonistic factors contend to inhibit or not inhibit function.

The mechanisms of induction and repression are illustrated in Figure 3-11. These are reasons to believe that positive mechanisms also exist. Some operons seem to function when in the presence of proteins that specifically promote their synthesis of mRNA, but this mechanism is not presently understood.

Still another means by which cellular metabolism may be controlled is called *end-product inhibition.* It is thought that inhibitors can bind to completed enzymes at a location other than the active site of the enzyme. The enzyme slightly changes its shape as a result of the binding of the inhibitor, so that its active site no longer accepts its substrate and is temporarily inactive. Proteins that change shape as a result of combination with another molecule are called *allosteric proteins.*

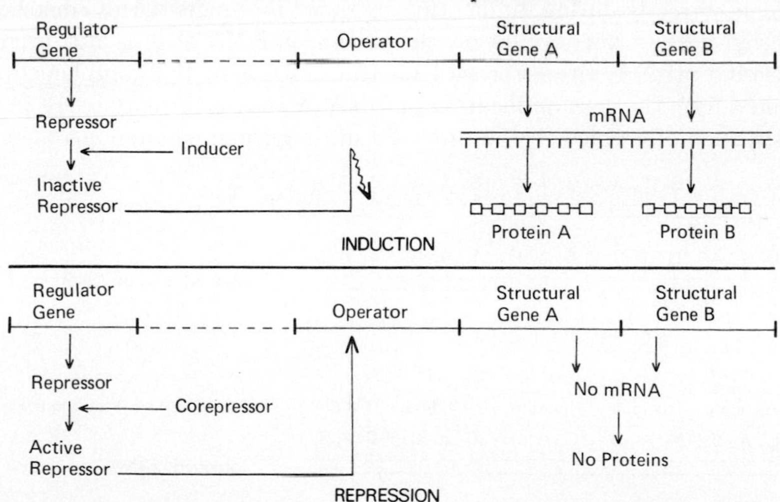

FIGURE 3-11. Induction and repression mechanisms. Induction takes place when the repressor is inactive. Repression occurs when an active repressor combines with the operon and halts the transcription of mRNA from the structural genes (DNA). The repressor is made active or inactive by the presence of a corepressor or inducer, respectively.

The Chemical Basis of Mutation

It is now appropriate to consider the phenomena associated with mutations as they are related to the chemical structure of genes and to their ultimate effects on proteins. A cursory examination of the genetic code in Table 3-1 is sufficient to observe that a change of a single letter of a codon might result in the incorporation of a different amino acid in a protein chain.

How serious is an error of one amino acid out of the hundreds that make up a protein? Consider a protein like hemoglobin, found in red blood cells. The two protein chains in hemoglobin contain a total of 287 amino acids. Normally the amino acid in the number 6 position from the amino terminal end of one chain is glutamic acid. A person whose hemoglobin has valine instead of glutamic acid at this position suffers from sickle cell anemia. The side chain on valine does not carry a charge as does glutamic acid, and consequently the sickle cell hemoglobin is less soluble than the normal variety and tends to form crystals easily. The result is fragility in the red blood cells and death for the bearer of the defective gene.

Sickle cell anemia originates from a mutation of a single nucleotide of a single gene. Mutations may occur in a number of ways, some of which are briefly considered here.

Copy errors result from the rare occasions when the bases on DNA do not pair correctly during duplication. A sometimes pairs with C instead of T. Even if the errors occurs on the strand of DNA that is not used to transcribe mRNA, the error will be transmitted to the complementary strand during the next duplication of DNA. A copy error may result in the substitution of an incorrect amino acid in the protein chain controlled by

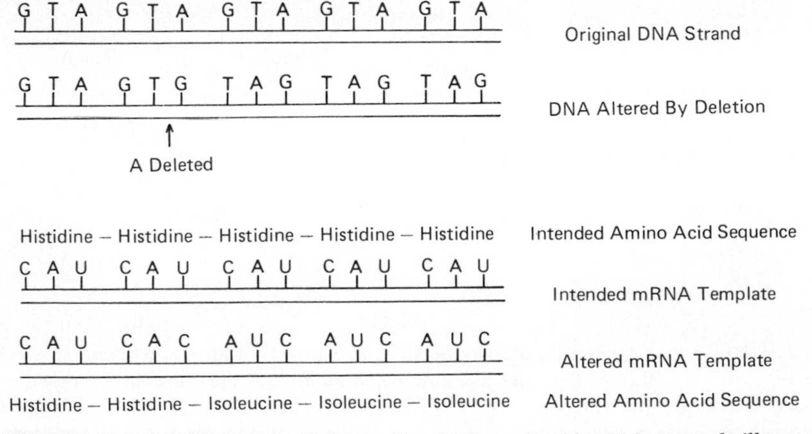

FIGURE 3-12. Mutation by deletion of a single nucleotide. This example illustrates how the deletion of one nucleotide from a DNA molecule can drastically change the amino acid sequence it produces from the sequence intended by the original DNA.

the DNA if the error occurs in a structural gene. It may completely eliminate the production of a protein if the error occurs in a regulator gene or operator.

Transition mutations are similar to copy errors, but transitions refer to the substitution of one purine for another or one pyrimidine for another. For example, if A is substituted in a DNA strand for G, an A-T pair results instead of G-C pair. This change, is, of course, transmitted to the mRNA that is transcripted from the altered DNA and results in an error in the amino acid chain.

Deletion or *addition* of a nucleotide in a DNA strand radically changes the information on the DNA. Figure 3-12 shows, as an example, how the deletion of a single nucleotide from a chain could effect an amino acid chain for which it is coding.

References and Suggested Reading

Asimov, I. (1963). "The Genetic Code." Grossman, New York.

Beadle, G. W., and Tatum, E. L. (1941). *Proc. Nat. Acad. Sci. U.S.* **27,** 499.

Crick, F. H. C. (1954). The structure of the hereditary material. *Sci. Amer.* **191,** 54 (Offprint 5).

Gardner, E. J. (1960). "Principles of Genetics." Wiley, New York.

Garen, A. (1968). Sense and nonsense in the genetic code. *Science* **160,** 149.

Levine, R. P. (1968). "Genetics," 2nd ed. Holt, New York.

Nomura, M. (1969). Ribosomes. *Sci. Amer.* **221,** 28 (Offprint 1157).

Rich, A. (1963). Polyribosomes. *Sci. Amer.* **209,** 44 (Offprint 171).

Watson, J. D. (1968). "The Double Helix." Atheneum, New York.

Watson, J. D. (1970). "Molecular Biology of the Gene," 2nd ed. Benjamin, New York.

Yanofsky, C. (1967). Gene structure and protein structure. *Sci. Amer.* **216,** 80 (Offprint 1074).

CHAPTER IV Molecular Transport in Living Systems

The importance of chemical reactions in living systems has been emphasized repeatedly up to this point. The stress that has been placed on enzymatic control of chemical processes has implied the significant role of reaction rates. However, in living systems it is often found that the rate at which chemical reactions take place is limited by purely physical constraints. One of the most common and most important physical limitations is the speed with which the reacting materials can be transported to the site of the reaction. This chapter is concerned with the transport of materials through fluids and across membranes and with the distribution of these materials in the fluids and along the membrane surfaces.

A number of physical phenomena can be identified that are especially applicable to molecular transport and distribution in living systems. These phenomena will be considered first. Then the relationship between these phenomena and specific chemical materials and cellular situations will be discussed.

Diffusion

The motion of molecules in cells is influenced by electrical attractions and repulsions, by van der Waals forces, and by mechanical barriers such as membranes. The most important factor in the motion of cellular molecules, though, is the random motion of the molecules due to their collisions with other molecules in the cytoplasm. This random motion and some of its consequences are now considered.

Random Motion

Kinetic theory relates the average kinetic energy $\bar{E}$ of particles of mass m whose instanteous velocity is v to the absolute temperature T by

$$\bar{E} = \tfrac{1}{2}m\overline{v^2} = \tfrac{3}{2}kT \tag{4-1}$$

in which $k = 1.38 \times 10^{-23}$ joules/°K is the Boltzmann constant. In three dimensions, a point particle or the center of mass of a larger body has three degrees of freedom associated with it. In other words, three coordinates like x, y, and z are required to specify its location with respect to an arbitrary origin. The translational motion of such a particle also requires specification by three velocities v_x, v_y, and v_z. Each degree of freedom of a particle is associated with an average kinetic energy $\bar{E}_i = \tfrac{1}{2}kT_i$, so the average kinetic energy $\bar{E}$ of a particle in three dimensions is $\bar{E} = 3\bar{E}_i = \tfrac{3}{2}kT$. The average kinetic energy associated with each degree of freedom, like the x direction, is

$$\bar{E}_x = \tfrac{1}{2}m\overline{v_x^2} = \tfrac{1}{2}kT \tag{4-2}$$

Of course the average velocity is not directly measurable, but the molecules are in motion and they collide with other molecules randomly. The average displacement of a particle in a particular direction, say the x direction, is zero because the motion is random. This means that a particle is equally likely to be displaced in the positive x direction as the negative. But the average of the square of the displacement in the x direction, $\overline{x^2}$, is not zero. The problem of determining $\overline{x^2}$ for a particle with a given root-mean-square velocity, usually called the "random walk" problem, will be considered here for cellular molecules.

For molecules in cells, the motion is more complicated than in gases because of the retarding frictional force of the liquid medium. It can be shown experimentally that for small particles at low velocities, the frictional force is proportional to the particle velocity. This is expressed by

$$F_x = -fv_x \tag{4-3}$$

in which only the x direction is considered, v_x is the velocity, and F_x is the frictional force. The negative sign indicates that the frictional force and velocity are in opposite directions. The constant of proportionality f is the coefficient of friction per unit velocity. Since we shall now only consider motion in one dimension, we shall omit the x subscript. For a molecule of mass m, Newton's second law becomes

$$m \frac{dv}{dt} = -fv \tag{4-4}$$

By separating the variables, we obtain

$$\frac{dv}{v} = -\frac{f}{m} dt \tag{4-5}$$

and integration gives

$$\ln v = -\frac{ft}{m} + C \tag{4-6}$$

The constant of integration can be expressed in terms of an assumed velocity v_0 at time $t = 0$. When $t = 0$ in Equation (4-6), C is given by

$$\ln v_0 = C \tag{4-7}$$

Then Equation (4-6) becomes

$$\ln v = -\frac{ft}{m} + \ln v_0 \tag{4-8}$$

or

$$v = v_0 \, e^{-ft/m} = v_0 \exp\left(-\frac{ft}{m}\right) \tag{4-9}$$

The exponential notation introduced as a convenience here is defined such that $\exp(x) = e^x$. This result can be substituted directly into the definition of kinetic energy to give an expression for the kinetic energy as a function of the time:

$$E = \tfrac{1}{2}mv^2 = \tfrac{1}{2}mv_0^2 \, e^{-2ft/m} = E_0 \, e^{-2ft/m} \tag{4-10}$$

Here, $E_0 = \frac{1}{2}mv_0^2$ is the kinetic energy at $t = 0$. Now the average value of E over a period of time t_1, is defined by

$$\bar{E} = \left(\int_0^{t_1} E \, dt\right) \Big/ \left(\int_0^{t_1} dt\right) = \frac{1}{t_1} \int_0^{t_1} E \, dt \tag{4-11}$$

and substitution of Equation (4-10) into Equation (4-11) gives

$$\bar{E} = \frac{1}{t_1}\int_0^{t_1} E_0\, e^{-2ft/m}\, dt = \frac{E_0 m}{2t_1 f}\left[1 - \exp\left(\frac{-2ft_1}{m}\right)\right] \tag{4-12}$$

For times t_1 sufficiently long, the exponential term is negligible compared to unity. The frictional coefficient f can be crudely estimated from Stoke's law,

$$f = 6\pi\eta a \tag{4-13}$$

in which a is the radius of a spherical particle and η is the viscosity ($= 10^{-2}$ cgs units for water) of the liquid medium. For a very large spherical molecule in which a may be taken as $1000\ A = 10^{-5}$ cm, f is about 2×10^{-6} cgs units. The largest macromolecules have molecular weights of about 6×10^7 or a mass of about 10^{-16} gm; so the exponent in Equation (4-12) is $(2ft_1)/m = 4 \times 10^{10} t_1$. Thus even for times t_1 as short as a microsecond, the exponential term is about $e^{-40,000}$, a value quite negligible compared to unity. Then Equation (4-12) is simplified for times t_1 greater than 10^{-6} sec to

$$\bar{E} = \frac{E_0 m}{2ft_1} \tag{4-14}$$

The displacement x can be related to the time interval t_1 by

$$x = \int_0^{t_1} v\, dt = v_0 \int_0^{t_1} e^{-ft/m}\, dt = \frac{v_0 m}{f}\left(1 - e^{-ft_1/m}\right) \tag{4-15}$$

which, for times $t_1 > 10^{-6}$ sec, becomes

$$x = \frac{v_0 m}{f} \quad \text{or} \quad x^2 = \frac{v_0^2 m^2}{f^2} \tag{4-16}$$

In terms of E_0, x^2 can then be expressed as

$$x^2 = \frac{2m}{f^2}\left(\tfrac{1}{2} m v_0^2\right) = \frac{2m}{f^2} E_0 \tag{4-17}$$

and using Equation (4-14), x^2 can be expressed in terms of $\bar{E}$ according to

$$x^2 = \frac{4\bar{E}t_1}{f} \tag{4-18}$$

Now we may take the average (over many molecules) of the square of the displacement, and by recalling Equation (4-2), we obtain

$$\overline{x^2} = \frac{2kT}{f} t_1 \tag{4-19}$$

Equation (4-19) is the solution to the one-dimensional "random walk." Similar relationships describe the y- and z-direction solutions, so that the average of the square of the displacement acquired by a molecule in time t_1 in three dimensions is given by

$$\overline{r^2} = \overline{x^2} + \overline{y^2} + \overline{z^2} = 6 \frac{kT}{f} t_1 \tag{4-20}$$

It is important to note that we have calculated an *average* squared displacement. Some molecules will be displaced more and some less than the average. The distribution of the molecules about the average is known as a gaussian distribution. For a mean-square displacement $\overline{x^2}$, the probability $P(x)\,dx$ of finding a molecule with displacement between x and $x + dx$ is given by

$$P(x)\,dx = \frac{1}{(2\overline{x^2})^{1/2}} \exp\left(-\frac{x^2}{2\overline{x^2}}\right) dx \tag{4-21}$$

The factor $1/(2\overline{x^2})^{1/2}$ is called a normalizing factor and is chosen to ensure that

$$\int_{-\infty}^{\infty} P(x)\,dx = 1 \tag{4-22}$$

that is, the probability of finding the molecule at some value of x between $-\infty$ and $+\infty$ is unity (a certainty). The symmetry of Equation (4-21) provides that positive and negative displacements are equally probable, i.e., that $P(x)\,dx = P(-x)\,dx$.

The results of the foregoing analysis will be useful when the diffusion process itself has been described and mathematically formulated.

The Diffusion Equation

Because molecules in a fluid are in random motion, they are equally likely to move in one direction as another. If molecules are concentrated in one region more than another, more will move out of the high-concentration region than move into it because of the greater number of particles in the region of high concentration. The net flow of particles from a region of higher concentration to one of lower concentration is called *diffusion.*

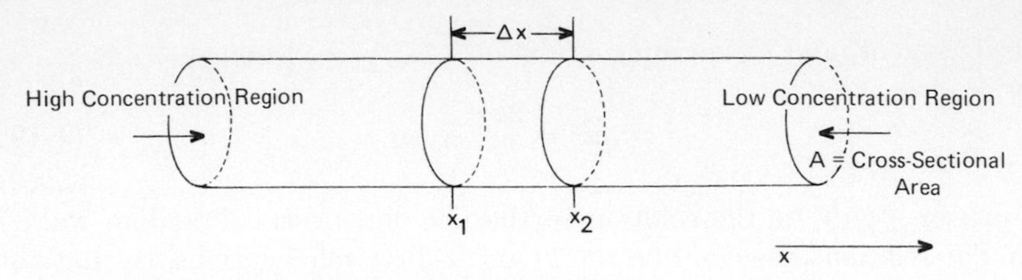

FIGURE 4-1. A one-dimensional concentration distribution. The particles in solution are in greater concentration at the left. The concentration is assumed to be constant across any cross-section of the tube.

An analytical description of diffusion can be obtained by consideration of the flow of particles in a pipe in one dimension from a region of high concentration to a region of lower concentration, as shown in Figure 4-1. It is assumed that the concentration of particles expressed in kg/m^3 is the same at every point along any cross section of the pipe, while the concentration varies from one cross section to another and may vary at any single cross section with time. All of these variations can be described by a single equation, which can be obtained by the following considerations.

At the planes x_1 and x_2, where the concentration c_1 at x_1 is higher than the concentration c_2 at x_2, the mass Δm that moves from x_1 toward x_2 in the time Δt is proportional to the concentration c_1. Similarly the mass that moves from x_2 toward x_1 in the time Δt is proportional to the concentration c_2. Both of the masses are proportional to the cross-sectional area A. These facts are stated mathematically by

$$\frac{\Delta m}{\Delta t} \propto Ac_1 - Ac_2 = A(c_1 - c_2) \tag{4-23}$$

This can be made into an equation by introducing a constant of proportionality k. Then Equation (4-23) becomes

$$\frac{\Delta m}{\Delta t} = kA(c_1 - c_2) = -kA\Delta c \tag{4-24}$$

in which the negative sign indicates that the change in concentration is negative in the positive x direction. In this notation, Δm is the net transfer of mass in the positive x direction across the plane x_2. If the planes x_1 and x_2 were far apart, one would expect the rate of mass transfer from x_1 to x_2 to be small; in other words, k would be small. Similarly if x_1 and x_2 are close together, k would be large. Thus, k is inversely proportional to $(x_2 - x_1)$, which we shall call Δx. Then we may relate k and Δx through another constant of proportionality by

$$k = \frac{D}{\Delta x} \tag{4-25}$$

where the proportionality constant D is called the *Fick diffusion constant.* Substituting Equation (4-25) into (4-24), we obtain

$$\frac{\Delta m}{\Delta t} = -DA \frac{\Delta c}{\Delta x} \tag{4-26}$$

In the limit as $\Delta t \to 0$ and $\Delta x \to 0$, Equation (4-26) becomes

$$\frac{\partial m}{\partial t} = -DA \frac{\partial c}{\partial x} \tag{4-27}$$

in which the derivatives are partial because concentration and mass are functions of both time and position. Equation (4-27) is the diffusion equation known as *Fick's law.* For particles of equal mass, i.e., a single species of molecule, Fick's law can be written in terms of the number of particles n:

$$\frac{\partial n}{\partial t} = -DA \frac{\partial c}{\partial x} \tag{4-28}$$

It is now useful to express the diffusion equation in an alternate form. This can be done by noting that the time rate of change of the mass (i.e., the mass per second) in the region between x_1 and x_2 is the mass per second going into the region at x_1 less the mass per second leaving the region at x_2. The mass in the region is the volume of the region times the concentration (mass per unit volume) within the region, or $A\Delta xc$. The mass per second going into the region at x_1 is given by Equation (4-27) as

$$\left(\frac{\partial m}{\partial t}\right)_{x_1} = -DA\left(\frac{\partial c}{\partial x}\right)_{x_1} \tag{4-29}$$

and the mass per second going out of the region at x_2 is

$$\left(\frac{\partial m}{\partial t}\right)_{x_2} = -DA\left(\frac{\partial c}{\partial x}\right)_{x_2} \tag{4-30}$$

The subscripts on the derivatives denote the points at which the derivatives are to be evaluated. We then have

$$\frac{\partial(A\Delta xc)}{\partial t} = A\Delta x \frac{\partial c}{\partial t} = -DA\left(\frac{\partial c}{\partial x}\right)_{x_1} + DA\left(\frac{\partial c}{\partial x}\right)_{x_2} \tag{4-31}$$

The concentration gradient at x_2 may be expressed as the gradient at x_1

plus the rate at which the gradient is changing with x multiplied by the distance Δx, i.e.,

$$\left(\frac{\partial c}{\partial x}\right)_{x_2} = \left(\frac{\partial c}{\partial x}\right)_{x_1} + \frac{\partial(\partial c/\partial x)}{\partial x}\Delta x = \left(\frac{\partial c}{\partial x}\right)_{x_1} + \frac{\partial^2 c}{\partial x^2}\Delta x \qquad (4\text{-}32)$$

Now we may put Equation (4-32) into Equation (4-31):

$$A\Delta x \frac{\partial c}{\partial t} = -DA\left(\frac{\partial c}{\partial x}\right)_{x_1} + DA\left(\frac{\partial c}{\partial x}\right)_{x_1} + DA\frac{\partial^2 c}{\partial x^2}\Delta x \qquad (4\text{-}33)$$

or

$$\frac{\partial c}{\partial t} = D\frac{\partial^2 c}{\partial x^2} \qquad (4\text{-}34)$$

The form of Fick's law in Equation (4-34) can be extended to three dimensions by a treatment similar to that given above in one dimension. The three-dimensional version is

$$\frac{\partial c}{\partial t} = D\left[\frac{\partial^2 c}{\partial x^2} + \frac{\partial^2 c}{\partial y^2} + \frac{\partial^2 c}{\partial z^2}\right] \qquad (4\text{-}35)$$

which, when written in vector notation, is the elegant form

$$\frac{\partial c}{\partial t} = D\nabla^2 c \qquad (4\text{-}36)$$

In deriving the diffusion equation, it has been tacitly assumed that the molecules that are diffusing are neither being produced nor consumed during the diffusion process. Quite frequently in cellular diffusion systems, the molecules of interest are being produced while being diffused (like oxygen in the photosynthesis process) or are being consumed while being diffused (like oxygen in the respiration process). If s is the rate at which the diffusing molecules are being produced (or $-s$ is the rate at which they are being consumed), Equation (4-36) becomes

$$\frac{\partial c}{\partial t} = D\nabla^2 c + s(x, y, z, t) \qquad (4\text{-}37)$$

This is called the inhomogeneous diffusion equation.

An explicit connection between random motion and diffusion can now be made. The one-dimensional diffusion equation, Equation (4-34), has the intrinsic feature that the concentration is changing, and consequently the concentration gradient is changing during the diffusion process. The solutions of the diffusion equation take different forms depending on the boundary conditions of the problem under consideration. It is convenient here to consider an initial condition in which all the particles are at the origin at $t = 0$. As time goes on, the particles diffuse away from $x = 0$ while executing their random walks. Figure 4-2 shows graphically the boundary conditions and distributions at arbitrary intermediate times. As $t \rightarrow \infty$, the concentration approaches zero for all values of x. Since a solution that satisfies the boundary conditions is the general solution for the problem under consideration, we may submit a solution that satisfies the diffusion equation and the boundary conditions:

$$c = \frac{1}{(4\pi\, Dt)^{1/2}}\, e^{-x^2/4\,Dt} \tag{4-38}$$

The numerical coefficient $1/\sqrt{4\pi}$ is chosen to normalize the concentration so that

$$\int_{-\infty}^{\infty} c\, dx = 1 \tag{4-39}$$

It then follows that the value of c at any x and t is the probability that a particle will have diffused to that position x in the time t. Then the solution given by Equation (4-38) should be identical to the random walk solution given in Equation (4-21). Comparison of the two equations shows them to be identical when we impose the relation

$$\overline{x^2} = 2\, Dt \tag{4-40}$$

The importance of this result in the transport of cellular materials can be seen from two simple examples. Suppose we take an enzyme like urease,

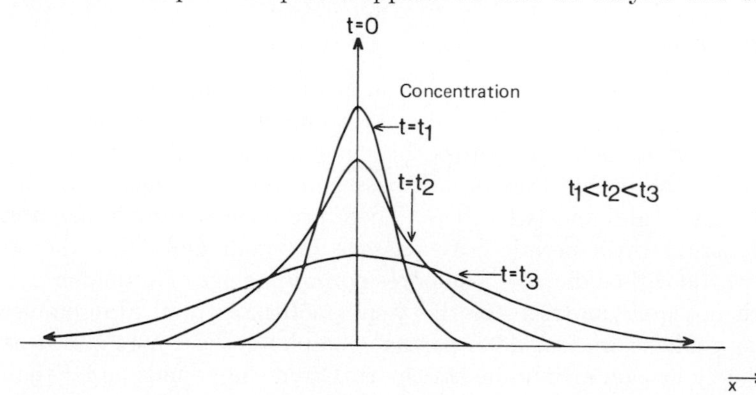

FIGURE 4-2. A graphical representation of the concentrations of particles as a function of x for different times when the initial concentration (at $t = 0$) is such that all the particles are at the origin.

whose molecular weight is about a half-million. The Fick diffusion constant D for urease is 3.5×10^{-11} m^2/sec. Suppose we carefully place a gram or so of urease on the bottom of a cup of water 4 cm high. The time it takes for the urease to diffuse to the surface of the water can be found from Equation (4-40):

$$t = \frac{x^2}{2D} \tag{4-41}$$

We have $x^2 = (4\text{ cm})^2 = (0.04\text{ m})^2 = 1.6 \times 10^{-3}$ m^2. Substituting the values for x^2 and D, we obtain

$$t = \frac{1.6 \times 10^{-3}\text{ m}^2}{2 \times 3.5 \times 10^{-11}\text{ m}^2/\text{sec}} = 2.3 \times 10^7\text{ sec} \tag{4-42}$$

or about 9 months required to diffuse throughout the cup of water. Compare this result to diffusion of urease across a cell's length, which might be typically about 5 μm (5×10^{-6} m) across. If we substitute again into Equation (4-40) or (4-41), we obtain

$$t = \frac{25 \times 10^{-12}\text{ m}^2}{2 \times 3.5 \times 10^{-11}\text{ m}^2/\text{sec}} = 0.36\text{ sec} \tag{4-43}$$

which means that even very heavy molecules can be transported across cellular dimensions in a fraction of a second. We may conclude then that in general no concentration of molecules can exist very long in a cell unless there is some force field to maintain the concentration. Furthermore, we may conclude that diffusion is an efficient mechanism for transporting molecular material throughout cells.

Membranes

The membranes of cells—the plasma membrane, the nuclear membrane, and the endoplasmic reticulum—have a common characteristic relating to the transport of molecules across these membranes. All of them are *selectively permeable* in that they permit passage through the membrane barrier of some molecules and bar others. There are openings in the membranes called *pores*, which permit easy passage of small molecules like water, oxygen, and carbon dioxide. The pores permit passage of a number of salts, though not quite so freely as the very small molecules. Macromolecules such as proteins are forbidden passage completely. Yet pore size does not completely explain membrane transport. Often compounds larger than the pores pass through, and often compounds smaller then the pores do not. A number of other factors affect the permeability of a membrane with respect to a particular diffusing molecule. The concentration and solubility of the diffusing molecule, as well as the electrical charge distribution on the membrane contribute to the permeability of the membrane. Two other phenomena are known to take place at membrane barriers. Cells can acquire certain materials through the expenditure of metabolic energy, regardless of the relative concentration of the materials inside and outside the cells. This process is called *active transport.* Certain kinds of cells have the ability to engulf materials into invaginations (inward folds) of the plasma membrane and reseal the membrane at the original surface, thus transporting relatively large quantities of materials across the membrane. This process is known as *pinocytosis.*

Many of the details of the membrane transport functions are not yet understood. Some basic principles that apply to membrane transport are understood, and a few models to explain some of the more complex mechanisms have been conjectured. Nevertheless, the functional details of membrane transport await further elucidation.

Osmosis

The process known as *osmosis* is a special case of diffusion that applies to the transport of water across semipermeable membranes. Osmotic processes are crucial in living systems because cell membranes surround a watery medium in which all of the cell's organelles, metabolites, and structural materials are immersed, and the cell is bathed in fluids that are principally water.

Osmosis is the phenomenon by which the concentration of a solute in water is equalized on both sides of a membrane that is permeable to water. It can be illustrated by a simple experimental situation (see Figure 4-3)

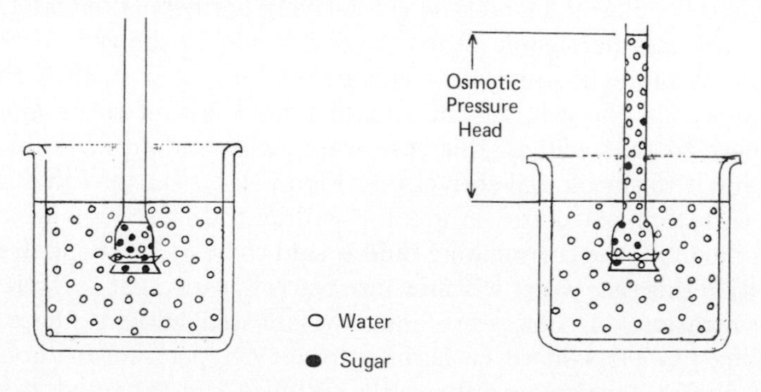

FIGURE 4-3. Experimental demonstration of the development of a pressure head by osmosis. A thistle tube is covered with cellophane, which encloses a concentrated sugar solution. On the left, the thistle tube is placed in the water. On the right, water has crossed the cellophane membrane and caused the solution to rise in the tube.

using cellophane as the semipermeable membrane that will easily pass water, but which is almost completely impermeable to sugar. A thistle tube with cellophane stretched securely over the funnel's outer opening and partially filled with a solution of highly concentrated sugar molecules is placed under the surface of the water. Water will cross the membrane toward the high sugar concentration until an equilibrium situation is obtained, i.e., until a pressure head is established in the thistle tube such that there is no longer a net gain or loss of water across the membrane. The pressure exerted by the height of the weakened solution in the thistle tube above the water surface is called the *osmotic pressure*. The osmotic pressure can be described analytically by the relation

$$\Pi = cRT \tag{4-44}$$

in which Π is the osmotic pressure in atmospheres, c is the concentration in moles of solute per liter of solution, R is the gas constant (0.082 atm-liter/°K-mole), and T is the absolute temperature.

Biochemists sometimes use the osmotic pressure relationship in a slightly different form to determine the molecular weight of biological macromolecules. Equation (4-44) can be written in the form

$$\Pi = \frac{g}{MV} RT \tag{4-45}$$

where g is the mass of the solute molecules in grams, V is the volume of solution in liters, and M is the molecular weight of the solute molecule.

A number of terms are used to describe the osmotic relationships that exist between the contents of cells and the media surrounding the cells. Of course, the cell's plasma membrane is relatively permeable to water, but it may be much less permeable to many solutes within the cell. If the concentration of solutes is greater in the surrounding medium than the concentration within the cell, the surrounding fluid is said to be *hypertonic* with respect to that cell. In this case water will tend to flow out of the cell, causing it to shrink and shrivel (see Figure 4-4). On the other hand, if the concentration of solutes is greater within the cell than in the surrounding medium, the surrounding fluid is said to be *hypotonic* with respect to the cell; in this case water will flow into the cell, causing it to swell. When the concentrations of solutes are equal within and without the cell, the surrounding fluid is said to be *isotonic*. Since "hyper" means greater or above, "hypo" means below or beneath, and "iso" means equal, it is often helpful to remember that the descriptive terms refer to the amount of material (solute) dissolved in the surrounding media.

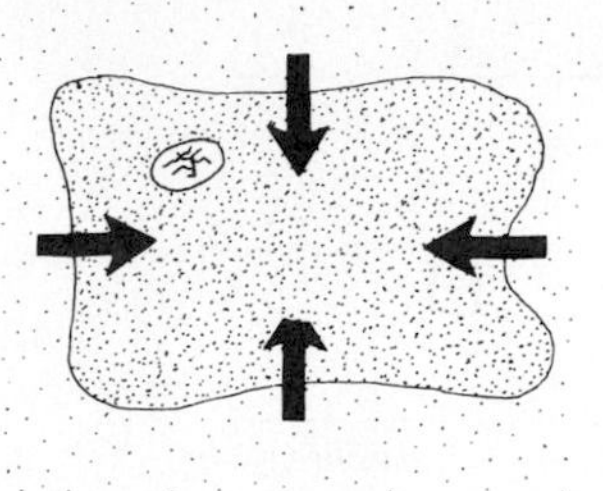

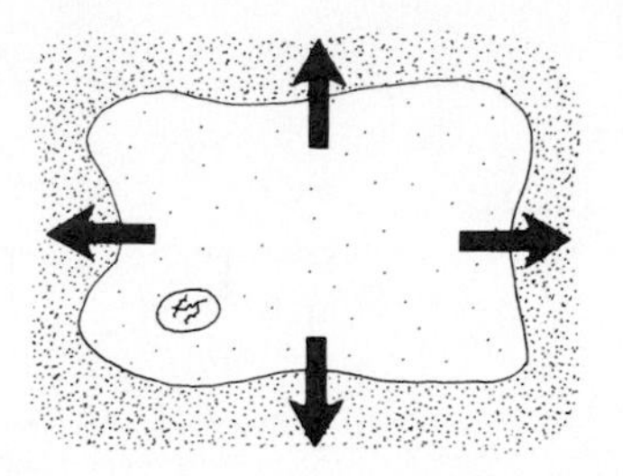

FIGURE 4-4. Water transport into and out of cells placed in solutions that are hypotonic and hypertonic with respect to the cells. The solute material under consideration is indicated by the dots in each illustration.

Simple osmotic considerations provide an explanation for a number of easily observable cellular phenomena. For example, when a plant is placed in water (which is hypotonic with respect to the plant cells), water enters the plant cells and builds up an osmotic pressure. In plants this pressure is outward on its cell walls and is called *turgor pressure.* Since plants have relatively rigid cellulose cell walls, the cells are not unduely swollen, but are firm (or *turgid*). Turgor pressure helps support the plant, and when insufficient water is available, the plant wilts from lack of this turgor pressure.

A simple means of emptying red blood cells of their contents for study is to place some blood into distilled water. The hypotonic medium causes water to enter the cells, swell them to a spherical shape and burst them. One can observe the shrinking of red blood cells with a microscope if the blood cells are placed in a solution that contains more than 0.9% NaCl (which is the concentration that is isotonic for cells of humans).

Diffusion across Membranes

The most general form of the diffusion equation that has been developed in this chapter is Equation (4-37), which is not in a suitable form for consideration of diffusion across membranes. There are often very marked differences in concentration of certain substances across cell membranes, and a mathematical description that is appropriate to this physical situation will be considered here.

Suppose a membrane and the fluids on either side of it are considered as three separate regions, as shown in Figure 4-5. Suppose further that each of the three regions has associated with it a concentration and a diffusion coefficient. In the case of cell membranes, the middle region is so thin that

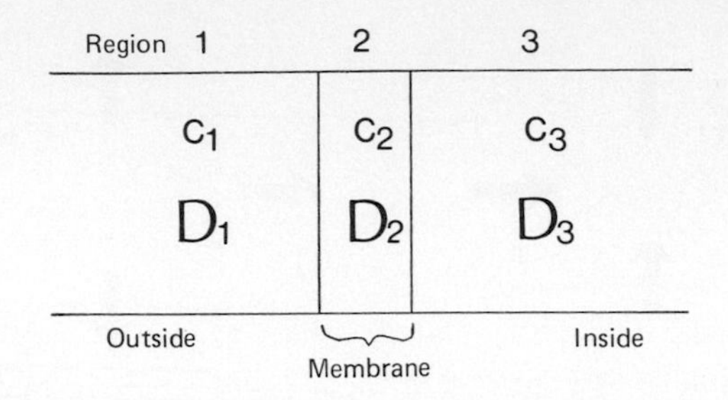

FIGURE 4-5. A model for characterizing diffusion across a membrane. Each region, including the membrane, is characterized by one of the concentrations c_1, c_2, or c_3 and by diffusion coefficients D_1, D_2, or D_3.

no realistic significance can be attached either to the concentration c_2 or to the diffusion coefficient D_2. It is convenient to characterize the membrane's transport properties by a parameter called the permeability k, which is a constant of proportionality between the mass per unit time per unit area passing through the membrane and the difference in the concentrations inside and outside the membrane. In terms of the same symbols used to develop the earlier diffusion equations, this relationship is expressed by

$$\frac{1}{A}\frac{\partial m}{\partial t} = k(c_1 - c_3) \tag{4-46}$$

Since the mass per unit time passing through a unit area of the membrane must be equal to the mass per unit time entering and leaving a unit area of the membrane (the membrane does not absorb or generate the diffusing molecule), we may recall Equations (4-29) and (4-30) and equate these quantities on either side of the membrane, which gives

$$\frac{1}{A}\frac{\partial m}{\partial t} = -D_1\frac{\partial c_1}{\partial x} = -D_3\frac{\partial c_3}{\partial x} = k(c_1 - c_3) \tag{4-47}$$

One other factor that has, up to this point, been tacitly assumed not to be present will now be considered. Electrical potential gradients may exist across membranes because of the presence of ions on the membranes. Such electrical gradients play an important role in a number of physiological situations. Electrical potential gradients in living systems exist only across membranes because the fluid media are good conductors of electricity and charges cannot accumulate in the fluids. If the number of electronic charges on an ion is n and V is the potential difference in volts across the membrane, the difference in the free energies (due to electrical charge distribution) inside and outside the membrane is given in joules per mole by

$$\Delta G = nFV \tag{4-48}$$

where F is the faraday (the number of coulombs in a mole of electrons—96,487). ΔG can be converted to kcal/mole by using the fact that 4.18 joule = 1 cal = 10^{-3} kcal. Then Equation (4-48) becomes

$$\Delta G = 23\, nV \text{ kcal/mole} \tag{4-49}$$

This value represents the free energy (available to do useful work) when the concentrations of ions are equal on each side of the membrane. Then the same factor that appears in the mass action formula of Chapter II appears in a modification of Equation (4-47) to account for charged particles:

$$D_1 \frac{\partial c_1}{\partial x} = D_3 \frac{\partial c_3}{\partial x} = -k(c_1\, e^{-\Delta G/RT} - c_3) \tag{4-50}$$

R is the gas constant and T is the absolute temperature. Note that if the electrical charges are absent, the available free energy due to ions, ΔG, is zero, and Equation (4-50) reverts to Equation (4-47).

From a mathematical point of view, Equations (4-47), (4-37), and (4-50) completely describe biological diffusion problems when the appropriate values of the diffusion coefficient D, the rate of production of the diffusing molecule s, and the permeability k are provided. In practice, the solutions may be very difficult to obtain. Many physical effects that have not been mentioned, like stirring the fluids or stretching the membranes, can be accounted for in the formalism developed here by adjusting the parameters D and k. Stirring or agitating the fluid medium effectively increases the diffusion coefficient; changing a physical characteristic of the membrane effectively changes the permeability k.

Membrane Models

The nature of the permeability of cell membranes is not understood. A few structural characteristics are known, but otherwise the functional mechanisms are at this time completely a matter of conjecture.

It is known, for example, that membranes are composed of layered lipids and proteins, arranged much like a sandwich in which the proteins are the "bread." It is also known that certain regions of the membrane have pores that are about 8 Å in diameter and permit easy passage to small molecules like water. Another fact that has been learned about the nature of membrane permeability concerns the solubility of the diffusing molecules in lipids. Substances with greater solubility in lipids are able to penetrate the membrane more readily than those less soluble in lipids, other factors being essentially equal. It is also known that electrolytes, substances that ionize in solution, penetrate the plasma membrane more slowly than nonelectrolytes of equal size.

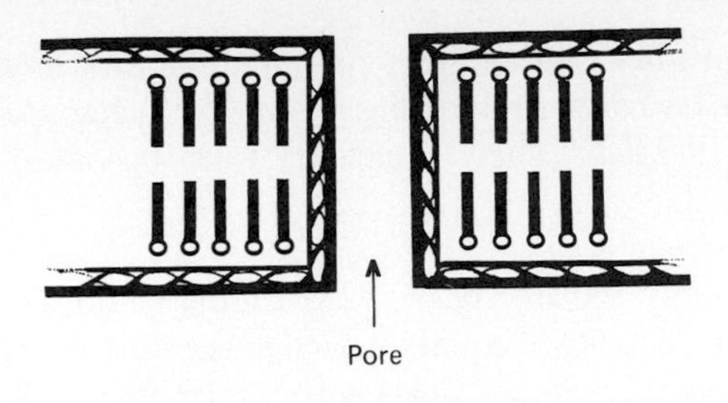

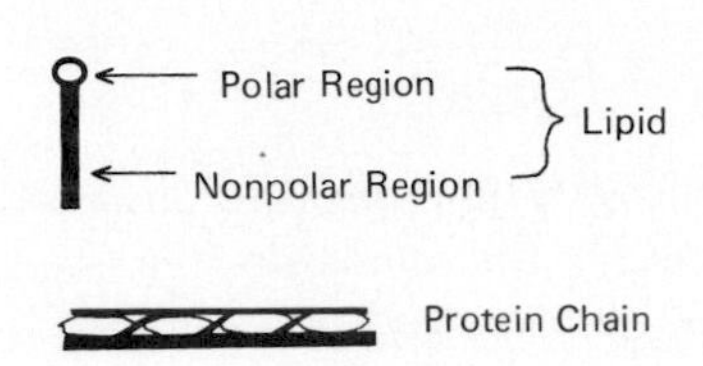

FIGURE 4-6. Schematic representation of one model of a pore in a plasma membrane. The membrane is depicted here in cross section.

A schematic representation of a possible model for a plasma membrane and pore is shown in Figure 4-6. Since lipids contain long, nonpolar chains attached to a polar region, it has been assumed that the nonpolar portions of the lipids are pointed toward the middle of the membrane and the polar portions toward the surfaces; the polar portions of the lipids are assumed to be covered by monolayers of unfolded proteins. At the pores, it is usually assumed that the protein layer also lines the pore region.

The means of transporting materials through a membrane like the one just described has been the subject of much recent speculation. One theory is based on the peculiar ability of proteins to acquire either a positive or negative charge, depending on whether they are in an acid or alkaline medium. In this model the pores that are lined with negatively charged protein attract positive ions into the pore region, where they presumably diffuse into the interior of the cell. Similarly, positively charged pore linings attract and permit entrance to negatively charged ions. The entire membrane surface is thought of as a "mosaic," composed of some positively lined pores and some negatively lined pores.

One of the most recent models for membrane transport suggests that the protein within a pore has a pair of polar active sites that act as "gates" to permit or refuse entry to ions. The negatively charged gates are swung closed when attached to a divalent positive ion such as Ca^{2+}, but are opened when bonded to a monovalent ion like Na^{+}. Thus, sodium might control the flow of potassium or of sodium itself, with calcium acting as a mediating material. All of these ions, Na^{+}, K^{+}, and Ca^{2+}, are present at the surfaces of nerve cells, and the roles of these ions in membrane transport will be of further interest when they are considered in Chapter V.

Active Transport

Many examples are known in which cells accumulate certain molecules, often in great quantity, against the concentration gradient between the cell and its surrounding medium. Some seaweeds, for example, accumulate iodine from the sea in concentrations in excess of two million times greater than that of iodine in the sea water. Similarly, the cells of the thyroid gland in humans concentrate iodine far above the level found in the bloodstream. The phenomenon of accumulating substances across a membrane in opposition to the concentration gradient is called *active transport,* and since it is difficult to imagine a passive system that would permit such transport, it is assumed that energy is expended in executing active transport across a cell membrane.

Although active transport is poorly understood, evidence exists that ATP is the energy source for this mechanism. The most interesting models proposed to explain active transport involve "carrier" molecules, which attach themselves, somewhat like enzymes do, to the material to be transported across the membrane. The carriers can then presumably be pumped across the membrane at the expense of metabolic energy. It is not at all clear how the carriers themselves are permitted such casual access across the membrane. Such theories require a number of different carriers—one for each variety of material transported across the membrane.

Obviously, many aspects of membrane transport function await further clarification. This area of research is one among those that are being vigorously pursued today.

References and Suggested Reading

Hokin, L. E., and Hokin, M. R. (1965). The chemistry of cell membranes. *Sci. Amer.* **213,** 78 (Offprint 1022).

Nash, L. K. (1962). "Elements of Chemical Thermodynamics." Addison-Wesley, Reading, Massachusetts.

Robertson, J. D. (1962). The membrane of the living cell. *Sci. Amer.* **206,** 64 (Offprint 151).

Solomon, A. K. (1960). The pores in the cell membrane. *Sci. Amer.* **203,** 146 (Offprint 76).

Szent-Györgyi, A. (1960). "Introduction to a Submolecular Biology." Academic Press, New York.

UNIT II

The Biophysical Approach to Special Organ Systems

CHAPTER V The Nervous System

The nervous system of an organism provides the means for rapid internal communication, sensory reception, and responsive reflex movement. In addition, the nervous system receives, processes, stores, and provides access to information from sensory organs. The complexity of nervous systems in the animal kingdom ranges from the simple nerve nets found in jellyfish to the complicated computerlike facility that we recognize in the mammals.

In this chapter we shall briefly identify the divisions of the nervous systems, discuss some physical and chemical properties of cells and membranes that are particularly applicable to the cellular units of the nervous system, and consider some details of the electrical phenomena that enable the nervous system to transmit information. Then, more in the realm of speculation, we shall consider some of the current approaches to a few of the problems associated with understanding brain function.

Anatomy and Physiology of the Nervous System

Anatomical Orientation

In discussions of animals, it is often necessary to refer to relative directions and planes within the body of the animal. A few descriptive terms make it easier to describe accurately the orientation of anatomical features. Since this text will be most frequently concerned with human structures, these terms will be defined with respect to the human body.

When the human body is in the *anatomical position*, i.e., standing with arms at the side and palms facing forward, three principal planes are defined. Figure 5-1 illustrates these three planes. The *median plane* passes through the center of the body and divides it into left and right halves. A *frontal plane* is any plane at right angles to the median plane that divides the body into front and back parts. A *horizontal plane* is any plane perpendicular to the median and frontal planes that divides the body into upper and lower parts. Structures that are referred to as *medial* are nearer to the median plane than those referred to as *lateral*. In referring to limbs, *proximal*

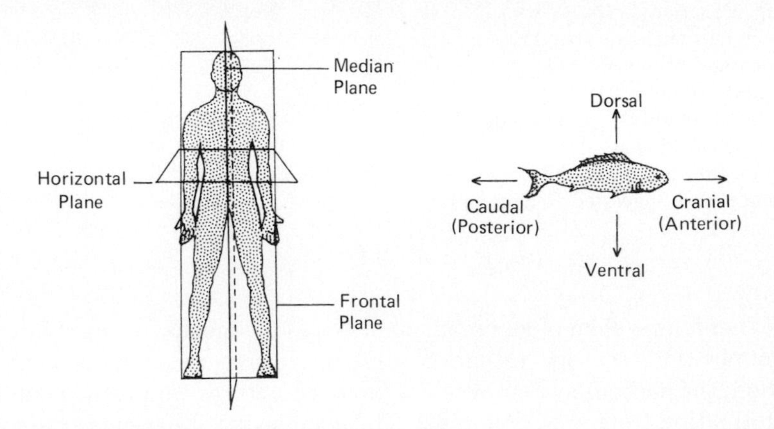

FIGURE 5-1. The planes and directions used in anatomical orientation. The human figure is in the anatomical position.

indicates a position nearer the trunk of the body or to the central axis (the intersection of the median and frontal planes); *distal* indicates a more peripheral or distant position.

In humans, the terms *anterior* and *posterior* refer to the directions toward the front and back of the body, respectively. In lower animals, like a fish for example, the direction of the head is called the anterior (or *cranial*) direction and the opposite direction is the posterior (or *caudal*) direction. The upper surface of a fish's body is called the *dorsal* surface; the under surface is called the *ventral* surface. Occasionally in humans, "dorsal" and "ventral" are used interchangeably with "posterior" and "anterior," respectively. Finally, the terms *superior* and *inferior* refer to upper and lower structures, respectively.

The Neuron

The functional units of the nervous system are cells called *neurons*. Like most cells, each neuron has a nucleus inside its cytoplasm. Unlike most cells, on the other hand, neurons have elongated parts and numerous *processes* (a process is anything that "sticks out"). The bulky portion of a neuron is called the *cell body*, which contains the nucleus, and from which extend a large number of short, highly branched processes called *dendrites*. Also extending from the cell body is a long, smooth process called the *axon*, which may occasionally branch. Axons have lengths of up to four meters in some mammals. At the end of the axon are the terminal branches. Figure 5-2 is a diagram of a neuron and its satellite cells, the Schwann cells.

Associated with neurons in the nervous system are two other varieties of cells, *Schwann cells* and *glial cells*. Schwann cells are found wrapped around the axons (but not around dendrites). Each Schwann cell contains a nucleus and encloses a section of the axon by a complicated series of invaginations that results in a number of thin layers of lipids wrapped

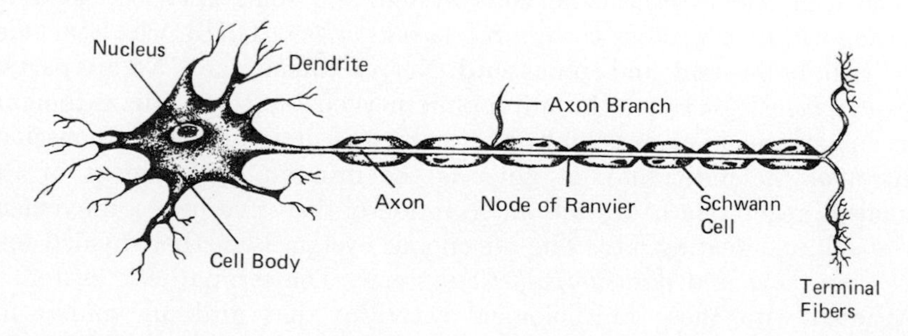

FIGURE 5-2. A neuron and its associated Schwann cells. The axon may be much longer, relative to the cell body size, than indicated here.

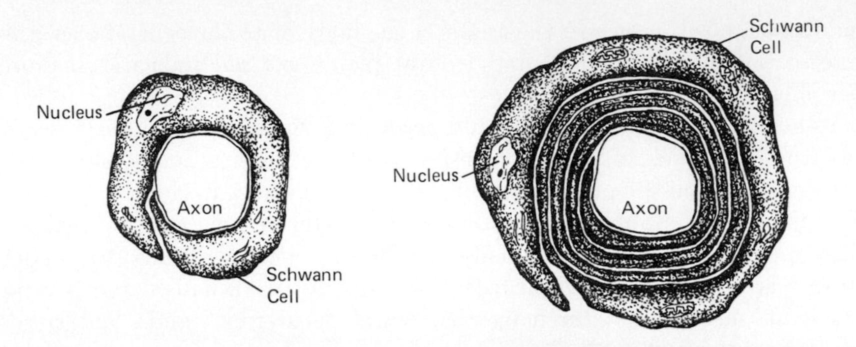

FIGURE 5-3. Cross section of an axon and one of its Schwann cells. As the Schwann cell grows, it develops a layered sheath (the myelin sheath) about the axon.

around the axon (see Figure 5-3). The region of the axon enclosed by the lipid layers is said to be *myelinated*, and the enclosing lipid layers are called a *myelin sheath*. The myelin sheath has a glistening white appearance and myelinated nervous tissue is often called *white matter* in contrast to non-myelinated *gray matter*. The myelin sheath serves to insulate axons from one another, and there is evidence that Schwann cells provide nutrients to damaged axons. The spaces between the Schwann cells along the length of the axon are called *nodes of Ranvier;* they are about one micron (10^{-4} cm) in width. The relatively few branchings of the axons occur at the nodes of Ranvier.

Glial cells, or *neuroglia*, are supportive cells found among the neurons in the brain and spinal cord. They are thought to be mechanical supports and have been implicated in the memory process of nervous tissue.

Classification in the Nervous System

The components of the nervous system are sometimes classified by location within or without the *central nervous system* (CNS), which includes the brain, brain stem, and spinal cord. Nerves outside the CNS are part of the *peripheral system*, or the autonomic nervous system. The autonomic nervous system is an involuntary system, i.e., it is not under conscious control of the individual. It governs the internal environment of the organism, regulating heart action, secretion of digestive juices, movement of intestines, sweating, etc. The autonomic system is further divided into the *sympathetic* and *parasympathetic* systems. The sympathetic system is associated with those physiological activities that gird an animal for action—activation of the adrenal gland that produces adrenaline and increases the heart and respiratory rates. The parasympathetic system is associated with the opposite, calming effects. Many physiological functions involve both the sympathetic and parasympathetic systems simultaneously in very complex ways.

Nerve tissue classification, besides making a distinction between white and gray matter, includes special names for groups of axons. A *funiculus* is a bundle of axons (including their coverings) covered by a tubular sheath of connective tissue called *perineurium*. The perineurium is penetrated by blood vessels that usually run parallel to the neuron fibers. A bundle of funiculi is covered by another sheath of connective tissue called *epineurium*, and it is this bundle that is commonly called a "nerve." When located within the CNS, bundles of axons are called *fiber tracts*.

A common feature of nervous tissue is the *ganglion* (plural, ganglia), an aggregate of neuron cell bodies outside the CNS. When a cluster of cell bodies occurs within the CNS, it is called a *nucleus* (not to be confused with the nucleus of a cell). There are other differences between neurons within and without the CNS. Within the CNS, a neuron degenerates and dies if it is injured; outside the CNS, injured axons and dendrites are replaced by new axons and dendrites that regenerate from the old cell body. Neurons within the CNS are more sensitive to oxygen deprivation than those outside the CNS. Those within the CNS can only exist a few minutes without an oxygen supply, whereas those outside the CNS might survive for several hours when deprived of oxygen.

Neurons themselves are usually classified according to their function. *Afferent neurons* carry signals from the peripheral system to the CNS, while *efferent neurons* carry signals from the CNS toward the peripheral system. The distinction is perhaps easier to remember if it is noted that the prefixes ("ad" = toward and "ex" = away from) refer to the CNS. Neurons are occasionally classified as *bipolar* or *multipolar*, names that refer to the number of processes on a particular neuron. Neurons are considered polarized in the sense that signals always travel into the neuron via dendrites and exit via the axon. Cells with only one dendrite and one axon are called bipolar; those with more than one dendrite are multipolar.

Nerve endings, or terminal branches, of nerve fibers may be simple fiber ends or these ends may be surrounded by accessory structures. The nerve endings of dendrites or axons are characterized either as *receptors* or *effectors*. The most common receptors are (1) *mechanoreceptors*, which respond to pressure as in the sense of touch, hearing receptors in the inner ear, and *kinesthesia receptors* that signal the position of joints and extemities; (2) *thermoreceptors*, some of which respond to heating, others to cooling; (3) *chemoreceptors*, which respond to chemicals, found in taste buds, olfactory (referring to the sense of smell) cells of the nose, and cells of the heart and aorta that respond to oxygen and CO_2 levels of the blood; and

(4) *photoreceptors*, which respond to light and are found in the retinal tissue of the eye. Free nerve endings in the skin, muscle, and joints are responsible for sensations of pain.

The Central Nervous System

The brain and spinal cord comprise the CNS. About 10^{10} neurons are contained in the human CNS. They are all present in the young child and do not undergo mitosis even to replace those destroyed by accident or disease. The neurons grow in size during maturation of an individual, but they do not increase in number.

Brain function is understandably complex, and most of its basic mechanisms of control and memory are not understood. How individual axons develop to contact their few specific terminal positions among millions of cells is a complete mystery.

The CNS has six major components shown in Figure 5-4. The *forebrain* (prosencephalon) and *midbrain* (mesencephalon) are collectively called the *cerebrum*. The *cerebellum*, *pons*, *medulla*, and *cord* complete the most prominent divisions of the CNS.

The cerebrum is the largest portion of the CNS. It is egg-shaped and fills the upper portion of the skull. The cerebral surface is called the *cortex* and

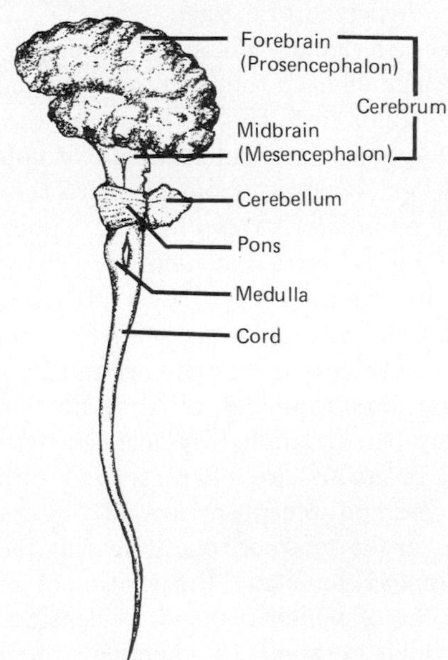

FIGURE 5-4. The central nervous system. The components are shown pulled apart somewhat for clarity.

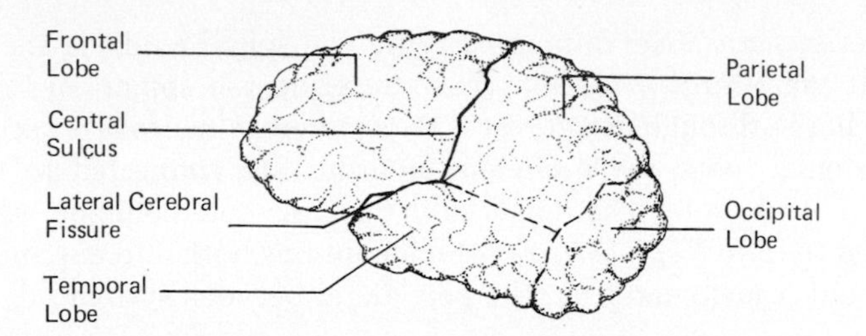

FIGURE 5-5. The geography of the brain. Each major area, or lobe, is delineated by the deeper creases in the folds of the brain called fissures or sulci.

is a highly convoluted, or folded, surface. The deepest furrows in the cortex are called *fissures;* the shallower ones are called *sulci*. The ridges between the sulci are called *gyri*. The cerebrum is divided into hemispheres, right and left, each of which is further divided into *lobes*. The lobes are geographical regions of the cerebrum and are shown in Figure 5-5.

The cerebellum is a deeply fissured structure behind the brain stem that functions primarily in control of various muscular activities. The pons is a bulbous body in the brain stem that joins the medulla, which in turn joins the spinal cord below. Both the pons and medulla contain a large number of fiber tracts. The details of the bodies within the brain and brain stem are beyond the scope of this book.

The spinal cord extends through most of the length of the spinal canal formed by the vertebrae. A typical cross section of the cord has several characteristic features. The white matter forms a shape (see Figure 5-6) that is symmetrical about the median plane and has an anterior median fissure and a posterior median septum (dividing plate). The gray matter has roughly the shape of a butterfly or the letter **H** with two anterior and

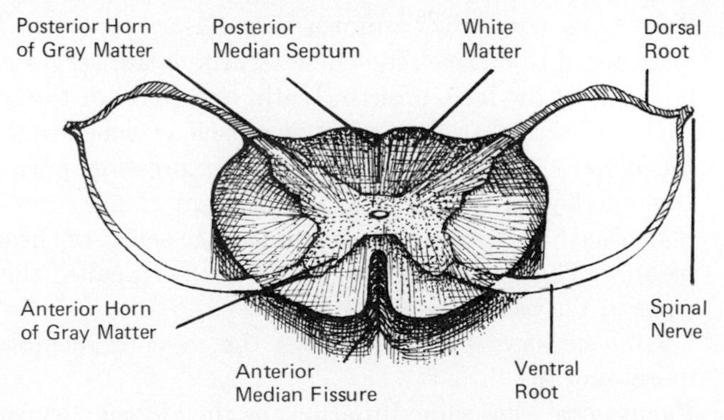

FIGURE 5-6. A cross section of the spinal cord. The outer region is white matter; the central area is gray matter, from which extend the dorsal and ventral roots that fuse to form spinal nerves.

two posterior *horns*. Nerve fibers leaving the spinal cord via the anterior horns exit via *ventral roots*, and those entering the spinal cord enter the posterior horns through *dorsal roots*. Thus, afferent neurons enter the CNS through dorsal roots, while efferent neurons are connected to the CNS through ventral roots. The dorsal and ventral roots combine outside the spinal cord to form a *spinal nerve*, which contains both afferent and efferent neurons that lead to and from the peripheral nervous system.

The Peripheral Nervous System

The peripheral nervous system includes the dorsal and ventral roots of spinal nerves, the spinal nerves themselves, and the *cranial nerves*. The cranial nerves are complex groups of nerves that are connected to the CNS above the spinal cord.

CRANIAL NERVES

There are twelve pairs of cranial nerves that are attached to the CNS at the base of the brain and at the brain stem. They serve specific peripheral structures, some of which will be considered in this text. The twelve cranial nerves are listed here with only a brief account of the function of each:

I. *Olfactory nerve*—arises in upper part of the nasal cavity and transmits information related to the sense of smell

II. *Optic nerve*—arises from retina of eye and transmits information related to vision

III. *Oculomotor nerve*—arises in muscles that attach to the eyeball and move it in various directions

IV. *Trochlear nerve*—arises in a single muscle attached to the eyeball

V. *Trigeminal nerve*—has a motor and sensory root. The former is distributed to muscles for chewing; the latter serves as sensors for some of the face, mouth, teeth, and parts of the scalp

VI. *Abducens nerve*—arises in a single muscle attached to the eyeball

VII. *Facial nerve*—serves muscles of facial expression, part of tongue (taste), the salivary glands, and the tear glands

VIII. *Statoacoustic nerve*—is associated with sense of hearing and balance. The statoacoustical nerve has been called the *auditory nerve* in the past.

IX. *Glossopharyngeal nerve*—supplies the mucous membranes and muscles of the throat

X. *Vagus nerve*—has same functions as the glossopharyngeal nerve and also has a complex distribution to internal organs of the chest and abdomen

XI. *Accessory nerve*—is divided into two parts, one of which serves the throat region. The other part serves two specific muscles in the neck, shoulders, and back

XII. *Hypoglossal nerve*—supplies the muscles of the tongue

SPINAL NERVES

There are thirty-one pairs of spinal nerves that join the CNS by dorsal and ventral roots. The spinal nerves interconnect with each other and with peripheral nerves by *rami* (singular, ramus) and form complex localized networks of nerves called *plexuses*. Each plexus then serves a number of specific organs or tissues in the peripheral system. The *solar plexus*, for example, is a complex of spinal nerves that innervates, i.e., provides the nerve supply for, the kidneys, liver, gonads, and some of the intestines.

The Nerve Impulse

A single axon can be teased out of a living animal under laboratory conditions and observed with relatively simple instruments. By touching the leads of an oscilloscope to the axon at two points near each other, one can observe the electrical changes that take place along the axon as information is passed to or from the animal's brain. A series of electrical "spikes" are observed, each exactly like the others. If the axon is part of the statoacoustic nerve, the rate at which the spikes occur, i.e., the repitition rate, depends on the sound entering the animal's ear. Further observation reveals that each spike corresponds to a change in potential of about 140 mV, which lasts about one millisecond. A careful observer would notice, in addition, that the pulses never overlap, i.e., one never begins while the previous pulse is still formed. Two pairs of electrodes placed at a known distance from the original pair could be used to measure the velocity at which the spike proceeds along the axon. In mammals the propagation velocity typically lies in the range between 50 m/second and 150 m/second. In general, it is found that the propagation velocity of the spike is directly proportional to the diameter of the axon. [Typical axon diameters in humans and most other vertebrates are about 10–20 microns (μm). Some invertebrates, like the lobster and giant squid have giant axons with diameters up to 200 μm; this characteristic makes these animals convenient subjects in nerve conduction studies.]

The above observations contain many of the known characteristics of the means by which information is transmitted along axons to or from the CNS. A few details can be added if an axon is probed with microelectrodes—one placed in the cytoplasm inside the axon (in the *axoplasm*), the other placed outside the neuron's plasma membrane. It is found that in the

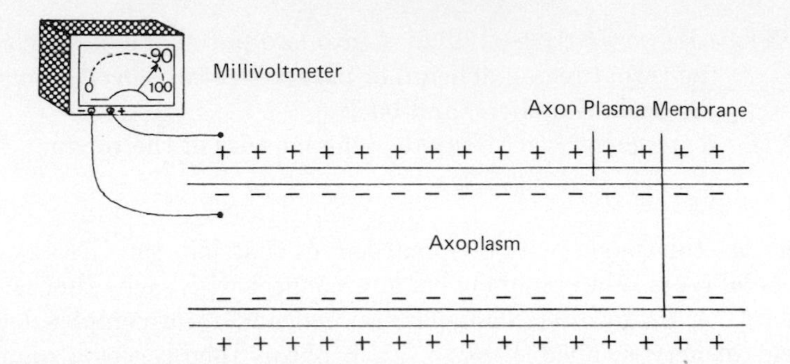

FIGURE 5-7. A schematic representation of an axon in the resting state. The dipole sheath across the plasma membrane, positive outside and negative inside, produces a difference in potential of about 90 mV across the plasma membrane.

resting state, i.e., when the axon is not transmitting a pulse, a difference in potential of about 90 mV exists, in which the outside is positive with respect to the inside. Figure 5-7 is a diagram of the axon in the resting state.

The Action Potential

When an axon is in the resting state, it is found that a stimulus is required to cause a change of state. The stimulus may be electrical, chemical, mechanical, or thermal in nature. In the laboratory, an electrical stimulus is the simplest to provide. The required stimulus is not simply how much difference in potential (voltage) is applied between two nearby points on the axon (or across the plasma membrane), but it also depends on the time it is applied. A stimulus of low voltage and long duration or a stimulus of high voltage and short duration can cause the resting axon to change from the resting state. In fact, the product of the voltage V and time t is a measure of stimulation in this case, and it can be shown experimentally that there is a minimum value of Vt required to "fire" a particular axon. The minimum stimulus required to change an axon from the resting state is called the *threshold* of that axon.

When an axon receives a stimulus that is equal to or in excess of its threshold, the axon undergoes a change in the resting state charge distribution across its plasma membrane. The change at a particular cross section is effectively a *depolarization* at first, in which the difference in potential across the membrane disappears (when the charge is the same on both sides of the membrane, the membrane is no longer polarized). However, the depolarization actually overshoots, and the outside of the membrane becomes negative with respect to the inside for a brief time. After a few milliseconds, the original resting state (polarized) is restored at a particular cross section. A graphical representation of the entire process that results from stimulation of the axon is shown in Figure 5-8. The "pulse" that appears, along with the ripples seen as it recovers, are called the *action potential* of the axon. The largest pulse is called the *spike potential*. Meanwhile, the depolarization has progressed along the length of the axon. The propagation process is like that of wave propagation, in that the disturbance at one point stimulates a neighboring point, and so on down the axon—*in both directions*. There are no directional characteristics inherent in the axon.

The pulse shown in Figure 5-8 is the form of the potential difference across the membrane of the axon at a single point as time proceeds. It is called a *monophasic* pulse. Experimentally, a monophasic pulse can be obtained by placing microelectrodes inside and outside the axon's membrane at one plane perpendicular to its length. If the two electrodes connected to the voltmeter are placed so that they touch the outside of the axon at two points close to each other along the length of the axon, a *diphasic* pulse is observed as a pulse that travels down the axon. This situation is seen in Figure 5-9, which shows the charge distribution at different times as the pulse passes the electrodes.

The monophasic pulse of Figure 5-8 can be seen to have some structure during the recovery phase. The *negative after-potential* is so-called because during this portion of the pulse the outside of the axon is less positive than

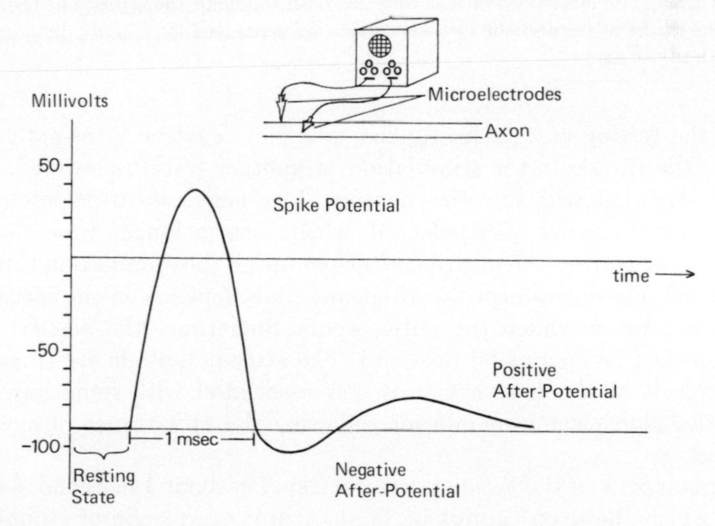

FIGURE 5-8. The action potential on an axon, observed in the form of a monophasic pulse. Two electrodes of an oscilloscope, one inside and the other outside of the axon, provide a trace of the action potential as an impulse moves along the axon. The spike potential is typically about 130 mV high.

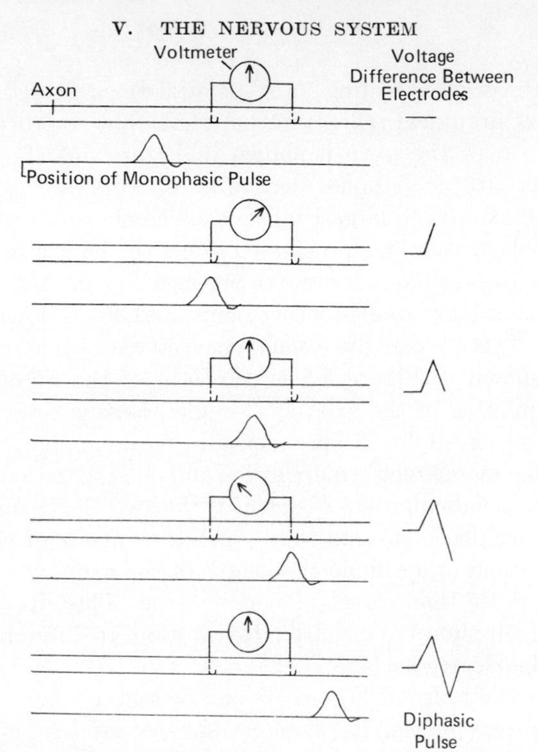

FIGURE 5-9. Formation of the diphasic pulse as an action potential propagates along an axon. The electrodes in this case are both touching the outside of the plasma membrane of the axon, and the two electrodes are separated by a short distance along the length of the axon.

during the resting state. During the time the negative after-potential is present, the threshold for stimulation of another pulse is lowered, i.e., a smaller stimulus will fire the neuron. The negative after-potential is followed by a *positive after-potential*, which lasts a longer time than the negative one. During the positive after-potential, the stimulation threshold is increased. The duration of the after-potentials depends on the species and type of neuron on which the pulse occurs. Sometimes the positive after-potentials last as long as 80 msecond. The after-potentials are thought to be the result of the recovery processes associated with ionic movement across the axon plasma membrane following the appearance of an action potential.

The main peak of the action potential lasts for about 1 msecond, a period during which the axon cannot be fired by any magnitude of stimulation. This period is called the *refractory period* of the neuron. This means that the maximum firing rate of a neuron is about 1000 pulses/second.

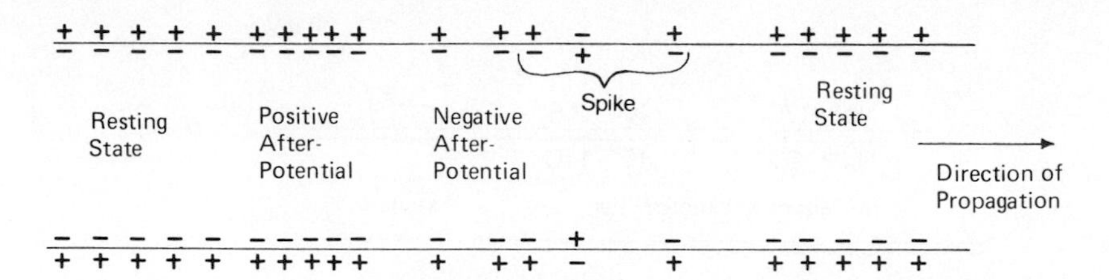

FIGURE 5-10. Details of the charge distribution inside and outside of an axon as an action potential is passing. The different regions of the charge distribution may be compared with the various portions of the monophasic action potential in Figure 5-8.

The details of the charge distribution inside and outside the axon plasma membrane at a passage of an action potential are seen in Figure 5-10.

Impulse Conduction

At the axon membrane, the concentration of sodium ions is about ten times greater in the external medium than in the axoplasm in the resting state. At the same time, there is a greater concentration of potassium ions inside than outside (by a factor of about 40). The relative numbers of ions, however, are such that the net charge is positive on the outside, a condition that results in a difference in potential of about 90 mV across the membrane. Figure 5-11a shows the ion concentration in the resting state. The concentrations of the Na^+ ions are maintained against the gradient by active transport, and it has been shown by experiment that ATP provides the energy required to maintain the concentration of sodium ions. The K^+ ions are relatively free to cross the membrane, and the electronegative condition in the axoplasm established by the exit of Na^+ ions causes the mobile K^+ ions to concentrate inside the axoplasm.

When a stimulus, electrical or otherwise, causes a disturbance at a point along an axon, the permeability of the membrane to Na^+ is altered so that the Na^+ ions rush into the axoplasm. Simultaneously there is a net outward flow of K^+ ions in the region of the disturbance. These ionic movements produce the spike potential, which is thus seen to be a *depolarization* phenomenon. After the initial depolarization, an immediate recovery process (*repolarization*) ensues, in which metabolic energy is expended to restore the original relative ion concentrations. Sodium is "pumped" back across the membrane to the outside, and the potassium diffuses back inside. The details of this process are not known, but the name *sodium pump* has been applied to the restorative mechanism. The gross features of the sodium pump are illustrated in Figure 5-11b.

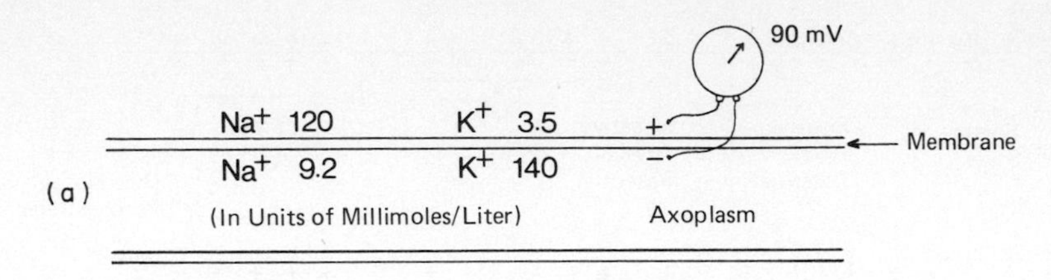

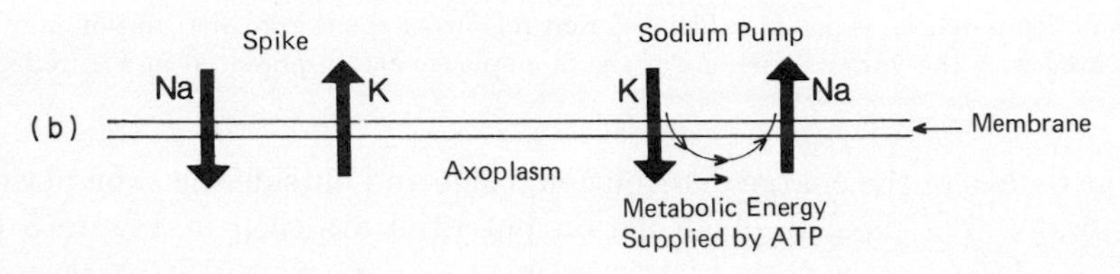

FIGURE 5-11. Ionic concentrations and movements of ions on either side and across the plasma membrane of an axon. (a) In the resting state the concentrations produce a difference in potential of 90 mV across the membrane. (b) As the spike passes a point along the membrane, sodium ions pass into the axon and potassium ions pass outward. After the spike has passed, the "sodium pump" restores the resting concentrations of the ions at the expense of metabolic energy within the axon.

The depolarization of a local region of an axon changes the permeability of the neighboring regions of the membrane to Na^+ ions so that the depolarization propagates in both directions away from the initial disturbance. Since the depolarization pulse has its energy supplied continuously along its course by the metabolism of the neuron, the pulse displays no attenuation, i.e., it does not suffer a decrease in amplitude as it propagates along the axon.

Synaptic Conduction

Although the action potential propagates in both directions along the length of an axon, the information conveyed by the pulse is meaningfully transmitted in only one direction in an intact animal. The unidirectional character of information flow along neurons is achieved at junctions between neurons called *synapses*. A synapse can be visualized (see Figure 5-12) as a specialized knob at the end of an axon that fits near, but separated by about one micron from, a receptor surface on a dendrite. The axon is referred to as the *presynaptic fiber*, and the dendrite is called the *postsynaptic fiber*. The knob at the end of the presynaptic fiber has small compartments containing *transmitter molecules*, which are released when a spike reaches the synapse. The receptor surface has areas sensitive to the transmitter molecules, which means that the permeability to Na^+ is altered on the postsynaptic fiber so that an action potential appears on the postsynaptic fiber and propagates away from the synapse.

The case considered here is chemical transmission in a synapse. In a few cases, it has been demonstrated that an electrical mode of transmission is possible at synapses, but electrical transmission is imperfectly understood and will not be considered here. The only substance that has been definitely identified as a synapse transmitter is *acetylcholine* (ACh), although others may be involved.

When ACh is released on arrival of the spike potential at the presynaptic knob, it diffuses across the synapse gap. The diffusion process requires less than a millisecond. The presence of ACh on the postsynaptic fiber changes the membrane permeability of Na^+ ions and generates a new action potential on the postsynaptic fiber. It is, of course, necessary that the ACh be inactivated immediately after initiating the depolarization pulse on the postsynaptic fiber so that the synapse is ready to conduct another pulse. The inactivation of ACh is accomplished by an enzyme, *acetylcholinesterase*. Some of the very efficient and lethal nerve gases function by joining onto the active site of the enzyme acetylcholinesterase and disabling the enzyme. The result is immediate dysfunction of the synapses with consequent disruption of communication in the nervous system and death.

The ACh released at the presynaptic knob appears in very small, apparently quantized, amounts—about 10^{-20} mole. This means that about 1000 molecules of ACh are released at each crossing of a pulse at a synapse.

It is clear then that because of the structure and chemistry of the presynaptic and postsynaptic components of a synapse, depolarization pulses can propagate in only one direction. The synapses serve as the

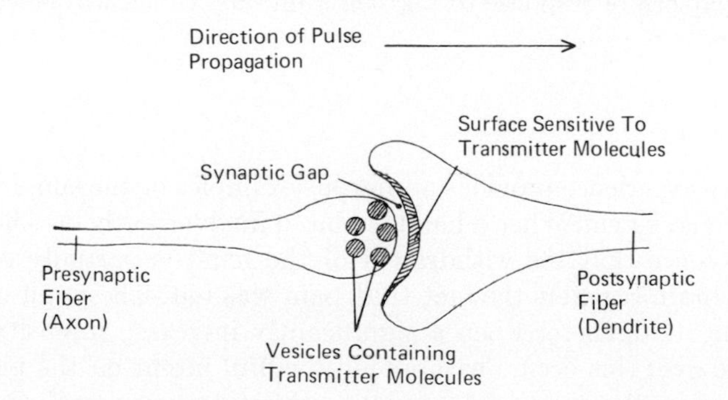

FIGURE 5-12. A synapse (chemical mode of transmission). A pulse along the axon on the left releases the transmitter molecules (ACh) when it reaches the knob at the end of the axon. These transmitter molecules diffuse across the synaptic gap and recreate the pulse on the end of the dendrite. The recreated pulse proceeds to the right.

"one-way valves" of the nervous system's electrical circuit, blocking flow in the unwanted direction.

Nervous System Function

The human nervous system performs a myriad of very complex functions, very few of which are understood in even a qualitative way. The interpretive capabilities of the brain will be discussed in later chapters, when the visual and auditory senses are considered. The mechanisms by which the brain encodes information from sensory organs and stores it for future use are only beginning to be studied. The means by which the brain retrieves information, makes comparisons, and performs its seemingly creative manipulation of information is not understood in any quantitative or qualitative way. There are many interesting phenomena of brain function that are neither understood nor even adequately classified in a scientific manner. The nature of sleep and the electrical phenomena observed in brain activity are being intensively studied. More exotic phenomena, like hypnosis and amnesia, apparently must await progress before the details of such functions can be explained in any meaningful and satisfying scientific way. The possibility exists, of course, that some types of phenomena are associated with brain function that will require understanding physical principles that are not presently known. These include the phenomena such as extrasensory perception, telekinesis, clairvoyance, and a number of other so-called psychic manifestations of the brain. It is perhaps fair to say that no conclusive scientific evidence presently exists that confirms the existence of psychic phenomena.

Among the simpler functions of the central nervous system is a very primitive pattern of response to a given stimulus, which will be considered briefly here.

The Reflex Arc

Everyday experience provides numerous examples of one simple function of the nervous system. When a hand is placed inadvertently on a hot object, one can observe that the withdrawal of the hand is partially completed before the brain registers the fact that pain was felt. The pupil of the eye decreases in its diameter when a significantly increased intensity of light falls on the eye; this occurs without any willful intent on the part of the person viewing the light and occurs without his ever realizing that a physiological change has taken place. A more informative example can be observed when a weak acid is placed on the skin of a frog. The animal tries to remove the irritation by a coordinated series of leg movements. If the animal's brain is destroyed and acid is applied to its skin, the same response occurs. But if the animal's spinal cord is destroyed or the nerves from the cord to the legs are severed, the movements do not take place. The pathway of information apparently consists of receptors in the skin, afferent neurons to the spinal cord, where synaptic connections transmit the signal to efferent neurons that are connected to the appropriate skeletal muscles. The total pathway of information in such a system is called a *reflex arc*.

All reflex arcs are not limited to the peripheral system and spinal cord but may include the brain stem as well. Most reflex arcs involve three or more neurons, though the simplest involve only two. Figure 5-13 illustrates two- and three-neuron reflex arcs. The knee-jerk reflex is an example of a two-neuron arc, in which the receptors are in a muscle, and the motor fibers (i.e., effector terminal fibers) end on muscle fibers of the muscle. In three-neuron arcs, of which withdrawal of a limb in response to pain is an example, an intermediate neuron (called an *interneuron*) is synapsed to both the afferent and efferent neurons of the arc. Usually the afferent neuron is also synapsed to neurons that transmit the information through ascending fiber tracts of the CNS to the brain, where the pain is registered.

Many reflexes are important in diagnostic medicine. The clinical examination of reflexes may enable one to determine the nature and location of certain neurological disorders.

Brain Function

The functions of the brain in higher animals are so complex and so poorly understood that it is impossible to describe them systematically. The technique of electrically stimulating certain specific areas of the brain and observing the reactions of the subject or listening to the subject's description of sensations has enabled neurologists to map the motor areas of the brain and associate certain sensations with specific portions of the brain. Such a process is, of course, useful in localizing brain *lesions* (changes in tissue due to the disease or damage) but does not give any information on brain function at the cellular level. The details of brain function at the cellular and molecular levels must await further progress.

Electroencephalography

In a living vertebrate, it is possible to detect changing electrical potentials on the surface of the brain. In humans, electrodes can be attached at various locations on the scalp, and the potential variations at the brain surface can be detected at the skin surface because the skin and intervening

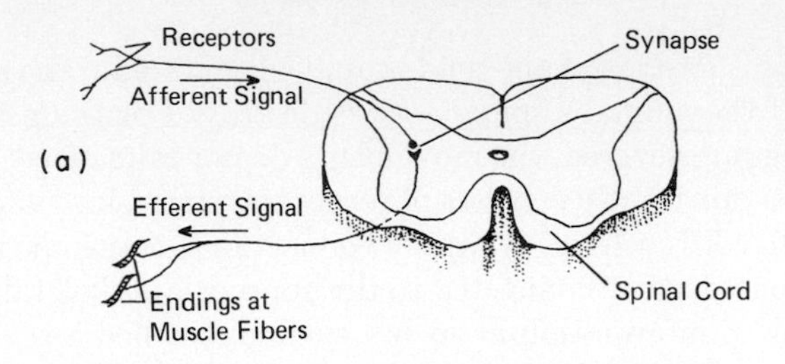

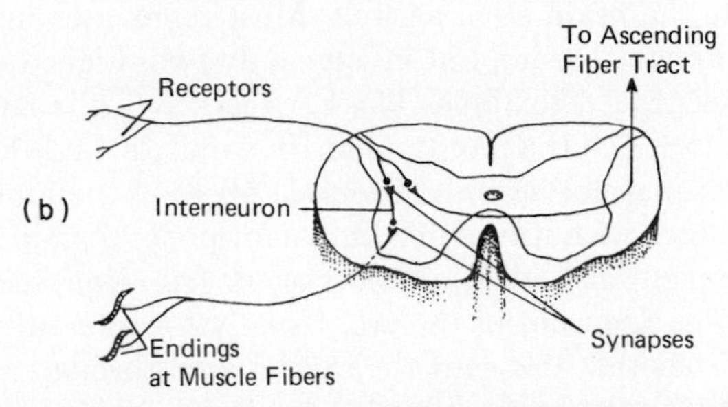

FIGURE 5-13. Reflex arcs. (a) A two-neuron arc, synapsed in the gray matter of the spinal cord. (b) A three-neuron arc, synapsed through an intermediate neuron (interneuron) to a neuron leading to the ascending fiber tract to the brain.

tissues are conductors of electricity. A record of the brain's electrical potential variations sensed on the scalp is called an *electroencephalogram* (EEG).

The potential differences between the surface of the cortex of the brain and a neutral point (such as the nose or ear) are as large as 10 mV, but when observed at the scalp, most potential differences are less than 100 μV. This difference is due to the relatively high electrical resistance of the intervening bone.

Normal humans show a group of characteristic rhythms in their EEG's. The most obvious rhythms are a wave form of amplitude less than 20 μV and of frequency near 10 Hz. This pattern has been termed the *α-rhythm* or simply *α-waves*. There are a number of identifiable EEG patterns that occur in humans, all of which are labeled with Greek letters—β-waves, δ-waves, θ-waves, etc. Many EEG patterns have been correlated with various physiological phenomena such as sleep, visual perception, maturation of the subject, and alarm reactions. More importantly, the EEG can be used to identify certain abnormalities like epilepsy and brain tumors.

At the level of understanding brain function, however, the EEG gives no information about neuron activity or the interrelation of the activities of groups of neurons, except that some electrical activity is present. Since the EEG is a means of observing the overall electrical activity of the brain, it is tempting to associate the EEG with fundamental brain function. Such an association is not possible at this time.

It seems then that the complex functions of the brain are not susceptible to direct measurement and analysis at the present time. Nevertheless, some of the fundamental brain functions are currently under study. A reasonable starting point seems to be to inquire into how the information in the electrical pulses on afferent neurons are stored in the brain, i.e., how the memory and learning processes initially confirm themselves in the CNS. Only then can one reasonably inquire into the mechanisms of access to information, the comparison functions of the brain, and the many functions of the brain that are familiar to everyone.

Learning and Memory

Learning processes in animals produce behavioral changes that are often subject to objective measurement. When this is the case, one can measure quantitatively the behavioral changes resulting from learning experiences that have altered the CNS in an experimental animal. There has been considerable speculation as to how a learning experience changes the CNS. Is it an ordering of synaptic connections? Is it an electrical phenomenon that forms "resonant circuits?" Or is it a chemical change that occurs in specific "memory molecules," or perhaps a more general chemical change throughout the cells of the CNS? These questions will be briefly considered here.

Memory Models

The electrical nature of the information transfer mechanism along neurons is suggestive of electrical mechanisms in the process of storing information in the brain. The complex interconnections that occur between the billions of neurons of the CNS suggest that an ordering of specific pathways by making (or breaking) synaptic connections is a mechanism capable of storing vast amounts of information. The presence and nearly continuous production of very large molecules (some of which, like DNA and RNA, are already known to store genetic information), suggests that chemical storage of learned information might be feasible.

ELECTRICAL AND SYNAPTIC MODELS

The concept that the electrical pulses transmitted by neurons to the brain either establish "circuits" or cause existing circuits to resonate has been under consideration for many decades. The notion that remembering something corresponds to exciting a current in one of the resonant circuits has apparently made this model appealing because it provides what appears to be a relatively simple mechanism for access (finding the stored information). Unfortunately, the more fundamental question of how the resonant circuits are established is not so obvious, and this seems to be the principal defect in the model.

One of the means that has been postulated for establishing electrical resonant circuits in the CNS is the activation of specific neuronal paths through synaptic connections. In such models, a particular set of input pulses in some way causes certain synaptic junctions to be established or to become active. Not all synaptic models assume that the pathways established in this manner are electrically resonant circuits. Some models simply assert that the synaptic connections form "memory traces": paths that can be retraced by signals at a later time.

The electrical and synaptic models suffer from a common drawback—neither have been (and perhaps cannot be) subjected to experimental verification. For that reason they will not be considered further here.

CHEMICAL MODELS

Holger Hydén (1961) performed a series of experiments in which he clearly demonstrated that extraordinarily large amounts of RNA were produced in brain cells of rats while the animals were undergoing learning experiences. The implication of RNA in the memory process suggests that the sequence of nucleotides might be used to store information, at least initially, in the brain in much the same way that it stores genetic information. For two reasons this mechanism has been rejected. First, it was not clear how the presence or absence of spike potentials could control the ordering of the sequence of nucleotides in RNA chains. The other reason is based on energy considerations. Considering the large amount of information sent to the brain per unit time by sensory organs, a large rate of energy expenditure would be required to arrange in proper order the vast numbers of nucleotides necessary to record information at such a rate. It is, however, a physiological fact that the human brain uses energy at a very low rate, even when the subject is deeply involved in mental activity.

An unusual mechanism for recording information in RNA was suggested by Fong (1969). He supposed that RNA might have an unusual property that permits molecules to become polarized in response to an electric field and maintain that polarization when the field is removed. Materials with this property are called *ferroelectric* materials. The ferroelectric model for recording information in the brain assumes that "tapes" of RNA are produced by the usual transcription process from DNA, and as the tapes are formed, each nucleotide of the RNA is either polarized or not polarized by spike potentials reaching the cell body at the time of its attachment to the tape. The result of such a recording mechanism would be a binary code, like dots and dashes, on the RNA. In this model, the RNA is the molecule of initial storage, i.e., the means of recording short-term memory. Since RNA is the template from which protein is made (see Chapter II), it is assumed that the memory information on the RNA is ultimately transferred to proteins (or polypeptides) to be confirmed into permanent memory storage. It is not clear how such a storage mechanism provides for later access or recall. It is likewise not clear how the polarization coding on RNA controls or changes the production of the polypeptides.

One of the more interesting and provocative aspects of a chemical model, like the one described here, is the possibility of experimentally testing at least some of the aspects of the models. Some recent attempts to test the chemical model will be discussed briefly.

Memory Experiments

A crucial point in the proposed model discussed in the preceding section hinges on the ferroelectric character of RNA. Since the model was proposed, it has been demonstrated experimentally that RNA is indeed a ferroelectric material.

A series of controversial experiments have been performed in the past decade that strongly support the chemical model of memory. McConnell (1962) reported having trained planarians (flat worms) that were ground up and fed to untrained planarians, which then showed measurable behavorial changes. These changes suggested they had acquired the information originally learned by the cannibalized worms. These experiments were subjected to considerable criticism and were not generally accepted as convincing evidence of chemical transfer of information.

Impressive evidence in support of a chemical memory model was the result of experiments by Ungar and his associates (1965). These experiments, in which rats were the experimental animals, demonstrated interanimal transfer of information by removing chemical extracts from the brains of trained rats, injecting the chemicals into untrained rats, and testing the recipient animals to demonstrate their acquired information from the donor animals. Such transfer experiments have since been performed in a number of laboratories. Certain specific polypeptide sequences have been reported as being the agent of transfer of a particular behavioral change. Many of Ungar's experiments have been verified by other workers; still, the area of memory transfer has been the object of considerable skepticism, and the ultimate contributions of work in this area are yet to be evaluated.

It seems that the chemical models of memory represent a start toward understanding some of the most fundamental mechanisms associated with memory and learning processes. It further seems obvious that a great deal of research remains to be done to understand the functions of the nervous system.

References and Suggested Reading

DiCara, L. V. (1970). Learning and the autonomic nervous system. *Sci. Amer.* **222,** 30 (Offprint 525).

Eccles, J. C. (1957). "The Physiology of Nerve Cells." Johns Hopkins Press, Baltimore, Maryland.

Eccles, J. C. (1965). The synapse. *Sci. Amer.* **212,** 56 (Offprint 1001).

Fong, P. (1969). *Physiol. Chem. Phys.* **1,** 24.

Hodgkin, A. L. (1964). *Science* **145,** 1148.

Huxley, A. F. (1964). Excitation and conduction in nerve. *Science* **145,** 1154.

Hydén, H. (1961). Satellite cells in the nervous system. *Sci. Amer.* **205,** 62.

Kandel, E. R. (1970). Nerve cells and behavior. *Sci. Amer.* **223,** 57 (Offprint 1182).

Katz, B. (1966). "Nerve, Muscle, and Synapse." McGraw-Hill, New York.

McConnell, J. V. (1962). *J. Neuropyschiatry* **3,** 542.

Miller, W. H., Ratliff, F., and Hartline, H. K. (1961). How cells receive stimuli. *Sci. Amer.* **205,** 222 (Offprint 99).

Penfield, W., and Rasmussen, T. (1960). "The Cerebral Cortex of Man." Macmillan, New York.

Ranson, S. W., and Clark, S. L. (1959). "The Anatomy of the Nervous System," 10th ed. Saunders, Philadelphia, Pennsylvania.

Sperry, R. W. (1969). The growth of nerve circuits. *Sci. Amer.* **201,** 68 (Offprint 72).

Unger, G., and Oceguero-Navarro, C. (1965). *Nature* (*London*) **217,** 1259.

CHAPTER VI Sound and Hearing

To most animals, including humans, the ability to perceive sounds is an important survival function. The production and detection of sound are among the important means by which animals communicate, locate food, and sense danger. In humans, and perhaps in other higher animals, the sense of hearing provides, in addition to the essential survival functions, esthetic pleasure in the form of music and rhythm.

In this chapter, the physical nature of sound will be considered with emphasis on those properties of sound that affect the sense of hearing. In examining the anatomical and physiological characteristics of hearing mechanisms, the human auditory system will be used as an example of a highly developed, yet rather typical, sound detection system. Finally, the close association of the auditory and nervous systems is considered by an examination of various models that have been suggested to explain the considerable capabilities of the human sense of hearing.

Some Physical Aspects of Sound

Everyday experience tells us that vibrating objects produce sounds. We can often detect the mechanical effects of vibration without using our ears at all; occasionally we can simultaneously detect the effects of vibration using the senses of touch and hearing. It therefore seems natural to associate sound with vibration.

Vibrations are back-and-forth motions of objects. When a vibrating object executes its cyclic motion over and over with equal periods of time required for each cycle, the motion is called an oscillation, which is said to be *periodic*. If the vibrating object is not in a vacuum, its oscillatory movement will impart energy to the medium surrounding the object. This energy as it spreads away from the vibrating source is *sonic energy*, or sound.

The ancient natural philosophers conjectured that sound was a flow of particles away from a source. We know this cannot be the case, because we can observe that sounds from two or more sources (or even a single sound and its echo) freely pass through one another, neither interfering permanently with the other.

Another simple experiment may help to characterize the physical nature of sound. If the rate at which energy is conveyed away from a vibrating sound source is changed by factors of millions or more, it is found that the speed at which the sound travels in a particular medium is a constant.

On the basis of these and many other considerations, it is known that sound is transmitted by *waves*, disturbances in the medium that propagate without bodily translating the particles of the medium along with the disturbance. Waves on a taut rope or on the surface of water are familiar examples in which a disturbance moves along while the particles in the rope or water simply jiggle locally. In both these examples as well as in sound waves, one can observe that the speed of propagation depends only on the physical characteristics of the medium and not on the magnitude of the disturbance. One may also observe that more than one wave can propagate along the rope or water surface without permanently altering each other.

Sound Wave Characteristics

Sound waves in air, in contrast to waves on a rope or on the surface of water, are *longitudinal waves*. The molecules of air oscillate along a line that lies in the direction the wave is moving. A vibrating source of sound exerts pressure on the air molecules in its immediate neighborhood, and the air molecules in turn transmit the pressure to their neighbor molecules. Thus, the sound wave is a pressure wave. The time required for the source to execute a single cycle of its vibration is called the *period* of oscillation, and the time required for an air molecule to oscillate through one cycle about its equilibrium position is the period T of the wave. The frequency f of the wave is the reciprocal of the period ($f = 1/T$) and is the number of cycles executed per unit time by each air molecule as the wave passes. A cycle per second is called a *Hertz* (Hz). Each particle in the medium oscillates about its equilibrium position in a manner that is quite accurately described by a sinusoidal expression,

$$y = y_0 \sin 2\pi ft \tag{6-1}$$

in which y is the displacement of a particle from equilibrium, t is time, and y_0 is the *amplitude*, or maximum displacement from equilibrium of the particle.

At any point along the wave, the pressure p varies in the same way:

$$p = p_0 \sin 2\pi ft \tag{6-2}$$

In this case, p_0 is the maximum pressure, or pressure amplitude, at any point along the wave. Equation (6-2) formulates the wave as a function of time, i.e., it describes the motion of a single particle or group of particles at a particular location x as time passes. Figure 6-1 shows a wave as a function of position; this is how the wave would look if it were photographed at a particular time. In this representation it can be seen that the crests

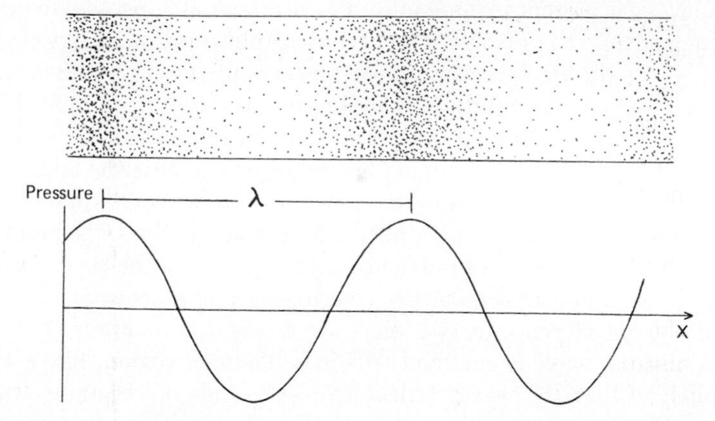

FIGURE 6-1. A pressure wave in a gaseous medium. The wave is shown here at a particular instant in time. The particles shown indicate the compressions and rarefactions of the gas as a function of position. The separation λ between successive maxima (or minima) of pressure is the wavelength of the pressure wave.

and troughs of the wave (the compressed and rarified regions in the case of a pressure wave) are repeated at equal spatial intervals in the x direction. The distance from one pressure maximum to the next is the *wavelength* λ of the wave. The wavelength and frequency are related to the speed of propagation v of any wave by

$$v = f\lambda \tag{6-3}$$

The waves we have been describing are called *traveling waves,* in which energy is transported by a wave from a source in a direction toward a point arbitrarily far away from the source. The displacement from equilibrium of any particle in the medium can be expressed in terms of the wave parameters we have already defined by the equation

$$y = y_0 \sin 2\pi \left(\frac{t}{T} \pm \frac{x}{\lambda} \right) \tag{6-4}$$

This equation, when the positive sign is used, represents a wave traveling to the left (or negative x direction); when the negative sign is used, it represents a wave traveling in the positive x direction. In a longitudinal wave, the displacement y is in the same direction as x, the direction in which the wave is propagating. The amplitude of the wave is y_0. Thus, a sound wave may be thought of as either a displacement wave or a pressure wave.

STANDING WAVES AND RESONANCE

Two sound waves in a medium can *interfere* with each other. When more than one wave is present in a medium, the *principle of superposition* applies, which means that the displacement from equilibrium of a particle in the medium is just the algebraic sum of displacements due to each wave. Thus if two waves of equal amplitude both happen to have maximum displacement in the same direction at a certain point in the medium, the displacement of a particle in the medium at that point will be twice as large as when only one wave is present. Similarly, if the two waves happen to have maximum displacement, but in opposite directions, at the same point in the medium, the displacement of particles at that point will be zero. These two cases are illustrations of *constructive interference* and *destructive interference,* respectively. Interference of two waves is illustrated in Figure 6-2.

When a sound wave is confined within a bounded region, like a tube or organ pipe, so that the waves reflect from the ends of the tube, waves of equal amplitudes propagate in opposite directions within the tube. If the ends of the tube are closed, the particles in the medium cannot move longitudinally at the end points, and boundary conditions are imposed such that the ends of the tube must have zero displacement. A point at which the

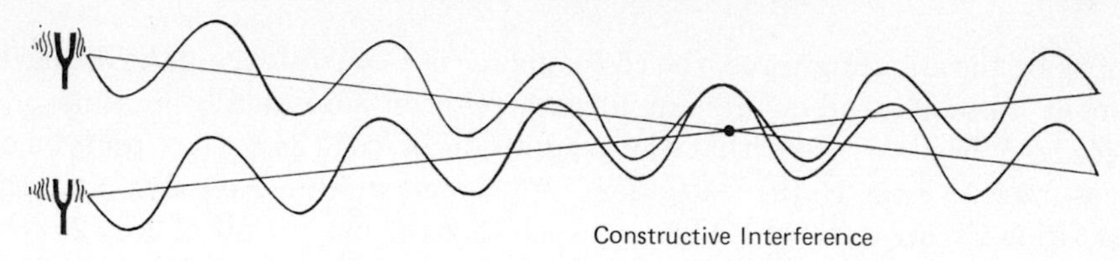

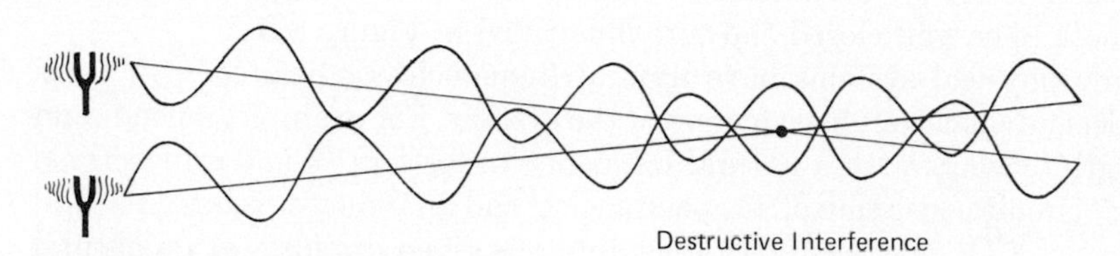

FIGURE 6-2. Interference of sound waves. In the upper illustration two sound waves constructively interfere at the point of crossing, where both waves are at maximum pressure simultaneously. In the lower diagram, the waves destructively interfere; one is at maximum pressure, the other at a minimum.

displacement is zero is called a *node*. Interference between the two waves traveling in opposite directions produces regions of maximum displacement called *antinodes* separated by nodes. The resulting wave inside the tube is called a *standing wave*, in contrast to the two traveling waves that combine to produce the standing wave. At nodes in the standing wave, no motion occurs in the medium, while at antinodes the particles are oscillating about an equilibrium position with displacements that are twice the amplitude of each of the two traveling waves that comprise the standing wave.

Since the displacements of particles in a standing wave can be described as the algebraic sum of displacements of the two waves traveling in opposite directions, the displacements in the standing wave can be represented by

$$y = y_0 \sin 2\pi \left(\frac{t}{T} - \frac{x}{\lambda}\right) + y_0 \sin 2\pi \left(\frac{t}{T} + \frac{x}{\lambda}\right) \qquad (6\text{-}5)$$

in which the first term on the right represents a wave of amplitude y_0 traveling in the positive x direction, and the second term represents a wave of the same amplitude traveling in the negative x direction. If we recall the trigonometric identity, $\sin(A + B) + \sin(A - B) = 2 \sin A \cos B$, Equation (6-5) becomes

$$y = 2y_0 \sin \frac{2\pi t}{T} \cos \frac{2\pi x}{\lambda} \qquad (6\text{-}6)$$

which is an equation representing a standing wave. At a particular position

x along the standing wave, the cosine factor is a constant, and the displacement y oscillates about its equilibrium position sinusoidally as time proceeds. When x is such that the magnitude of $\cos(2\pi x/\lambda)$ is unity, i.e., $\cos(2\pi x/\lambda) = \pm 1$, the excursion of y is maximum, i.e., $2y_0$, and the position x is an antinode. When x is such that the magnitude of $\cos(2\pi x/\lambda)$ is a minimum, or zero, y is always zero and that position of x is a node.

Tubes or pipes with open ends may also support standing waves. In this case, there is always an antinode at an open end. Standing waves in pipes with both open and closed ends are illustrated in Figure 6-3.

Many physical systems have natural frequencies associated with them that depend on certain parameters of the system. For example, a pendulum of length l swings with a natural frequency of $2\pi(l/g)^{1/2}$, and an electrical "tank" circuit consisting of a capacitance C and an inductor L has a natural frequency of $(1/2\pi)(L/C)^{1/2}$. If a pendulum is given impulses at its natural frequency (or a person in a swing is pushed with impulses at the natural frequency), the pendulum acquires a large arc of swing (amplitude). The pendulum is then said to be pumped at its *resonant frequency,* and the phenomenon (acquiring large amplitudes when pumped at the natural

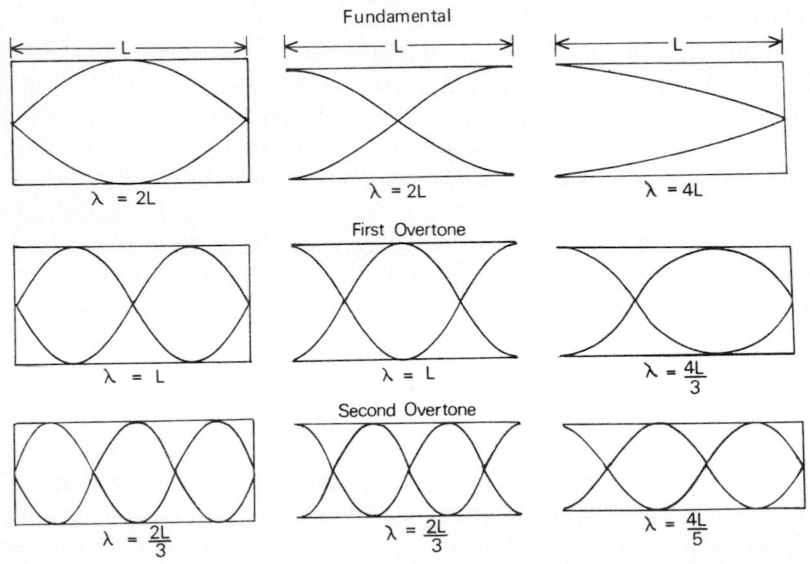

FIGURE 6-3. Standing waves in pipes. The wavelengths of the fundamental and first two overtones are illustrated for pipes of length L that are closed at both ends, open at both ends, and open at one end but closed at the other. The boundary conditions require a node at a closed end and an antinode at an open end. The vertical extent of each curve inside a pipe represents the maximum displacement of particles in the gas at various locations along the pipe (though the actual particle displacements are longitudinal).

frequency) is called *resonance.* A pipe or tube when filled with a medium, air, for example, has several frequencies at which it will resonate.

Before considering resonant sound waves in pipes, let us note that the most common medium in which we observe sound waves is air. The velocity of sound in air depends on the atmospheric pressure and temperature, but for our purposes, we may assume that it has a typical value of about 330 m/second (roughly 1100 ft/second). When a pipe is filled with air, its resonant frequencies can be calculated using the boundary conditions (closed ends are nodes; open ends are antinodes). Figure 6-3 shows the three cases: (1) both ends closed, (2) both ends open, and (3) one end closed with the other open. For a pipe of length L, the longest wavelength that satisfies the boundary conditions in each case is seen to be

$$1. \quad \lambda = 2L \qquad 2. \quad \lambda = 2L \qquad 3. \quad \lambda = 4L$$

The frequencies corresponding to each of these wavelengths are obtained from Equation (6-3):

$$1. \quad f = \frac{v}{\lambda} = \frac{v}{2L} \qquad 2. \quad f = \frac{v}{2L} \qquad 3. \quad f = \frac{v}{4L}$$

The frequency corresponding to the longest wavelength that will resonate in a pipe is called its *fundamental frequency.* The next higher frequency above the fundamental to which a pipe is resonant is called the *first overtone,* or *second harmonic.* Figure 6-3 shows the positions of nodes and antinodes in each of the cases cited above for the fundamental and the first two overtones. For the nth overtone in each case, the corresponding frequencies are

$$1. \quad f = \frac{(n+1)v}{2L} \qquad 2. \quad f = \frac{(n+1)v}{2L} \qquad 3. \quad f = \frac{(2n+1)v}{4L}$$

DIFFRACTION OF SOUND WAVES

Diffraction is a property of waves that permits them to bend around the edge of an obstacle. All of us have heard sounds that came from around corners of buildings, and since the sound could not have traveled in a straight line, it must have been diffracted at the edge of the building. Experiments show that diffraction is pronounced only when the dimensions of an obstacle are comparable to the wavelength of the waves striking the edge of the obstacle.

These facts about diffraction help us to understand the physical reasons for the localization of sounds. Long wavelength audible sound waves might have wavelengths of several meters, comparable to the size of a house. The source of these low-frequency waves is therefore difficult to locate by hearing the wave. On the other hand, high-frequency sound waves, say at

10 kHz, have wavelengths of less than 3 cm. When passing obstacles, these waves produce relatively sharp "shadows," and the source of such waves is relatively easy to locate.

Subjective Characteristics of Sound

In describing various qualities of sound when sensed by the human ear, we use terms like *pitch, timbre,* and *loudness*. Each can be related to physical properties of sound waves.

Pitch is the quality of sound that we casually describe by phrases like "high note" or "low tones." Pitch is directly related to frequency of a sound wave—the higher the frequency, the higher the pitch. Subjective judgments by humans in discrimination of pitch (deciding whether one tone is higher than another) vary somewhat with the frequency itself. At higher frequencies, the ear is slightly more discriminating to a relative change in frequency. The frequency range of human hearing extends from about 20 to 20,000 Hz and varies considerably between individuals. In general, as a person ages, the upper limit of frequency to which his ear responds decreases. The human ear is capable of distinguishing a difference in pitch of about 3 Hz at 1000 Hz. That is to say, the ear has a frequency discrimination considerably better than 1%.

We have been discussing sound waves as if they were always pure sinusoidal waves. Such pure tones rarely occur in nature. Most sounds are complex wave forms, as illustrated in Figure 6-4. These complex forms can be thought of as being composed of a number of pure sine waves, each of which is a harmonic (or multiple) of the fundamental frequency of the complex wave. The quality of sound called timbre is that quality that enables a listener to distinguish between a trumpet, violin, or an oboe, each of which is playing a "middle C." Each instrument plays the same fundamental frequency (middle C is 262 Hz), but each instrument emits waves with different harmonic content, i.e., different amounts of various overtones of the fundamental. The relative harmonic content in a complex sound wave is the physical basis of timbre.

Loudness is a quality of sound that is related to the physical property of *intensity*. Sound intensity is the quantity of energy in a sound wave passing through a unit area in a unit time. The human ear responds to intensity changes in a nonlinear manner, e.g., doubling the intensity does not correspond to a subjective doubling of loudness. Further, sounds at different frequencies require widely different intensities in order to be subjectively judged of equal loudness. The measure of loudness is usually expressed in terms of sound pressure level L, defined by

$$L = 20 \log_{10} (p/p_0) \quad \text{decibels} \tag{6-7}$$

In this equation, p is the pressure amplitude of the sound wave being measured and p_0 is the threshold pressure amplitude of the "average ear," which has been arbitrarily set at 2×10^{-5} newtons/m^2. The unit of sound pressure level, the decibel (db), is convenient to describe a wide range of pressure values. From Equation (6-7), it can be seen that a change in

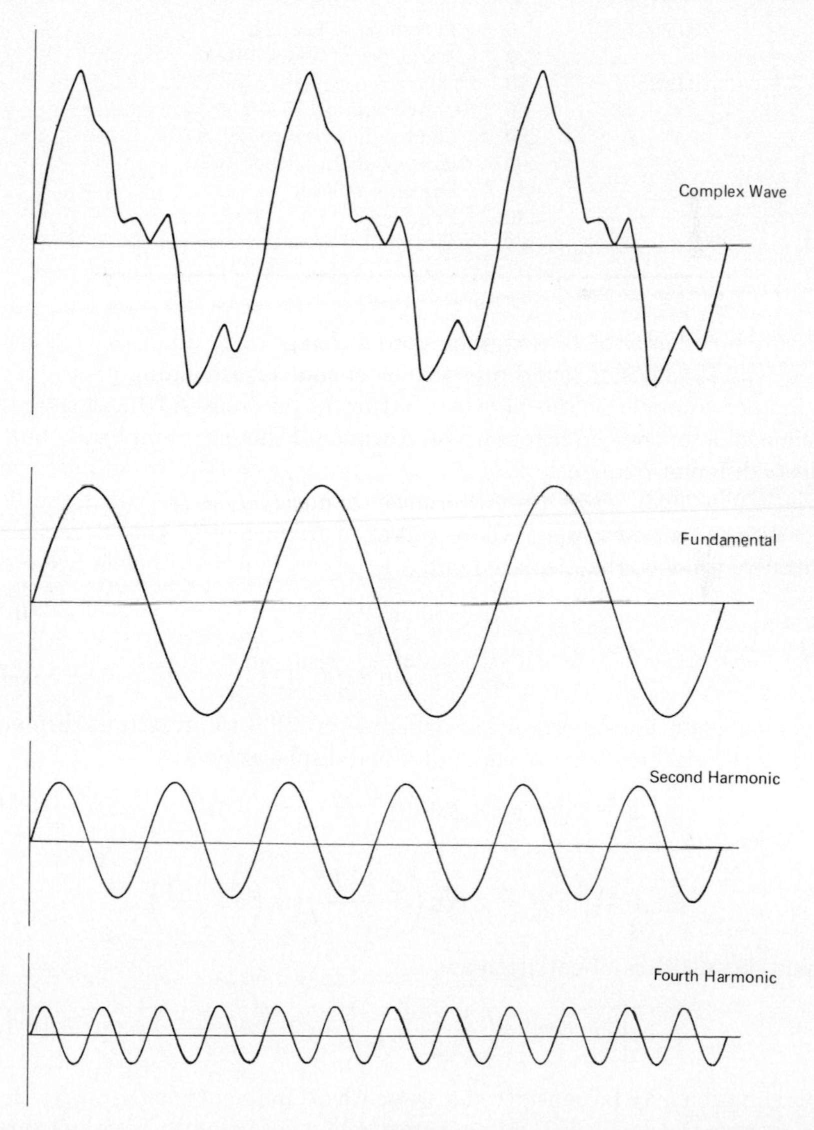

FIGURE 6-4. A complex wave composed of a fundamental and two harmonic frequencies. In accordance with the principle of superposition, the amplitude at any position along the complex wave is the algebraic sum of the amplitudes of its components.

Table 6-1 A Scale of Sound Pressure Levels

Pressure (newtons/m²)	Pressure level (db)	Example
0.00002	0	Threshold of hearing
	20	A whisper at about 4 feet
0.002	40	City noise at night
	60	Conversational speech
0.2	80	Loud radio: very heavy traffic
	100	Riveter about 35 feet away
20	120	Discomfort level
	140	Pain level
2000	160	Mechanical damage to human ear

pressure by a factor of 10 corresponds to a change of 20 db. Table 6-1 gives some typical values of sound pressure level and corresponding pressures.

Another example of the physical nature of our sense of hearing is the phenomenon of *beats*. When two sound waves of the same amplitude but of *slightly* different frequencies, say f_1 and f_2, impinge on the ear, we perceive a variation in loudness at the difference frequency $f_1 - f_2$, called the *beat frequency*. At a point in air where waves of frequency f_1 and f_2 are simultaneously present, the displacements of the medium due to each wave are

$$y_1 = y_0 \sin 2\pi f_1 t \tag{6-8}$$

and

$$y_2 = y_0 \sin 2\pi f_2 t \tag{6-9}$$

Invoking again the superposition principle, we find the resultant displacement is the algebraic sum of the individual displacements:

$$y = y_1 + y_2 = y_0(\sin 2\pi f_1 t + \sin 2\pi f_2 t) \tag{6-10}$$

and since

$$\sin A + \sin B = 2 \cos\left(\frac{A - B}{2}\right) \sin\left(\frac{A + B}{2}\right)$$

Equation (6-10) can be written as

$$y = \left[2y_0 \cos 2\pi\left(\frac{f_1 - f_2}{2}\right) t\right] \sin 2\pi\left(\frac{f_1 + f_2}{2}\right) t \tag{6-11}$$

This equation may be considered a wave whose frequency is $(f_1 + f_2)/2$, or the average frequency of the two waves, and whose amplitude (the expression in brackets) varies slowly with time at a frequency $(f_1 - f_2)/2$. Since a maximum in amplitude, i.e., a beat, occurs whenever $\cos 2\pi[(f_1 - f_2)/2]$ is 1 or -1, the number of beats per second f_b is twice the frequency $(f_1 - f_2)/2$, or

$$f_b = f_1 - f_2 \tag{6-12}$$

Anatomy and Physiology of the Ear

The ear, the organ of hearing, is a *transducer*—a system for changing one form of energy into another. We have seen in Chapter V that the "language" of the brain is electrical impulses. The ear changes the sound language of the outside world, pressure waves, into the language of the brain. The ear therefore converts mechanical energy into electrical energy. It is one of the objects of biophysical study to determine how the ear accomplishes this function. A great deal can be learned by asking cogent questions, performing meaningful experiments, and constructing reasonable models that provide guides to further informative experimentation. But it might as well be stated at the outset: We do not understand how the ear actually accomplishes the conversion from mechanical energy to meaningful electrical impulses on the auditory nerve. Nonetheless, a "great deal" has been learned about the nature of the hearing process by questioning, experimenting, and devising and testing models.

One reasonable approach to understanding the function of the ear seems to be first learning how it is constructed. In this section the construction of the human ear is described. We shall apply the physical principles of the earlier sections when they seem to clarify the nature or purpose of particular components of the ear. The human ear is pictured in Figure 6-5 in a frontal section. The spatial compartments are clearly divided into three regions—

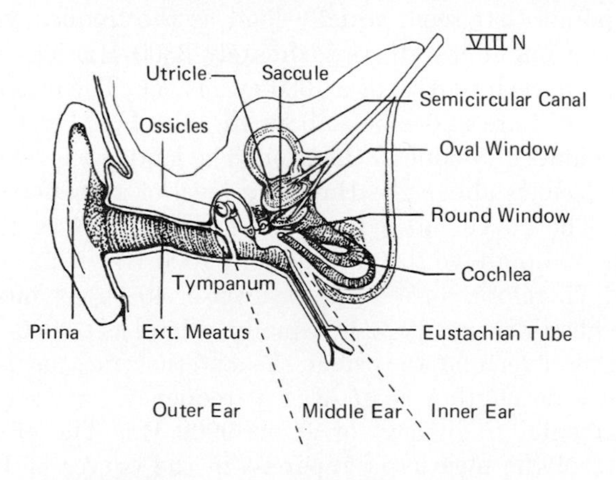

FIGURE 6-5. A frontal section of the human ear.

the outer ear, the middle ear, and the inner ear. They will be considered here in that order.

The Outer Ear

The most obvious feature of the outer ear is the *auricle* (or *pinna*), the curiously wrinkled appendage to the head. In many animals the pinnae are useful in collecting sound energy and can often be moved to optimize reception, and they perhaps aid in locating directions from which sounds emanate. In humans, the auricles serve little known purpose, for sounds can be heard about as well without them. The canal leading into the head from the auricle is called the *external meatus* of the ear (a meatus is any natural opening in the body). The meatus is, of course, open to the air and serves as a pipe to conduct sound waves to the *tympanic membrane* or *tympanum* (eardrum), which separates the outer ear from the middle ear. Around the edge of the conically shaped tympanic membrane is a gland that secretes a protective wax called *cerumen*. The tympanic membrane has an area of about 65 mm^2 and is about 0.1 mm thick. It moves in and out in response to pressure variations in sound waves impinging on the ear.

The external meatus of the ear is about 2.5 cm in length, open at one end and closed at the other. As we have seen, a pipe of length L with these boundary conditions has a fundamental resonant frequency given by $f = v/4L$. Substituting $v = 330$ m/second and $L = 2.5 \times 10^{-2}$ m, we obtain a frequency of 3300 Hz. This frequency, according to experiments, is the frequency to which the ear is most sensitive.

Experiments in which subjects judge sounds of different frequencies to be equally loud provide the results shown in Figure 6-6, in which each curve is the locus of points that seem equally loud as the frequency is varied. Every curve has a minimum at approximately 3300 Hz. The curves are labeled according to the intensity of a 1000-cycle tone. The intensities (the ordinate of the graph) are in decibels above the threshold level. The labels on the curves are units of loudness, the *phon*. The loudness level in phons is the intensity in decibels above the threshold level of a 1000-cycle tone of equal loudness. The curve labeled 40, for example, passes through an intensity level of 40 db above the threshold at 1000 Hz; it passes through 70 db at 60 Hz. Therefore, for a loudness level of 40 phons, 60-cycle tone must be 30 db higher in intensity to be equally as loud as a 1000-cycle tone.

The next higher overtone to which the external meatus is resonant corresponds to a wavelength $\lambda = 4L/3$ or a frequency $f = 3v/(4L)$, three times the fundamental frequency, or about 9900 Hz. The effect of this second resonance of the meatus is apparent in the curves of Figure 6-6.

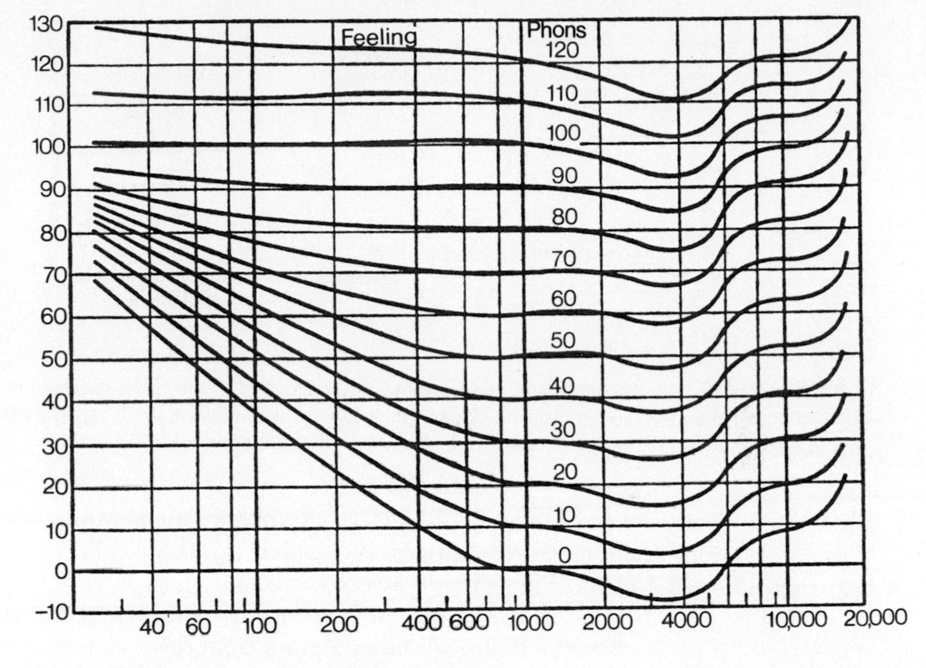

FIGURE 6-6. The response of the human ear to sound intensity. Each curve represents the subjective judgement of an average listener that the intensity seems constant as the frequency varies. The reference level for each curve is the actual intensity of a sound wave whose frequency is 1000 Hz measured in decibels above the threshold of hearing. See text for definition of the unit of subjective loudness, the phon. (Adapted from Lüscher and Zwilocki, 1947).

The upward trend of each curve is flattened at about 10,000 cps and then resumes the upward direction for higher frequencies.

The Middle Ear

The middle ear, shown in Figure 6-7, is a small cavity in one of the bones of the skull on the medial side of the tympanic membrane. Like the outer ear, the middle ear is filled with air. At the medial end of the cavity are two separate membranes, the *oval window* and the *round window*. Through the inferior floor of the cavity is an opening into the *Eustachian tube*, which connects the middle ear to the *pharynx* (the region that connects the mouth cavity to the esophagus). The Eustachian tube serves to maintain equal average pressure on either side of the tympanic membrane. This tube, which usually is closed, can be opened by yawning or swallowing when

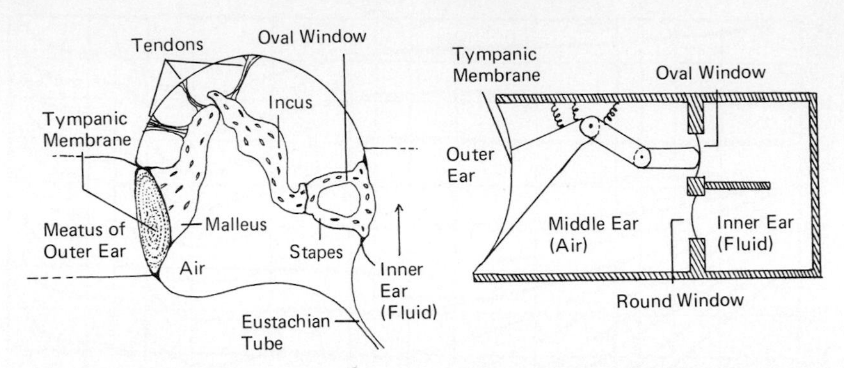

FIGURE 6-7. A cross section of the human middle ear and a schematic diagram of the ear. The schematic representation shows the relationship between the lever system of the middle ear and the membranes and chambers of the outer and inner ear.

pressure is decreased or increased outside the tympanic membrane. The discomfort from such pressure changes is familiar to those who have experienced rapid ascents or descents in aircraft.

There are three tiny bones in the middle ear that are collectively called *auditory ossicles*. They derive their individual names from their shapes: the *malleus* (hammer), the *incus* (anvil), and the *stapes* (stirrup). Together, these bones form a lever system, the long arm of which is attached to the tympanic membrane, the short arm to the oval window. Figure 6-7 shows a cross section of the middle ear and a schematic model of the lever system. The lever system reduces the amplitude of the vibration on the tympanic membrane that is transmitted to the oval window, while increasing the force of the vibration. This effectively increases the pressure on the oval window while decreasing its displacement for a given amplitude of vibration on the tympanic membrane. The increase of pressure and decrease in displacement is an "impedance matching" system, similar to matching a loudspeaker to an amplifier, the purpose of which in each case is to maximize the transfer of power. The medium on the medial side of the oval window is a liquid, and the matching system facilitates transfer of power from air, which is light and compliant, to the fluid, which is heavy and stiff.

The pressure amplification associated with the transfer of power from outer ear to inner ear can be computed theoretically (ignoring friction). The area of the tympanic membrane is 56 mm^2, of which about 50 mm^2 is in contact with the malleus. Using areas expressed in mm^2, we can express the force on the malleus F_{mal} as a product of pressure on the tympanic membrane P_{tym} due to a sound wave incident on the ear and the area of contact between the malleus and the tympanic membrane:

$$F_{\mathrm{mal}} = 50P_{\mathrm{tym}} \tag{6-13}$$

Studies of models indicate that the theoretical mechanical advantage of the lever system is about 1.3. This means that the force on the stapes F_{st} is 1.3 times F_{mal}, or

$$F_{st} = 1.3F_{mal} \tag{6-14}$$

Then P_{tym} of Equation (6-13) can be expressed in terms of F_{st} by

$$P_{tym} = \frac{F_{mal}}{50} = \frac{F_{st}}{(1.3)(50)} = \frac{F_{st}}{65} \tag{6-15}$$

The stapes is in contact with the oval window over an area of about 3 mm^2, so the pressure on the oval window P_{ow} is given by

$$P_{ow} = \frac{F_{st}}{3} \tag{6-16}$$

The theoretical pressure amplification P_{ow}/P_{tym} is then found to be

$$\frac{P_{ow}}{P_{tym}} = \frac{F_{st}/3}{F_{st}/65} = 21.7 \tag{6-17}$$

a value that compares favorably to actual experimental measurements, in which the pressure amplification has been reported to be 17. In the domestic cat, the pressure amplification is 60.

The Inner Ear

The inner ear, like the middle ear, is a cavity in the temporal bone of the skull. The cavity of the middle ear is, however, multichambered, and it contains a single anatomical organ that has a number of parts: the *cochlea*, the *saccule*, the *utricle*, and the *semicircular canals*. Of these parts, only the cochlea functions in the hearing process. The function of the saccules in man is not known with any certainty. The utricles serve to keep the brain informed of the position that the head occupies in space. This is accomplished by cells attached to nerves that contain small masses of calcium carbonate called *otoliths* (ear stones) that increase the effectiveness of gravity. Since these cells are most effective when the head is upside down, the weight of the otoliths apparently produces tension in the special cells that trigger impulses along nerves to the brain. The semicircular canals are a series of small fluid-filled tubes, shaped like semicircular arcs, that function as organs of balance. The specialized sensing cells of the semicircular canals respond to motion of the fluid. When a person spins around rapidly for a few seconds, the fluid acquires some rotational motion. When the person stops spinning, the rotational motion continues because of its inertia

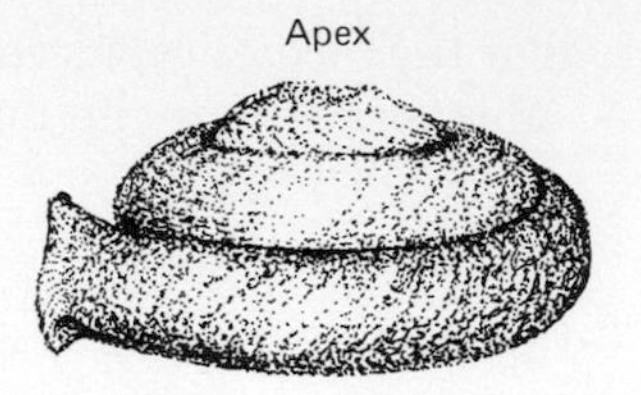

FIGURE 6-8. The intact cochlea.

and dizziness results because the motion of fluid continues to excite the sensory cells that transmit a sense of rotation to the brain.

The cochlea is a tube that is coiled into a helix whose two and a half turns decrease in size as it progresses away from the middle ear. The shape of the cochlea, seen in Figure 6-8, resembles a snail shell. Internally, the cochlea has a septum, called the basilar membrane, that divides the primary helical tube into roughly equal parts. A cross section of the cochlea, seen in Figure 6-9, shows that the coiled tube contains three tubular regions, two of which are above the basilar membrane, and one below. Above the basilar membrane are the *scala vestibuli* and the *cochlear duct*; below is the *scala tympani*. The stapes fits against the oval window, which is the end of the scala vestibuli. The scala vestibuli communicates with the scala tympani at the small end (apex) of the cochlea through an opening called the *helicotrema,* and the scala tympani meets the middle ear at another membranous window, the *round window*. The fluid in the scalae is *perilymph*. Somewhat sandwiched between the two scalae is the coclear duct, which is filled with a different fluid, the *endolymph*. The two fluids are anatomically and electrically different. The membrane separating the cochlear duct from the scala vestibuli is called *Reisner's membrane*.

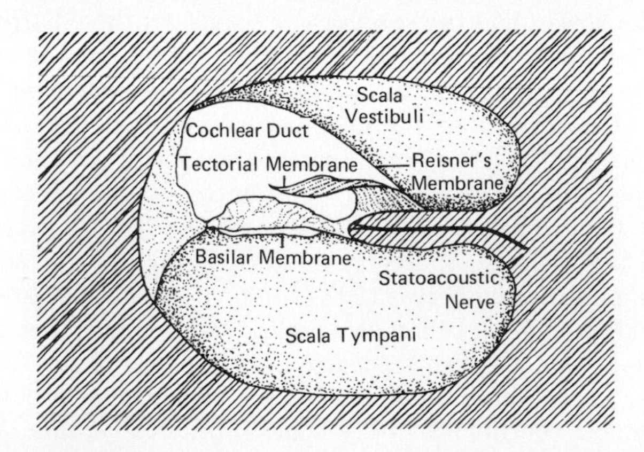

FIGURE 6-9. A cross section of one turn of the cochlea.

Seated on the basilar membrane is the apparatus that transduces mechanical energy into electrical impulses. This is the *organ of Corti,* seen in Figure 6-10. Here is the site of the sensory cells, or *hair cells,* which are connected to fibers of the statoacoustic (VIII) nerve. The hair cells have projecting "hairs," or fibers, that extend upward and are imbedded in a tonguelike, stiff tissue, the *tectorial membrane.* It is an important anatomical feature that the tectorial membrane is hinged at a point not on the basilar membrane so that motion of the basilar membrane causes the hair cells to bend.

BASILAR MEMBRANE CHARACTERISTICS

Although the cochlea becomes smaller as it spirals away from the middle ear, the basilar membrane thickens and widens as it nears the helicotrema. The membrane consists of supporting tissues through which run fibers arranged crosswise to the long dimension of the cochlea. The individual fibers are up to 1 or 2 microns (μm) in diameter and range in length from about 75 μm at the oval window end to about 475 μm at the helicotrema end.

Von Békèsy (1949) made some very informative observations on human basilar membranes of fresh cadavers after grinding away the bone over the cochlea. He observed the displacements of the basilar membrane as a function of sound frequency (at constant amplitude) and as a function of distance from the oval window. The actual movements of the basilar membrane are in the form of waves traveling from the oval window. Figure 6-11 shows the amplitudes of the displacements of the basilar membrane as a function of position (distance from oval window) for a number of frequencies of sound waves. As the frequency increases, the maximum of the amplitude moves toward the oval window. At high frequencies, those portions of the basilar membrane far from the oval window do not move at

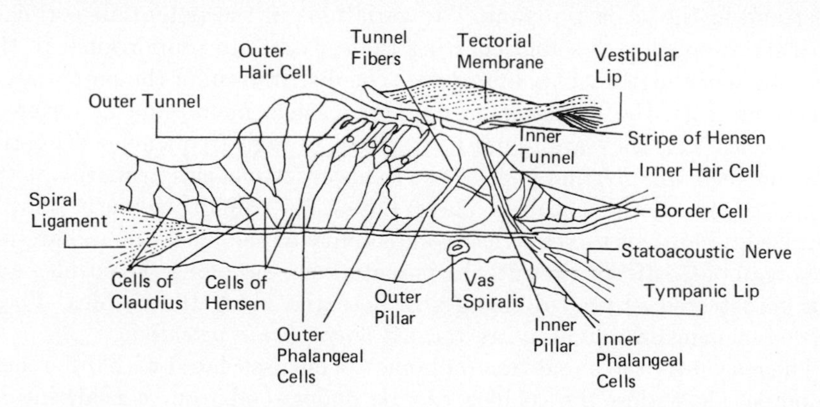

FIGURE 6-10. A cross section through the organ of Corti, which is situated on the basilar membrane of the cochlea.

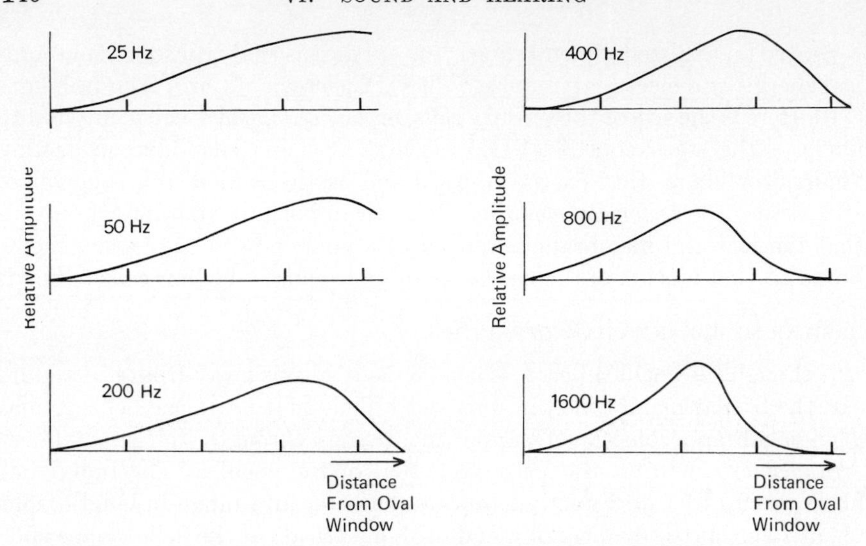

FIGURE 6-11. Displacement amplitudes of the basilar membrane as a function of position for various frequencies of sound incident on the ear (after von Békèsy, 1953).

all. It appears, therefore, that the sensing of pitch by the inner ear depends on where the hair cells are that trigger the auditory fibers.

COCHLEAR MICROPHONICS

Microelectrodes can be placed into the cochlear spaces and into tissues of living animals to record differences in electrical potential, if they exist. Experiments such as these have shown that potential differences are developed across the organ of Corti, i.e., between the fluid in the cochlear duct and in the scala tympani. The variations in the potentials correlate with the frequency of sound entering the ear and are proportional to the amplitude of the incident sound wave. The distribution of the potentials is consistent with the displacements of the basilar membrane at different frequencies, i.e., they are found in response to high frequencies primarily near the oval window and in response to lower frequencies primarily at the opposite end of the cochlea. These potentials cannot be very sharply localized because of the electrical conductivity of the media in and around the organ of Corti. As a result, the potentials spread along the cochlea and can be detected as graded changes in potential along the cochlea. These graded changes in potential are termed *microphonic potentials.*

The reason for the name "microphonic" being associated with differences in potentials within the cochlea can be understood from considering an experiment that can be performed on live experimental animals. If one electrode is placed on the oval window and another on the skin of the neck, the potential differences of all frequencies can be picked up. If this output is amplified and put into a loudspeaker, it is found that sounds incident on the animal's ear, like voice or music, are rather faithfully reproduced. In other words, the cochlea used in this way acts exactly as a microphone.

The greatest differences in electrical potential within the cochlea have been found in the immediate region of the hair cells and their associated nerve fibers. In addition, it has been found that action potentials occur on the auditory nerve on each cycle of the microphonic potential, provided the frequency is not too high. [Recall that the refractory period on an axon is about 1 millisecond (msecond), so that a maximum frequency of spike potentials on the axon is limited to 1000/second.] These observations suggest that there is an association between microphonic potentials and nerve impulses, even though the potentials may not be the immediate cause of the initiation of the impulse.

As was pointed out earlier, the detailed mechanism that fires the nerves in the ear is not understood. The hair cells of the organ of Corti are the sites of the fiber endings of the statoacoustic nerve, and we know that the hairs are bent by stresses between the tectorial membrane and the hair cells as the basilar membrane vibrates in response to sound waves. We also know that the microphonic potentials seem to be precursors of nerve impulse formation at least at low frequencies. With these facts available, it is now reasonable to consider the models and theories that have been proposed to explain the neural mechanisms of hearing.

Theoretical Mechanisms of Hearing

An adequate explanation of hearing function must explain how the mechanical energy of sound is converted into meaningful impulses to the brain. It must do this within the constraints imposed by the anatomy and physiology of the ear, yet it must account for the capabilities of the ear. This problem has been recognized and pursued for over a century—and by some impressive physicists. Helmholtz, whose contributions to thermodynamics have been emphasized in the naming of "Helmholtz free energy," and Lord Rayleigh, a master of physical acoustics, both made extensive studies of the hearing process. Both are responsible for models intended to relate the physical and physiological phenomena of sound and hearing to the results that are recognized by the brain. In the light of modern measurements, the models proposed by Helmholtz and Lord Rayleigh are no longer tenable. Still, even the most modern hypotheses leave a great many questions unanswered. It is perhaps informative to consider briefly some of the theories in historical order, if only to emphasize those aspects of the hearing sense that must be explained by an adequate model.

The Resonator Theory

An early concept of the hearing mechanism, developed by Helmholtz (1862), was based on knowledge of the physical role of the basilar membrane in the cochlea and the fibrous anatomy of that membrane. Helmholtz was familiar with mechanical resonators, and he had used these to analyze harmonic content in tones of musical instruments. He therefore sought mechanical resonant systems in the ear and recognized in the fibers of the basilar membrane a set of mechanical vibrators that varied in length along the cochlea. This system of fibers is much like the wires of a harp or piano, each length corresponding to a resonant frequency. This model assumes that each fiber of the basilar membrane is under tension (which is known to be the case) and is sharply tuned to a particular frequency. The basilar membrane is then, according to this theory, a mechanical analyzer of incoming sound waves.

When the measured ability of the human ear to discriminate pitch is considered, it must be recognized that no realizable mechanical resonating system can satisfactorily compare with the ear. A very sharply tuned resonator is difficult to excite into vibration, and once set into vibration, continues long after the excitation has ceased. Fibers in the basilar membrane do not demonstrate this sharp-tuned characteristic.

A series of experiments by von Békèsy finally ruled out the resonator theory. His measurements of tension and length of fibers in the basilar membrane indicated that there was not sufficient range of length and tension to account for the frequency range of human hearing. The resonant frequency of a taut fiber of length L, under tension T, and with mass per unit length m_1 are expressed by

$$f = \frac{1}{2L}\left(\frac{T}{m_1}\right)^{1/2} \tag{6-18}$$

Since the frequency range of hearing, from 20 to 20,000 Hz, is a factor of 1000, the value of T/L^2m_1 would have to vary by a factor of about 10^6. The measurements of von Békèsy showed that the ear fails to meet this requirement by at least a factor of 20.

The fact that the basilar membrane vibrates with maxima at different locations or "places" corresponding to different frequencies makes it tempting to retain certain aspects of the resonator theory. Any theory of hearing based on a model that utilizes the postulate that frequency discrimination is a function of localization of frequency sensitivity along the basilar membrane is called a *place theory*.

The Telephone Theory

Since the telephone is a transducer that converts sound waves into electrical pulses for transmission, it is not surprising that early workers in auditory sciences, including Lord Rayleigh, should consider explaining the hearing function of the ear by a system like the telephone. In this model, it is assumed that the cochlea simply serves as a microphone and transmits along the auditory nerve fibers signals whose form is that of the pressure wave incident on the ear. Cochlear microphonics had not been experimentally observed at the time the telephone theory was suggested—and perhaps that is fortuitous, because such knowledge would have only fortified the theory.

The telephone theory cannot be correct, at least over most of the audible frequency range. Since no more than 1000 pulses/second can be transmitted along neurons, sounds with frequencies above 1 kHz could not be perceived. Indeed, the frequency limit of hearing would be far below 1 kHz, because many identical spikes are required to reproduce a single cycle of a wave form (see Figure 6-12).

Although the telephone theory is untenable, it suggests that the brain might integrate the incoming spikes on neurons into recognizable wave forms. This suggestion has become the assumption of more modern theories of hearing.

The Volley Theory—Interpretation by the Brain

A recent theory of hearing does not depend on the place concept of the resonator theory and the telephone theory. Because single neuron paths

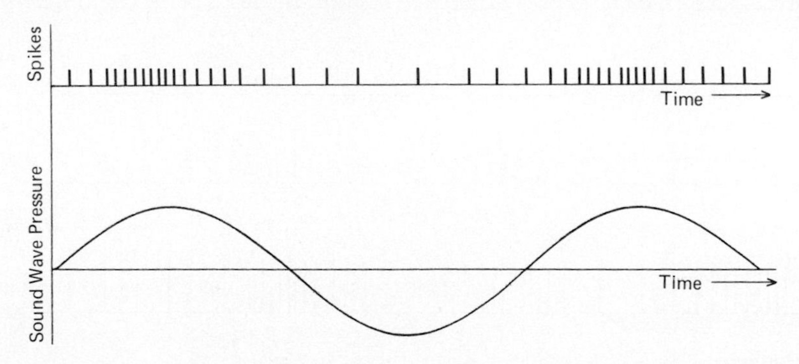

FIGURE 6-12. Diagram illustrating that a relatively large number of spikes (over 25 in this case) are necessary to approximately reproduce a sinusoidal wave. The spike repetition rate corresponds to the relative amplitude of the wave being reproduced.

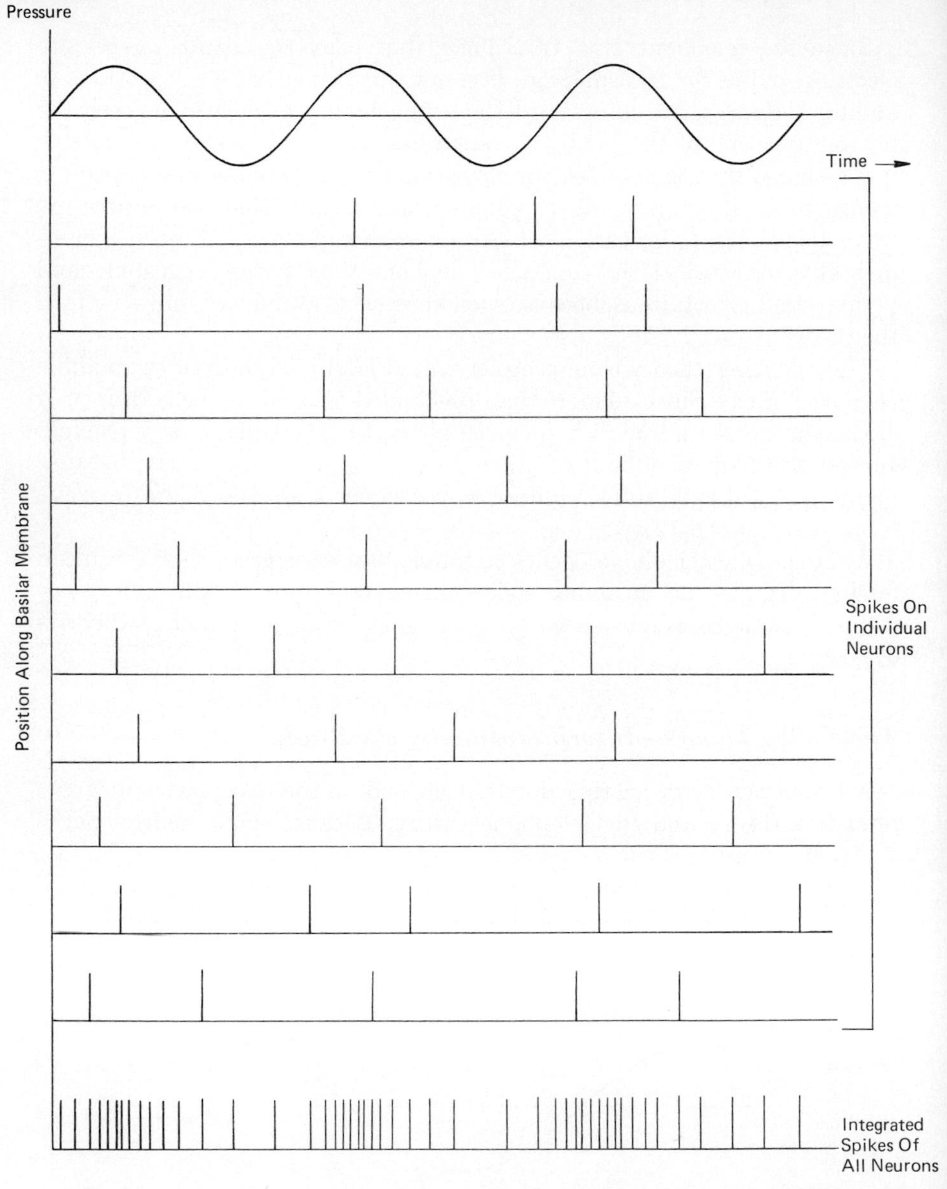

FIGURE 6-13. An indication of the manner in which a number of neurons could fire in volleys to reproduce a sinusoidal wave when the spikes of each neuron are integrated. The incident sound wave causes individual neurons at various positions along the basilar membrane to fire at relatively slow rates.

cannot carry more than 1000 pulses per second, it seems reasonable to assume that groups of neurons are involved in frequency discrimination. This is the basis of the *volley theory*. Pitch is associated with the grouping of pulses in time on a number of neurons in a group. Each neuron fires at a relatively slow rate, but the entire group of neurons reproduces the frequency of the sound by producing volleys of pulses. Figure 6-13 is an illustration of how a number of individual neurons, firing occasionally, but in a coordinated fashion, can reproduce a sine wave when all their pulses are accumulated. It is left to the interpretive abilities of the brain to integrate all the pulses of the volley into a meaningful pitch.

Experimental evidence has shown that numerous neurons do fire in synchronized volleys for sounds incident on the ear up to about 2000 Hz. Above that frequency, the nerve pulses are apparently more random. It has been suggested that the hearing mechanism of the ear uses the volley theory for the lower frequencies and uses a place theory to discriminate pitch for the higher frequencies.

The interpretation required of the brain by the volley theory in sensing frequency is not unreasonable. A large number of phenomena that are observed cannot be explained without requiring automatic interpretive processes of the brain. *Binaural beats* is an example of such a phenomenon. When two slightly different frequencies of sound waves are presented to each ear independently, the hearer senses a beat frequency. Such beats cannot be explained by mechanical motions of the cochlea, and the phenomenon implies that the brain interprets the two sounds in such a way that the hearer senses the beat frequency. The sensing of the localization of sound by a slight time delay between the two ears is another example of interpretation by the brain.

It seems, therefore, that any successful theory of hearing is going to be dependent on understanding brain function as well as on the understanding of physical and physiological processes of the mechanisms of the ear.

References and Suggested Reading

Fletcher, H. (1951). On the dynamics of the cochlea. *J. Acoust. Soc. Amer.* **23,** 637.

Lüscher, E., and Zwilocki, J. (1947). The decay of sensation and remainder of adaptation after short pure-tone impulses on the ear. *Acta oto-laryng. Stockholm* **35,** 428–445.

Mills, J. (1943). Sound. *In* "Medical Physics" (O. Glasser, ed.), Vol. I, p. 1438. Yearbook Publ., Chicago, Illinois.

Morse, P. M. (1948). "Vibration and Sound." McGraw-Hill, New York.

Van Bergeijk, W. A., Pierce, J. R., and David, E. E., Jr. (1960). "Waves and the Ear." Doubleday, Garden City, New York.

von Békèsy, G. (1949). The vibrations of the cochlear partition in anatomical preparations and in modus of the inner ear. *J. Acoust. Soc. Amer.* **21,** 233.

von Békèsy, G. (1953). Shearing microphonics produced by vibrations near the inner and outer hair cells. *J. Acoust. Soc. Amer.* **25,** 786.

von Békèsy, G. (1957). The ear. *Sci. Amer.* **197,** 66 (Offprint 44).

von Békèsy, G., and Rosenblith, W. A. (1951). The mechanical properties of the ear. *In* "Handbook of Experimental Psychology" (S. S. Stevens, ed.), p. 1075. Wiley, New York.

von Helmholtz, H. L. F. (1885). "Sensations of Tone." Longmans, Green, New York.

Weiner, F. M., and Ross, D. A. (1946). The pressure distribution in the auditory canal in a progressive sound field. *J. Acoust. Soc. Amer.* **18,** 401.

Wever, E. G. (1949). "Theory of Hearing." Wiley, New York.

Wever, E. G., and Lawrence, M. (1954). "Physiological Acoustics." Princeton Univ. Press, Princeton, New Jersey.

CHAPTER VII Light and Vision

Most animals have the ability to perceive light. In general, organs that detect light are called *photoreceptors*, regardless of their sophistication. Some of the lower forms of animal life have relatively simple photoreceptors that indicate only the presence of light and the general direction from which the light emanates. More complex animals often have sophisticated, image-forming visual systems that are perhaps typified by the human eye. The importance of a visual system as a survival aid to an organism is obvious. It is the primary means by which animals assess their environment, particularly their distant environment.

Our study of vision will concentrate on the human visual system as representative of relatively sophisticated image-forming systems in the animal kingdom. The study of vision in higher animals may be conveniently divided into (1) the physical phenomena associated with light and the formation of images using the optical properties of lenses and (2) the chemical reactions that convert the energy of light into nerve impulses to which the brain responds. In this chapter we shall review some of the essential characteristics of light that are important in the visual process and some of the principles of geometric optics that help in the understanding of image formation. Then we shall consider the anatomy of the human eye and some of its functional characteristics, including the chemistry of vision. Finally, we shall consider some examples of experimental and theoretical approaches to the understanding of vision.

Some Physical Aspects of Light

The physical nature of light was a topic of scientific debate and controversy until well into the twentieth century. The uncertainties concerning the fundamental nature of light stemmed from experimental observations that seemed contradictory. Sometimes light exhibited those phenomena, like interference and diffraction, that had been previously associated with waves. Sometimes, particularly when interacting with matter, light seemed to have characteristics that are associated with beams of individual particles. In recent years we have come to accept the dual nature associated with light. Indeed it does sometimes exhibit its wave nature and at other times its particulate nature. Both aspects of light will be considered here briefly.

The Wave Nature of Light

A great variety of experiments assure us that light is, in part, a wave phenomenon. Unlike some other waves, such as sound waves and waves on ropes or on water, light does not require the presence of matter in order to propagate. This fact is true because light is an *electromagnetic wave,* composed of electric fields and magnetic fields. Because of the unique physical basis of electromagnetic waves, it is perhaps worthwhile to review the basic characteristics of these waves and how they determine the physical properties of light.

ELECTROMAGNETIC WAVES

The presence of an electric field or a magnetic field in space changes the properties of that space in a specific and measurable way. The presence of an electric field in otherwise empty space alters that space in such a way that *if* a positive charge q were placed at a point where the electric field has a value $\mathbf{E}$, the charge *would* experience a force $\mathbf{F}$ in the same direction as $\mathbf{E}$. The magnitude of the force on the charge would be qE, where q is the magnitude of the charge and E is the magnitude of the electric field. Note that the electric field may be present at a point in space whether or not we choose to detect it or measure it by placing a charge at that point.

Magnetic fields also alter the space in which they occur. A magnetic field $\mathbf{B}$ is said to be present at a point in space if, when a wire of length ΔL carrying a current $\mathbf{I}$ in a direction perpendicular to $\mathbf{B}$ is placed at that point, it would experience a force $\mathbf{F}$ perpendicular to both $\mathbf{I}$ and $\mathbf{B}$. The magnitude of the force would be $BI\ \Delta L$. Again, a magnetic field may exist in space whether or not we choose to detect its presence or measure its magnitude and direction.

It should be clear from the foregoing descriptions that $\mathbf{E}$ and $\mathbf{B}$ are vector quantities, having both magnitude and direction.

The basic theory of electromagnetism, developed by Maxwell (1831–1879), elucidates the dynamic properties and interrelationships of electric and magnetic fields. In particular, this theory quantitatively describes the fashion in which a changing electric field produces a magnetic field and the converse, i.e., a changing magnetic field produces an electric field. Without, for the moment, considering the source of an electromagnetic wave, but assuming the presence of electric and magnetic fields that are changing in time, we can visualize how such a wave propagates through space. Figure 7-1 illustrates the geometric arrangements of the fields at a particular instant in time. The electric and magnetic fields are perpendicular to each other and are perpendicular to the direction in which the wave is moving. The electric and magnetic fields have maxima and minima at the same times and at the same places. The envelopes of the vectors representing the fields are sinusoids. If we look at either field at a point in space as time proceeds (Figure 7-2), we observe that the magnitude of each field varies sinusoidally, changing direction as the field becomes zero. Electromagnetic

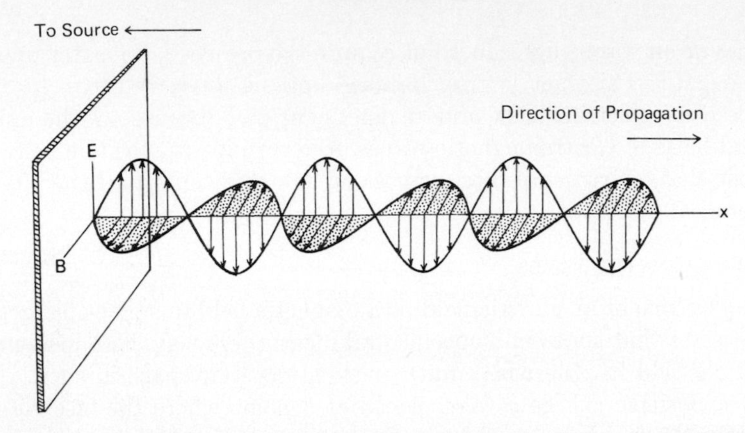

FIGURE 7-1. Geometric arrangement of the electric and magnetic fields that comprise an electromagnetic wave. This representation indicates the variation of either field along a line in space at a particular time as the wave passes.

waves are therefore *transverse waves*, in which the local motions of the disturbance (the fields in this case) are perpendicular to the direction of propagation of the overall disturbance.

Another result of the electromagnetic theory of Maxwell predicts that the velocity of propagation of an electromagnetic wave depends only on the electrical and magnetic properties of the medium in which it is traveling. Thus, the velocity is independent of the amplitudes of the field variations and of the history of the wave (i.e., whether or not the velocity of the wave has been slower in the past because of having been in a different medium). The velocity of light in free space is usually symbolized by c and has the magnitude 3×10^8 m/second, or 186,000 miles/second. In all other media, the velocity of light is less than c.

Like all waves, electromagnetic waves are characterized by frequency f and wavelength λ, both of which were defined in Chapter V. The relation-

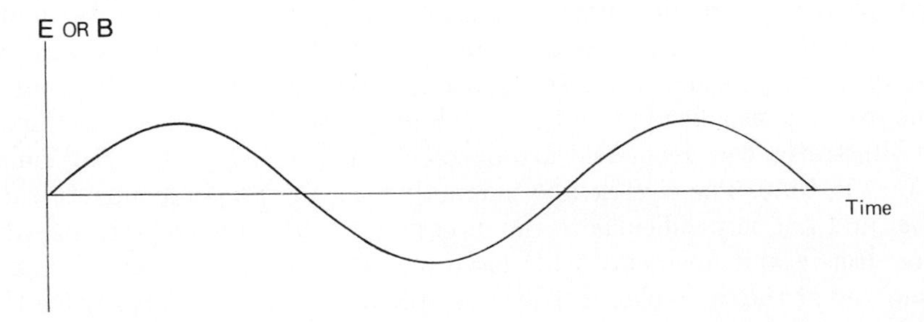

FIGURE 7-2. The sinusoidal variation of the field (either electric or magnetic) as a function of time. This representation indicates the variation of either field at a single point in space as the wave passes.

ship between frequency, wavelength, and velocity for electromagnetic waves is, as is the case for all waves, expressed by

$$v = f\lambda \tag{7-1}$$

Of course, for electromagnetic waves in free space, Equation (7-1) becomes

$$c = f\lambda \tag{7-2}$$

Visible light, i.e., electromagnetic waves visible to the human eye, is a term that specifies a particular band of frequencies (or wavelengths) within the total range, or *spectrum*, of all electromagnetic waves. Figure 7-3 shows the electromagnetic wave spectrum. It is seen that radio waves, X-rays, gamma rays, infrared light, and ultraviolet light are all electromagnetic waves like visible light. The only physical difference in these different varieties is frequency (or wavelength), though they often differ considerably in their interactions with matter. Visible light is usually considered to be the wavelengths between 380 and 760 nm.

The property of light that we discern as *color* is directly related to the frequency of the light wave. Within the visible portion of the spectrum,

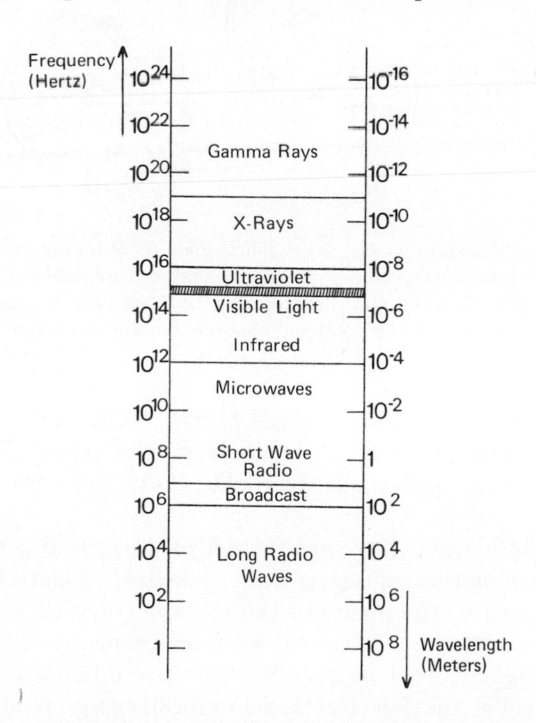

FIGURE 7-3. The electromagnetic spectrum.

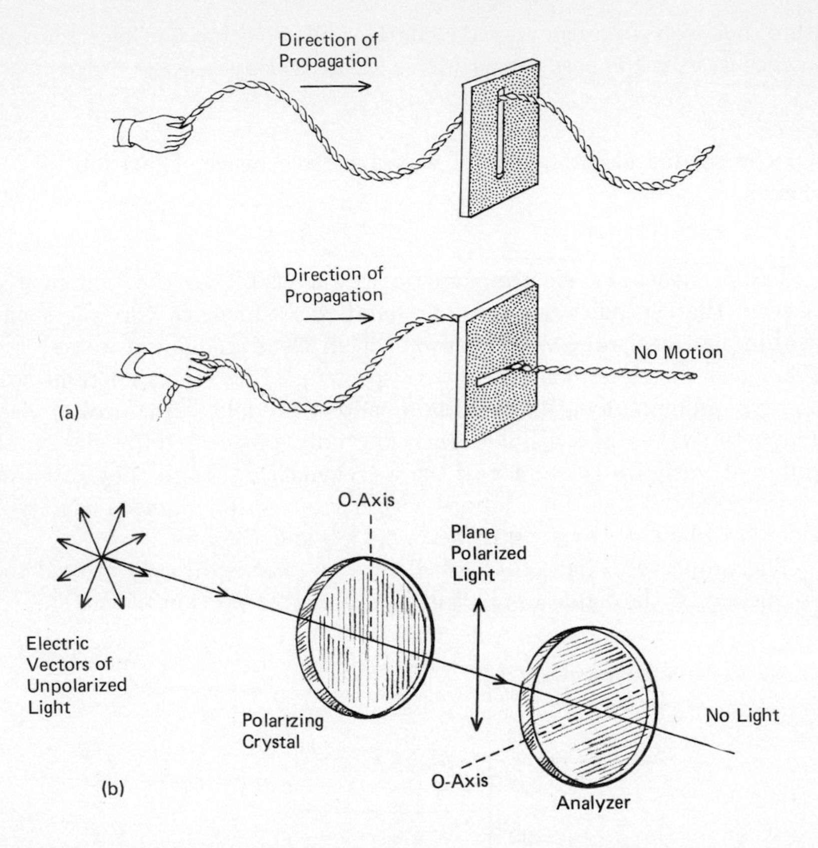

FIGURE 7-4. Polarized waves. (a) A mechanical wave on a rope may be polarized in a plane by shaking it in a plane; it may be analyzed by a board with a slot in it. (b) An electromagnetic wave may be polarized by passing it through a crystal that absorbs in all but one plane; the polarized electromagnetic wave may be analyzed by a similar crystal.

the longest wavelengths are perceived by the normal human as deep red. As the wavelength decreases, the perceived color passes through orange, yellow, green, violet, and finally deep blue at the short wavelength end of the spectrum.

Electromagnetic waves have one further property that is consistent with their transverse nature. They can be *polarized.* Polarization refers to restrictions placed on the planes of the vibratory motion within the wave. Figure 7-4a shows how a mechanical wave on a rope may be constrained to vibrate in a single plane. Figure 7-4b shows how a light wave can be similarly constrained so that its electric field vibrates only in one plane (which implies, of course, that the magnetic field is likewise constrained to one plane—one that is perpendicular to the plane of the electric field). The *polarizer* shown in this figure is a material that absorbs those waves in which the electric field is perpendicular to a special axis within the material called the O (for ordinary) axis. Those waves whose electric fields are parallel to the O axis are only very slightly absorbed and therefore pass through the material nearly unattenuated. The light that has passed through the polarizer can be shown to be polarized by placing a second polarizer (*analyzer*) in the beam. If the analyzer has its O axis 90° to the first polarizer, no light will pass through the analyzer. There are other means of polarizing light than using absorbing materials, but the effect is essentially the same.

PHOTOMETRY

One aspect of light that is important to visual studies concerns the measurement of parameters associated with brightness. The science of measurement of light intensity is *photometry*. The basic unit of light intensity, which refers to the energy passing through a unit area in a unit time, is the *candle*. Obviously, some standard of intensity must be chosen to which measurements are referenced, and the *standard candle* is defined as: the intensity at an opening 160 cm^2 in area above a cavity containing a white thorium oxide surface at the temperature of solidifying platinum.

Light flux is a measure of how much light energy is passing through a given solid angle from a "point source" of light. Assuming that a point source of light radiates isotropically, i.e., uniformly in all directions, the light flux is distributed evenly through 4π *steradians* (sr). (A steradian is the solid angle subtended by an area on the surface of a sphere of radius r divided by the square of the radius. Thus, the total solid angle subtended by a sphere is 4π sr.) The *lumen* (lm) is defined as the light flux from one candle through one steradian. Thus, the total flux from a light source of one candle is 4π lm.

Illumination is a measure of the light flux (lumens) falling on a unit area. The units of illumination are lumens/m^2. An engineering measure, *foot-candle* (fc) is often used to describe illumination. A foot-candle is a light flux of one lumen per square foot, or lm/ft^2.

For extended light sources (those other than point sources), the brightness of a source is characterized by its *luminance*. The unit of luminance is the *lambert*, defined as the luminance of a "perfectly diffusing surface" scattering or emitting one lumen of light per cm^2. In practice, magnesium oxide power is considered a perfectly diffusing surface, reflecting nearly all the light energy that falls upon it. The total light flux from an extended source is therefore the product of the area normal to the direction from source to observer and its luminance. Notice that if a surface reflects all

Table 7-1 Relations between Photometric Units

Unit	Quantity measured	Equivalent
1 candle	source intensity	4π lm
1 lumen	luminous flux	(intensity of 1 candle)/4π
1 candle	source intensity	1 lm/sr
1 fc	illumination	1 lm/ft^2
1 lm/cm^2	illumination	luminance/reflection coefficient
1 lambert	luminance	1 lm/cm^2 on perfectly diffusing surface

the light incident upon it, the luminance in lamberts is numerically equal to its illumination in lumens/cm^2. When all the light is not reflected from a surface, the ratio of its luminance to its illumination (lamberts/lumens/cm^2) is called the *coefficient of reflection* of that surface.

A brief summary of the relationships between the various photometric units used here is given in Table 7-1.

The Quantum Nature of Light

Around the beginnning of the twentieth centry, two types of experiments were being conducted that involved light. Both types of experiments produced results that could not be explained by the classical concepts of light as a continuous electromagnetic wave. One type of experiment involved precision measurements of electromagnetic energy emitted from a "blackbody" source as a function of wavelength and temperature. The observed distribution of energy throughout the electromagnetic spectrum could not be reconciled with classical theory. A second type of experiment demonstrated that electrons are ejected from metals when light of sufficiently high frequency falls upon the metal. These experiments included measurements of the energies of the ejected electrons as a function of the intensity and frequency of the light. The phenomenon observed in these experiments has become known as the *photoelectric effect*.

The blackbody radiation experiments and the photoelectric effect were explained by Max Planck (1858–1947) and Albert Einstein (1879–1955), respectively, by invoking a totally new concept. Instead of accepting the classical picture of electromagnetic radiation, in which the energy of a light beam is distributed continuously throughout the beam and may have any amount of energy within the beam, they proposed what has become known as the *quantum theory*:

1. The energy of electromagnetic radiation is not distributed continuously, but is localized in discrete packets of energy called *quanta*, or *photons*.

2. In electromagnetic radiation of frequency f, each proton has precisely the energy

$$E = hf \tag{7-3}$$

in which h is a universal constant of nature, now called Planck's constant, and has the value of 6.6262×10^{-34} joule seconds.

Thus, light is seen to have a particle-like nature, though this does not mean we can disregard its wave nature that is so convincingly expressed in such phenomena as interference and diffraction. In fact, we must now accept the dual nature of light—that it can in some experimental situations function as a wave and in others demonstrate its particulate nature.

Interaction of Light and Matter

EMISSION AND ABSORPTION

Emission and absorption of visible light usually is associated with changes in the electronic structure of atoms. The electrons that form the outer structure of atoms are capable of changing energy—but only in discrete (*quantized*) jumps. Thus, the electrons are said to jump between *energy levels* within the atom. When an electron jumps from a higher to a lower energy level, the difference in energy between those two levels is released from the atom and appears as a photon. The energy of the photon, $E = hf$, is exactly equal to the decrease in energy of the atom resulting from the electron jump. The photon is then either called light (visible, infrared, or ultraviolet) or an X-ray, depending on the frequency of the radiation. Atoms are capable of absorbing photons by a process similar to the emission process. The electrons in an atom may absorb a photon of frequency f, whose energy is $E = hf$, if that atom has electronic energy levels whose difference in energy is the same as the photon energy. An atom may also absorb a photon whose energy is sufficient to free an electron completely from an atom (this is the process that takes place in the photoelectric effect). Electrons freed from atoms by photon absorption may then heat up other matter by collisions. This sequence of events is one of the ways that light can be absorbed by matter resulting in the heating of matter.

Thus it is seen that light may have its beginning and its end in interactions with matter. Light can interact with matter in still other ways, however.

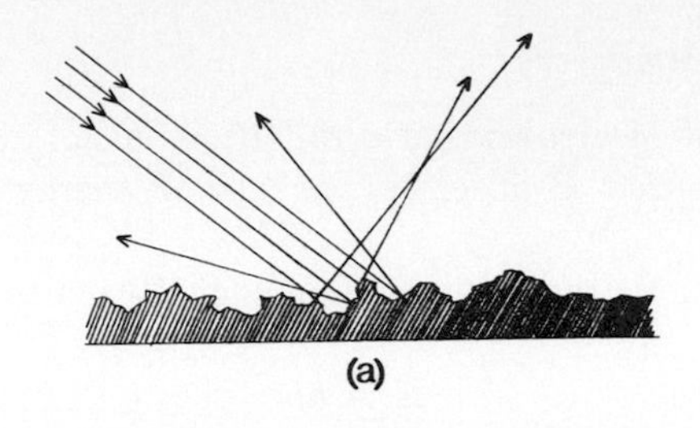

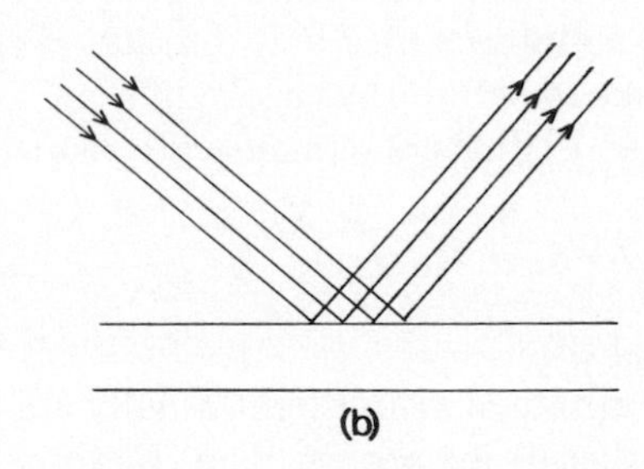

FIGURE 7-5. Reflection of light from a surface. (a) Diffuse reflection. (b) Specular (mirror) reflection.

REFLECTION AND REFRACTION

The phenomena that we recognize as *reflection* and *refraction* of light are manifestations of interactions between light and matter. Reflection takes place when light strikes a material surface on which it is not completely absorbed and which is not completely transparent to the light. Refraction is the bending of light as it traverses an interface between media in which its speeds of propagation are different.

When light is reflected from a surface, the quality of the reflected light depends on the smoothness of the surface. If the surface is relatively rough, i.e., if the irregularities are much larger than the wavelength of the light striking the surface, the light is diffused from the surface in many directions (see Figure 7-5a), and the light is said to have undergone *diffuse reflection.* If, on the other hand, light is reflected from a relatively smooth surface, i.e., the irregularities in the surface are small compared to the wavelength of the light, the surface is said to be *specular*, or mirror-like, and the light is reflected in a very regular manner (see Figure 7-5b). For specular surfaces the reflective properties of light take a simple and elegant form: the *angle of incidence* is equal to the *angle* of *reflection.* Figure 7-6 illustrates the law of reflection. The angles of incidence and reflection are measures of the incident and reflected rays from the normal to the specular suface.

The process of refraction, or bending of light, takes place when light crosses an interface between media with different *indices of refraction.* The index of refraction n of a medium is a measure of the ratio of the speed of light in free space c to the speed of light v in that medium. This relationship is expressed by

$$n = \frac{c}{v} \tag{7-4}$$

and since the speed of propagation of light in free space is greater than in any other medium, $n \geq 1$ is true for all media. (This is not strictly true for some X-rays in some media, in which case n can be less than unity. Nevertheless, in all cases energy is transported by the wave at a speed less than c.) When light passes from a medium whose index of refraction is n into one whose index of refraction is n', where $n' > n$, the path of the light is bent toward the normal to the interface between the media. This is illustrated in Figure 7-7. The degree to which bending takes place is predicted by *Snell's law*:

$$n \sin \theta = n' \sin \theta' \tag{7-5}$$

in which θ and θ' are the angles between the normal to the surface and the incident and refracted rays, respectively.

The arrows in both directions on the rays shown in Figure 7-7 indicate the reversible nature of a refraction process. Light emerging from the n' region into the n region is bent at the interface according to Snell's law.

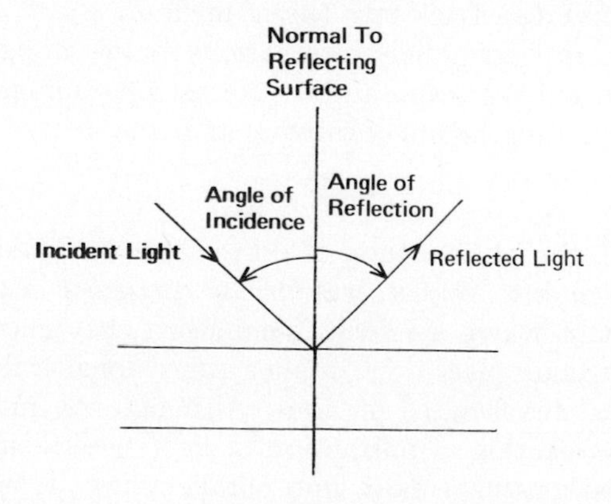

FIGURE 7-6. Illustration of the law of reflection: the angle of incidence is equal to the angle of reflection.

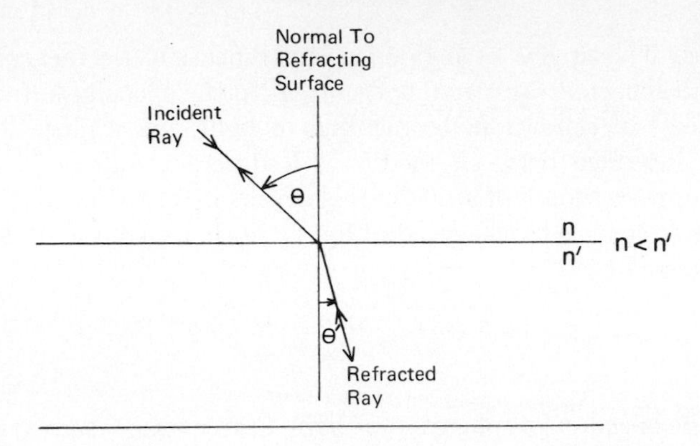

FIGURE 7-7. Illustration of the law of refraction: $n \sin \theta = n' \sin \theta'$. The indices of refraction of the media at whose interface refraction occurs are n and n'. The angles θ and θ' are measured from the normal of the interface to the incident and refracted rays, respectively.

It therefore follows that there must be some angle θ_c', called the *critical angle*, at which an emerging ray would lie parallel to the interface, i.e., such that $\theta = 90°$. By using $\theta = 90°$ in Snell's law, we find the critical angle in terms of the indices of refraction of the media to be

$$n = n' \sin \theta_c' \qquad \theta_c' = \sin^{-1} \frac{n}{n'} \tag{7-6}$$

For any angle θ' greater than θ_c', the light is said to undergo *total internal reflection*. In this case, no light escapes the n' medium, and all the light is reflected by the interface back into the n' medium.

The reflective and refractive interactions between light and matter permit the control of light beams and the formations of images by relatively simple geometric arrangements of mirrors and lenses.

DIFFRACTION

In Chapter VI, the phenomenon of diffraction, the bending of waves as they pass the edge of an object, was briefly discussed in association with sound waves. Light waves similarly bend when they encounter an edge, though the effect takes place on a smaller dimensional scale because of the relatively smaller wavelengths of light. Although the most common experimental demonstration of diffraction of light uses a single slit, double slit, or diffraction grating (many fine parallel slits), it is convenient for our purposes to consider the diffraction of light at a circular *aperture* (an opening admitting light).

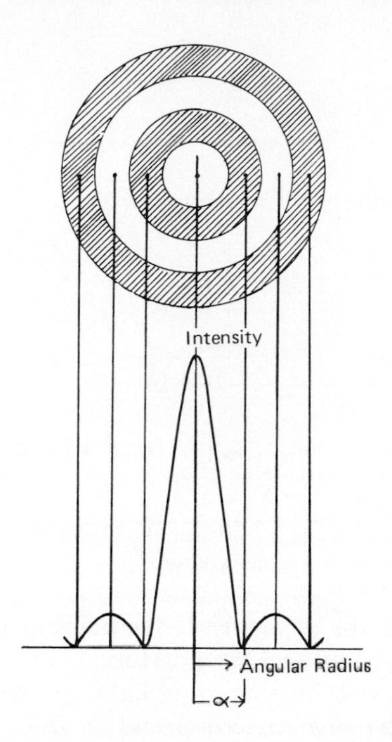

FIGURE 7-8. An Airy disk pattern and a plot of the intensity of light across the pattern. This pattern results when light from a point source passes through a circular aperture.

The light from a single point source, like a star, upon passing through a circular aperture, forms a pattern like that shown in Figure 7-8. This is called an *Airy disk pattern,* and it consists of a bright central disk surrounded by alternating dark and bright concentric rings that become dimmer as they proceed outward from the center. Beneath the pattern in Figure 7-8 is a plot of the relative intensity of the light that composes the pattern. The central maximum is far more intense than the subsidiary bright rings.

Analysis of the diffraction by a circular aperture shows that the diameter of the aperture d and the wavelength λ of the light being diffracted are related to the angular radius α (the angle subtended by a radius of the pattern from the center of the aperture) of the first dark ring. For small angles, which is the case in diffraction of light, this relationship is expressed by

$$\alpha = 1.22\,\frac{\lambda}{d} \tag{7-7}$$

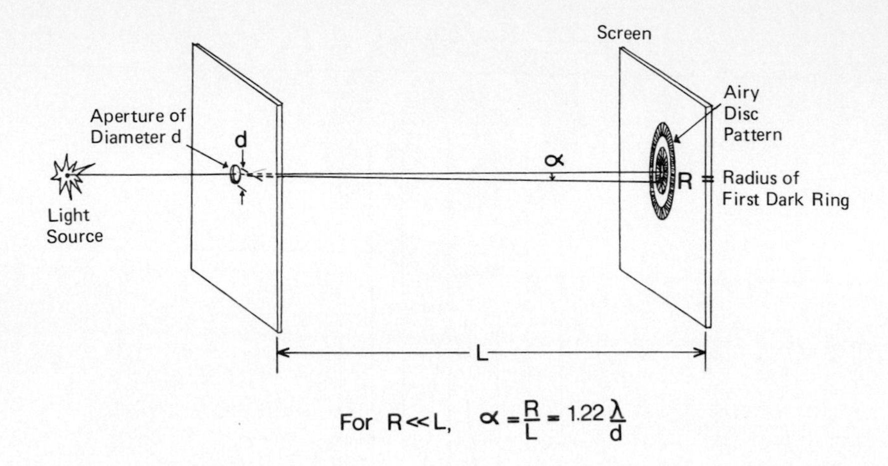

FIGURE 7-9. The relationships between the geometries of the Airy disk pattern and the circular aperture producing the pattern.

Figure 7-9 illustrates the geometrical relationship of the angular radius α to the diffraction pattern and the aperture.

If two closely spaced point sources of light are used, the question arises as to when the overlapping patterns resulting from them are distinguishable, or *resolved.* An arbitrary but useful criterion is usually accepted that determines when the images of two such point objects are resolved. This is the *Rayleigh criterion,* defined such that the central maximum of the diffraction pattern of one point source falls on the first minimum of the diffraction pattern of the other. Figure 7-10 shows examples of intensity plots of diffraction patterns of point sources at three different separations, one not resolved, one exactly meeting the Rayleigh criterion, and one clearly resolved.

Geometric Optics

Geometric optics concerns the treatment of light rays as they pass through or are reflected by a surface. In such treatments, it is assumed that light rays are *rectilinear,* i.e., they travel in straight lines until reflected or refracted at a surface. The surfaces of interest in geometric optics are either flat or curved so that all rays incident on a surface from a single point are brought together at another point.

Two categories of lenses distinguish between the geometrical treatments used in analyzing them. A *thin lens* is one that is sufficiently thin that it can be analyzed accurately by treating the lens as if it were in a single plane. A *thick lens* requires a somewhat more complex analysis.

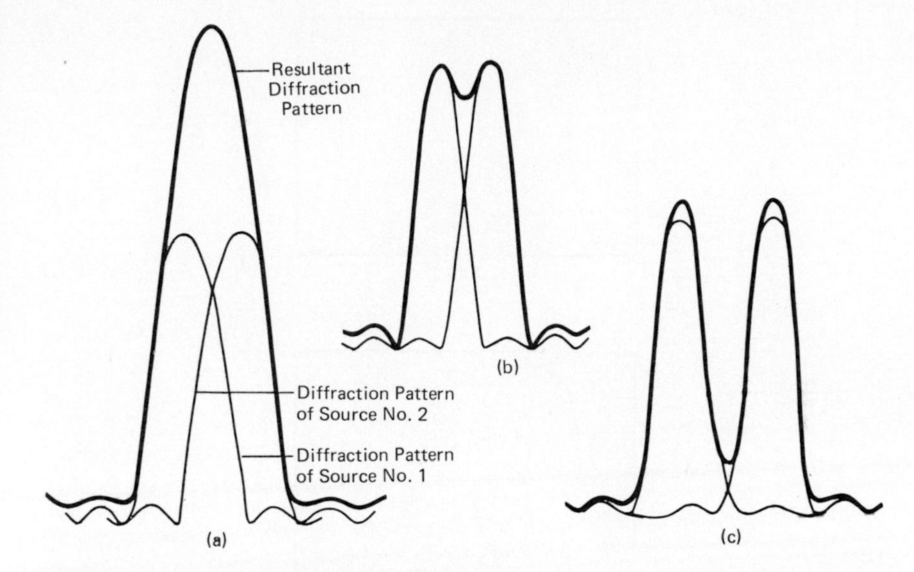

FIGURE 7-10. Intensity plots resulting from two closely spaced point sources of light whose rays have passed through a circular aperture. According to the Rayleigh criterion for resolution, (a) is unresolved, (b) exactly meets the Rayleigh criterion and is barely resolved, and (c) is completely resolved.

THIN LENSES

Figure 7-11 shows some of the common forms of thin lenses that can be classified either as *converging* or *diverging* lenses. A converging lens is thicker at the center than at the edge and causes parallel rays of light incident on the lens from one side to converge to a point, called the *focal point* (see Figure 7-12), on the opposite side. A diverging lens causes the incident parallel light to diverge apparently from a focal point on the same side of the lens as the incident light. Converging lenses are said to be positive, diverging lenses negative. The distance from either kind of lens to its focal point is called the *focal length* of the lens, and the line through the center of a lens perpendicular to the plane of the lens is called the *optical axis* of the lens.

Figure 7-13 illustrates the construction of an *image* using light rays coming from an *object* and passing through a thin lens. O and I are the *object distance* and the *image distance*; they are related to the focal length f of the lens by the *lens formula*:

$$\frac{1}{O} + \frac{1}{I} = \frac{1}{f} \qquad (7\text{-}8)$$

The construction in Figure 7-13 is based on two characteristics of thin

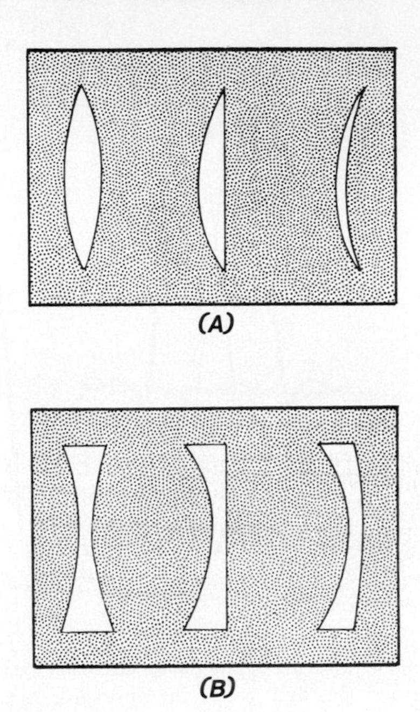

FIGURE 7-11. Some common forms of lenses. (A) Positive (convergent) lenses; (B) negative (divergent) lenses.

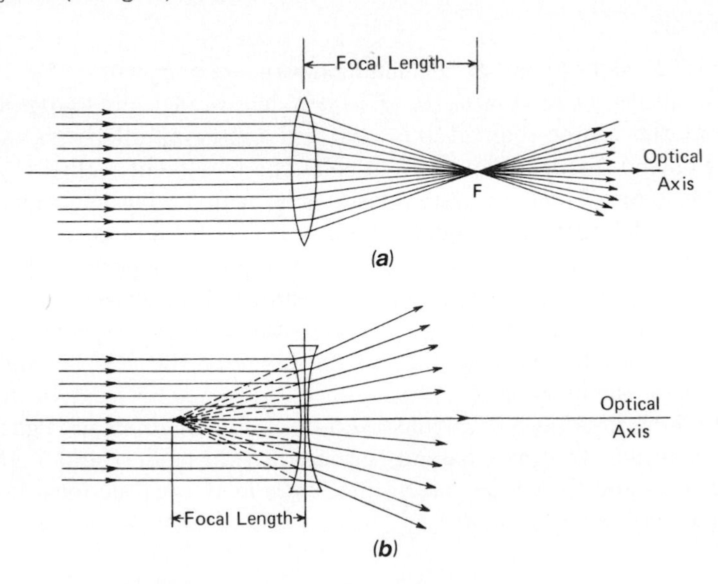

FIGURE 7-12. Illustrating the determination of focal length of (a) a convergent lens and (b) a divergent lens.

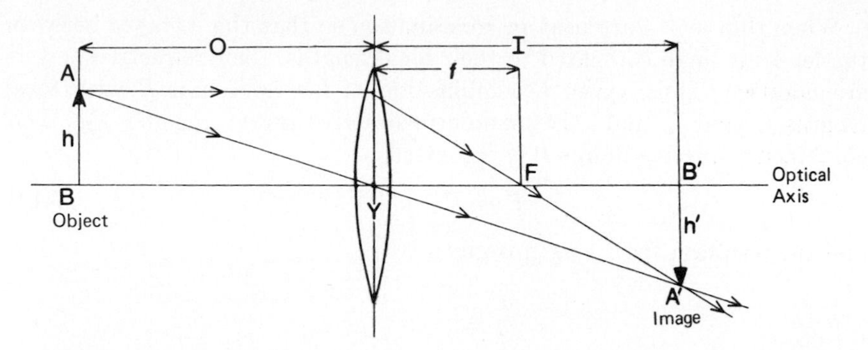

FIGURE 7-13. Construction of an image using a convergent lens. One ray from point A on the object is drawn parallel to the optical axis until it reaches the lens, where it is bent to pass through the focal point of the lens. A second ray from the same point on the object is drawn so that it passes through the center of the lens; this ray is undeviated by the lens. The two rays cross at A', the point on the image corresponding to point A on the object.

lenses: (1) rays parallel to the optical axis pass through the focal point of the lens, and (2) rays passing through the center of the lens are undeviated. In Figure 7-13, the triangles ABY and $A'B'Y$ are similar, and the sides of these triangles are related by

$$\frac{B'A'}{BA} = \frac{B'Y}{BY} \tag{7-9}$$

The *lateral magnification* m for this placement of object and lens is then obtained, in terms of object height h above the optical axis and the image height h', to be

$$m = \frac{h'}{h} = -\frac{I}{O} \tag{7-10}$$

The negative sign indicates that the image is inverted. Corresponding object and image points, like A and A' or B and B', are known as *conjugate points.*

The "strength" of a lens, or its *refractive power,* is related to its focal length. The customary unit of refractive power is the diopter, defined as the reciprocal of the focal length in meters, that is,

$$\text{refractive power (in diopters)} = \frac{1}{\text{focal length (in meters)}} \tag{7-11}$$

A convergent lens with a focal length of 25 cm has a refractive power of $+4$ diopters. Divergent lenses have negative refractive powers.

When thin lenses are used in combination so that the distance between the lenses is small compared to their focal lengths, their refractive powers are additive. Thus, given a combination of lenses A and B with focal lengths f_A and f_B and corresponding refractive power P_A and P_B, their combined refractive power P is expressed by

$$P = P_A + P_B \tag{7-12}$$

and the resultant focal length is given by

$$\frac{1}{f} = \frac{1}{f_A} + \frac{1}{f_B} \tag{7-13}$$

THICK LENSES

The optical properties of thick lenses require a larger number of parameters to specify the properties of the lenses. These parameters may be described in terms of the paths of particular rays. If an incident ray parallel to the optical axis is refracted through a thick lens, as shown in Figure 7-14a, the point at which the exiting ray crosses the opitcal axis is one of the focal points of the lens. A second focal point is located similarly by sending a ray parallel to the optical axis, but from the opposite side of the lens as depicted in Figure 7-14b. The two focal points are therefore on opposite sides of the lens. In Figure 7-15 both the incident and exiting rays are projected until they intersect. A line extended from the intersection perpendicular to the optical axis defines a *principal plane* of the lens. The intersection of a principal plane with the optical axis is called a *principal point* of the lens.

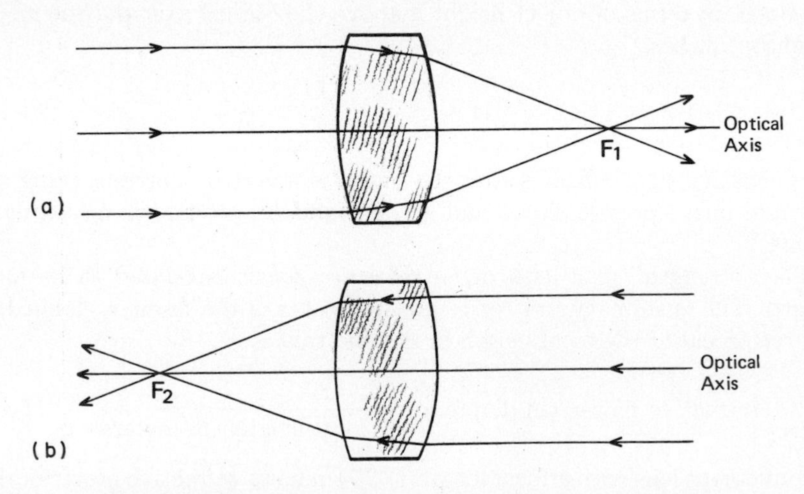

FIGURE 7-14. Locating the two focal points of a thick lens.

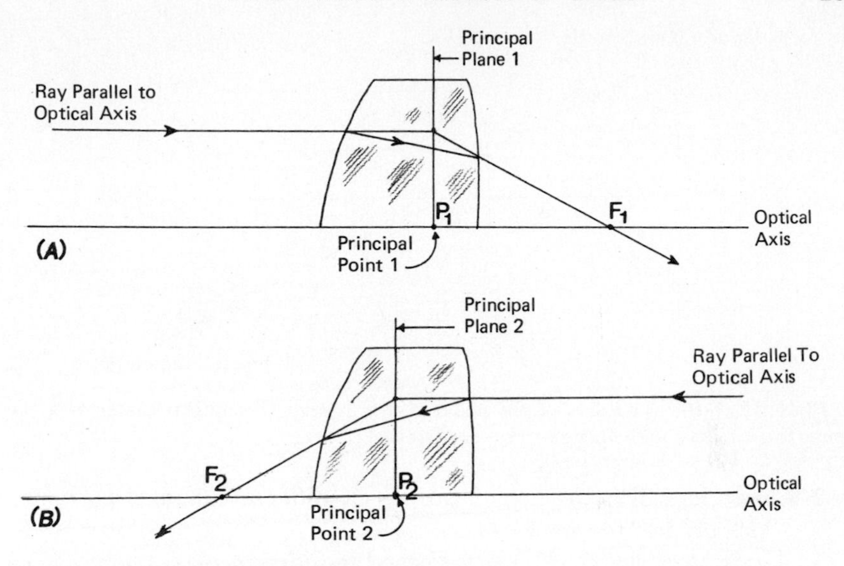

FIGURE 7-15. Location of principal points and principal planes of a thick lens.

Experimentally, one can determine the path of a ray from a point off the optical axis of a lens that results in an emerging ray parallel to the incident ray. Such an arrangement is shown in Figure 7-16. Projections of these rays to the optical axis locate two *nodal points* of the lens. The *optical center* of the lens is the intersection of the actual ray, as it passes through the lens, with the optical axis. The double arrows indicate that the nodal points and optical center can be located using light from either side of the lens.

A thick lens is described, therefore, by six cardinal points (two focal points, two principal points, and two nodal points) and its optical center. In the analysis of thick lenses, the distances to images and objects are measured from these points and not from the surfaces of the lenses.

LENS ABERRATIONS

In the detailed treatment of thin lenses, it is usual to consider only *paraxial rays*, i.e., those rays that are very near the optical axis. In this case, all of the angles involved in refraction are small. When the angles are not small, a number of aberrations, or defects in image formation, become apparent. Using *monochromatic light* (light of a single color, or frequency), several aberrations may occur.

1. *Spherical aberration*, in which all rays parallel to the optical axis do not cross the optical axis at a common point (see Figure 7-17)

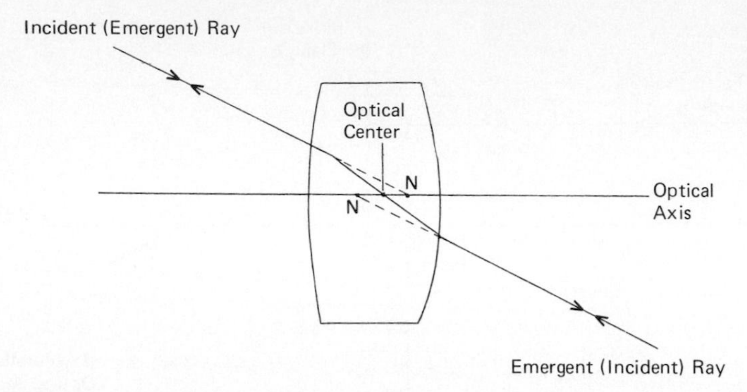

FIGURE 7-16. Location of the nodal points N and the optical center of a thick lens. The incident and emergent rays are parallel.

2. *Coma*, in which the magnifications of different annular portions of a lens are not the same
3. *Astigmatism*, in which images formed in different planes *perpendicular* to the optical axis are formed at different image distances
4. *Distortion*, in which the lens does not have uniform lateral magnification over its entire field

Another aberration of lenses occurs in connection with colors, or different wavelengths of light. *Chromatic aberration* refers to the inability of a lens to focus different colors of light at the same point. This defect occurs because the index of refraction (and therefore the ability to bend light) of a given lens material is slightly different for different wavelengths of light. The variation of index of refraction as a function of wavelength is called *dispersion.*

Anatomy and Physiology of the Eye

The human visual apparatus includes the *eyeball,* the *optic nerve,* and the *visual centers* of the brain. Understanding the visual process requires a

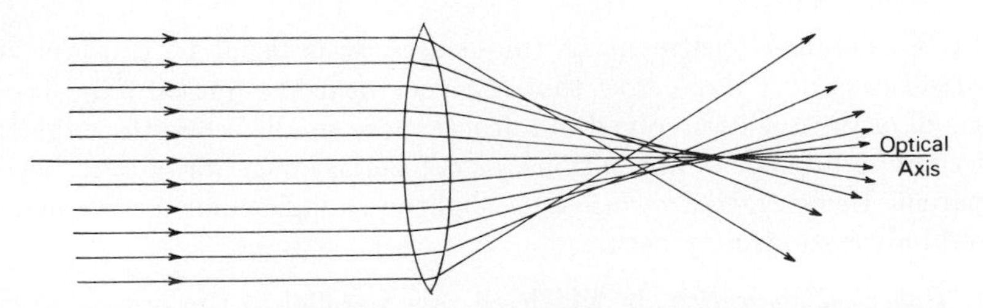

FIGURE 7-17. Spherical aberration in a lens.

familiarity with some of the structures involved as well as their functions. Here we shall consider some of the anatomical features of the human eye; then we shall develop a schematic version of the eye that will, it is hoped, serve as an aid in understanding the optical characteristics of the eye.

Anatomy of the Eye

The human eyeballs are couched within bony cavities of the head called the *orbits*. The eyeballs are further protected by eyebrows, eyelids, and eyelashes. The *conjunctiva* is a thin mucous membrane that lines the eyelid and folds back over the anterior portion of the eyeball, forming a surface covering. The lacrimal apparatus around the eyeball consists of glands, ducts, and sacs for producing, transferring, and storing tears, which are secretions of a dilute solution of various salts in water. Tears serve to keep the eyeball moist and help remove dust and microorganisms.

The eyeball is shown in horizontal section of Figure 7-18. The diameter of the eyeball is about 25 mm in the normal eye. It is composed of three tunics, or coats: the *fibrous tunic*, the *vascular tunic*, and the *nervous tunic* (the *retina*). Inside and on the anterior surface of the eyeball are the four *refracting media*: the *cornea*, the *aqueous humor*, the *crystalline lens*, and the *vitreous humor*.

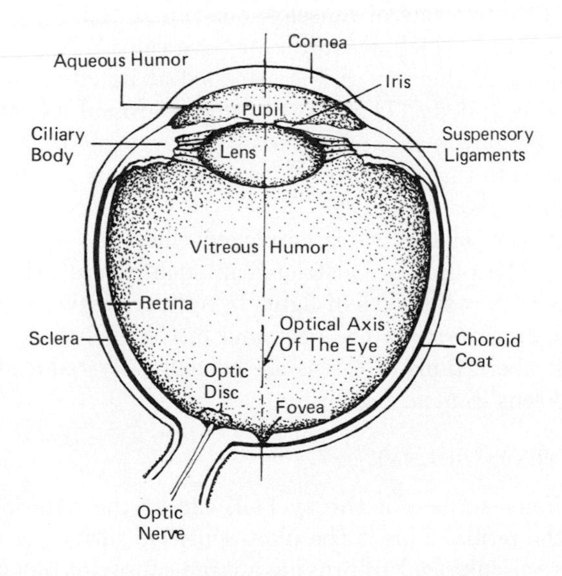

FIGURE 7-18. A horizontal section of the human eyeball.

THE FIBROUS TUNIC

The outer layer of the eyeball wall is the fibrous tunic. As the name implies, the fibrous tunic is a firm, unyielding, fibrous membrane. It serves to maintain the shape of the eyeball. The fibrous tunic consists of the *sclera*, or the white of the eyeball, which covers the posterior five-sixths of the eyeball, and the *cornea*, which covers the anterior sixth of the eyeball. Although the sclera and cornea are continuous, the cornea is transparent. The cornea is well supplied with nerves (so that a scratch on the surface of of the eyeball may be quite painful). However, it is devoid of blood supply and must depend on diffusion of oxygen from the atmosphere outside it and glucose from the aqueous humor inside it for its metabolic fuel.

THE VASCULAR TUNIC

The middle layer of the eyeball wall, the vascular tunic, consists of the *choroid coat*, the *ciliary body*, and the *iris*. As the name "vascular" suggests, the components of this tunic have a rich blood supply. The choroid coat is a dark brown sheath that darkens the eye chamber by preventing reflection of light.

The ciliary body is a specialized modification of the choroid coat and is a continuation of the choroid coat. The ciliary body contains a muscle whose contraction and relaxation changes the shape of the lens of the eye.

The iris is shaped like a very thin doughnut, and its anterior surface area can be changed by two sets of muscles, one radial and the other circular. The iris is pigmented, which accounts for "eye color." It is suspended between the cornea and the lens and is attached at its circumference to the margin of the ciliary body. The iris is arranged around a central opening, the *pupil*. By the movements of the iris, the eye regulates the amount of light that passes into the interior of the eyeball. This is a reflex response of the muscles in the iris.

A second function of the iris is employed when the eye is focused on nearby objects. The pupil is constricted in such a case, which limits the field of the lens to paraxial rays of light. In optics, restricting the field of a lens is referred to as *stopping* the lens, and in the eye, the iris serves as a *field stop*. Most aberrations of a lens are considerably reduced by limiting the rays to the lens' central region.

THE NERVOUS TUNIC (RETINA)

The entire inner surface of the eyeball, except the anterior portion, is covered with the retina. This is the photosensitive surface of the eye. The retina is highly vascularized to provide a large supply of blood. It is interesting to note that the retina has the greatest oxygen consumption per unit weight of any tissue in the body. The retina is called the nervous tunic because all the nerves that transmit the signals of visual perception to the brain have their endings in the retina. The nerve fibers all converge and pass out of the eyeball as the *optic nerve.* At the geometric center of the retina, in line with the optical axis of the lens, is the *fovea,* a depression in the retina about 1 mm in diameter.

An interruption in the retinal surface occurs on the nasal side of the central axis through the eyeball from cornea to fovea. This is the *optic disk,* often called the "*blind spot,*" through which the nerve fibers from the retina pass as a bundle that are collectively called the optic nerve.

The retina will be considered in further detail subsequently.

THE REFRACTING MEDIA

Several tissues in the eyeball are called refracting media because light entering the eye passes through these media in the process of forming an image on the retina. The cornea, at the anterior surface of the eye, is the first such medium. The aqueous humor, the medium between the cornea and lens, is a clear solution containing minute quantities of salts in water. Behind the lens is the vitreous humor, a semiliquid medium containing proteins, and whose refractive power is only slightly greater than water. The vitreous humor also helps to support the spheroidal shape of the eyeball and to support the retina.

The crystalline lens is a thick lens, with convex anterior and posterior surfaces. Its index of refraction is greater than those of the vitreous humor and aqueous humor. The lens has an outer covering of cells that enclose an elastic protein capsule that in turn encloses a clear, viscous gel. This gel is composed of about 25% proteins and 10% lipids. A metabolic defect sometimes causes the proteins in the lens to form fibrous aggregates that cause an opacity of the lens called *cataract.*

The lens of the eye can undergo changes in curvature of its refracting surfaces by the action of ciliary muscles that surround and attach to the lens. This change of shape permits the lens to change its focal length, thereby allowing the eye to focus on objects at various distances from the eye. This ability of the eye is called *accommodation.* Accommodation also includes the reduction of the useful volume of the lens by the iris as well as the adjustment of the orientation of the two eyes so that their visual axes converge on the object.

Formation of the Retinal Image

In order to set up patterns of nerve impulses that are related in a meaningful way to the light stimulus, it is necessary to distribute the incoming light in a suitable way among the receptors on the retina. This is accomplished in vertebrate eyes by forming a focused image on the retina. Such eyes are sometimes called "camera eyes" to distinguish them from "compound eyes" of, for example, some insects. In human eyes the image is inverted on the retina, just as the image in a camera is inverted on the film. A simple experiment demonstrates the inversion of the image on the retina as well as the fact that the brain interprets the inverted image as upright. Punch a small hole through a piece of paper or cardboard, hold it about 4 inches from one eye, and with one eye closed, focus on a distant lamp or window. Then place the pencil point as close to the eye as you safely can (within an inch—but be careful). Since the pencil is so close to the eye, the lens cannot accommodate for the pencil and will not change curvature. As you move the pencil upward, its shadows on your retina will also move upward, yet you will see the shadow move downward.

An even more impressive, though less generally accessible, proof of the inverted image on the retina is the observation of the image on the retina of the excised eyeball of an albino rabbit. This is possible because the sclera of an albino is unpigmented, and the image is clearly visible on the retina.

OPTICAL PARAMETERS OF THE EYE

The optical parameters of the eye that form the image on the retina can be characterized by treating all the refracting surfaces and media of the eye as if they collectively comprise a single thick lens. Such a model is called the *schematic eye*. The cardinal points of the schematic eye are usually measured from the corneal surface. Table 7-2 lists the optical constants of the normal eye under the two conditions that are the limits of accommodation, i.e., when the eye is relaxed and focused at infinity (on parallel rays) and when focused at the *near point*, about 10 cm from the eyeball. The diameter of the eyeball along the optical axis is about 25 mm in the normal eye.

DEFECTS OF IMAGE FORMATION IN THE EYE

Deviations in the shape of the eyeball from the *emmetropic*, or normal, eye account for a large variety of imperfectly formed images on the retina. The most common eye defects are *myopia* (nearsightedness) and *hypermetropia* (farsightedness). In myopia, the eyeball is extended beyond normal along the optical axis of the eye; the resulting optical defect is the formation of the image in front of the retina (see Figure 7-19A). Myopia is corrected by a negative lens in front of the eyeball (Figure 7-19B). Hypermetropia, a shortening of the dimension of the eyeball along the optical axis, seems to be often part of the normal aging process. In this

Table 7-2 Optical Constants of the Normal Eye[a]

	Relaxed (mm)	Accommodated to near point (mm)
1st Principal point	1.35	1.77
2nd Principal point	1.60	2.09
1st Focal point	−15.71[b]	−12.40
2nd Focal point	24.39	21.02
1st Nodal point	7.08	
2nd Nodal point	7.33	
1st Focal length	17.06	14.17
2nd Focal length	22.79	18.93
Distance to fovea	24.39	
	Relaxed (diopters)	**Accommodated to near point (diopters)**
Refractive power	42.9	48.4
	Indices of refraction	
Cornea	1.376	
Aqueous humor	1.336	
Vitreous humor	1.336	
Crystalline lens	1.424	

[a] Distances measured from the surface of the cornea.

[b] Negative sign means that the focal point lies outside the eyeball.

case, the image forms behind the retina (Figure 7-19C), and the defect is corrected by a positive lens (Figure 7-19D).

Astigmatism is relatively common among human eye defects. This occurs when the normally spherical surfaces of the cornea or lens become imperfectly shaped. The surfaces take on a cylindrical shape along some axis. A cylindrical lens, like the one shown in Figure 7-20, is completely astigmatic, i.e., a point object is transformed into a line image. A cylindrical aspect on an eyeball surface can be corrected by choosing a corrective lens in which a counteracting cylindrical surface, oriented in the proper direction, forms a combination of lens and eyeball that restores the point-to-point relationship of object and image.

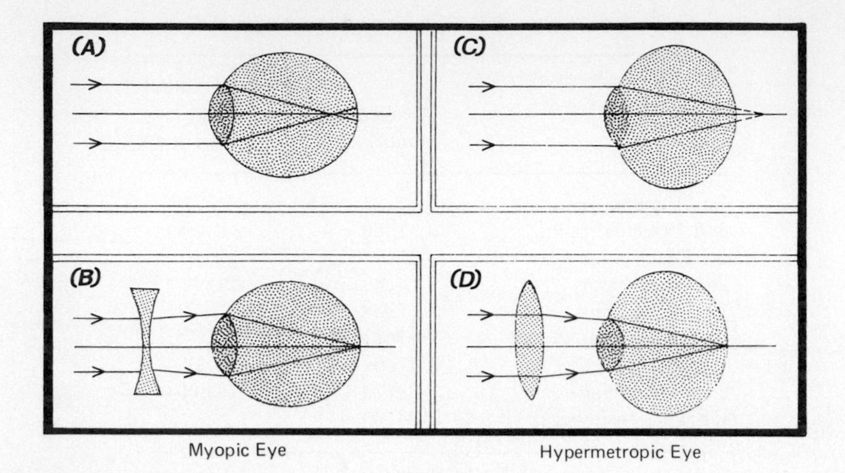

FIGURE 7-19. Correction of myopia (nearsightedness) and hypermetropia (farsightedness) with lenses. The myopic eye (A) focuses an image in front of the retina; correction is accomplished (B) by interposing a diverging lens. The hypermetropic eye (C) focuses an image behind the retina; in this case, correction (D) is accomplished by interposing a convergent lens.

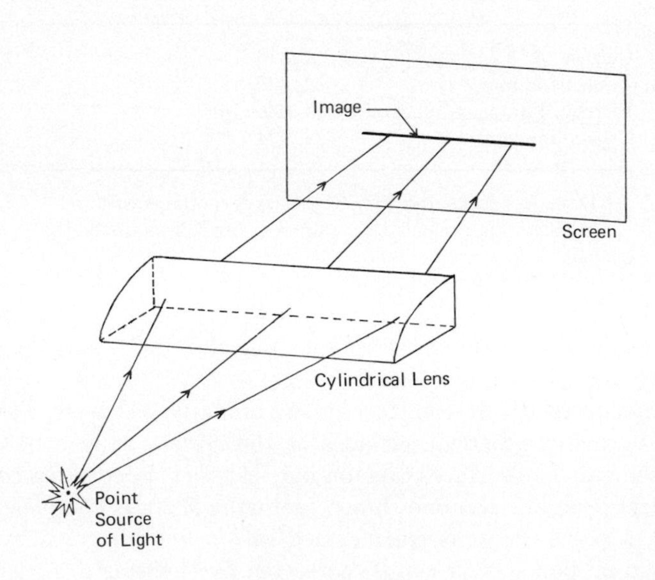

FIGURE 7-20. Illustrating the complete astigmatism of a cylindrical lens. A point source of light is extended into a line by such a lens.

The Retina

The light signals received by the eye, after passing through the refractive media of the eye, reach the retina, where this tissue converts the light into action potentials on neurons of the optic nerve. Since the retina plays such a crucial role in the visual process, we shall consider further detail of its anatomy and physiology.

ANATOMICAL DETAIL

The sheath that forms the retina is multilayered, consisting of three levels of neurons, two layers of granular materials, some membranes, and a a layer of pigmented material. The arrangement of the layers is shown in Figure 7-21. The neuron layers consist of ganglion cells on the innermost surface, which connect to bipolar neurons, which in turn are attached to the *rods and cones*. Notice the rather surprising spatial arrangement, in which the light impinges upon the retina so that it must pass through almost all the layers before it reaches the rods and cones. As we shall see, it is at the rods and cones that the absorption occurs that is effective in producing the visual sensation.

The fovea, previously described as a depression in the retina on the optical axis of the eye, is formed by a local thinning of the layers in the retina. In the fovea no rods, only cones, are found. Because of the thinning of the retina, light impinging on the fovea falls almost directly on the cone cells. Elsewhere on the retina beyond the fovea, both rods and cones are present. The entire retina contains approximately 7×10^6 cones and 125×10^6 rods; except at the fovea, the rods are roughly 18 times more numerous

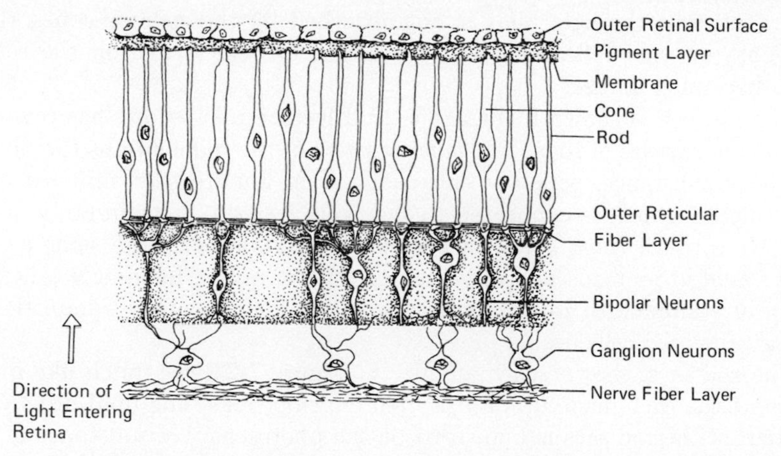

FIGURE 7-21. A schematic drawing of a section through the retina.

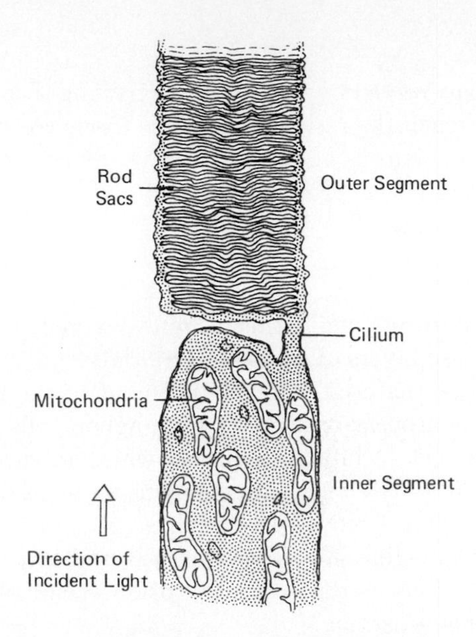

FIGURE 7-22. Part of a rod cell.

than cones. Histological examinations show that the cone cells of the fovea each have a "direct line" to the brain, i.e., one ganglion cell axon leading to the optic nerve serves one cone cell. Outside the fovea, one ganglion cell serves about 10 cones and one ganglion cell serves as many as 250 rods.

The pigmented layer of the retina, immediately behind the rods and cones, absorbs the light that is not absorbed by the rods or cones. This prevents internal reflections and thereby keeps the image on the retina from becoming diffuse.

The rod cell is shown in Figure 7-22. Electron microscopy has revealed the ultrastructure of rods. The *outer segment* of the rod contains the photosensitive substance, which is encapsulated in hundreds of thin *rod sacs.* The outer segment is connected to the *inner segment,* or main body of the cell, by a thin *cilium,* a fiber whose structure is exactly the same as the cilia found in protozoa and in vertebrate sperm tails. The inner segment is literally stuffed with mitochondria, which supply the large quantities of energy that rod cells use.

The rod sacs, shown schematically in Figure 7-23, are much like membrane disks, each made up of a protein envelope enclosing lipid disks. The protein of the rod sacs is known to be the photosensitive substance of the human eye. We will now consider its function further.

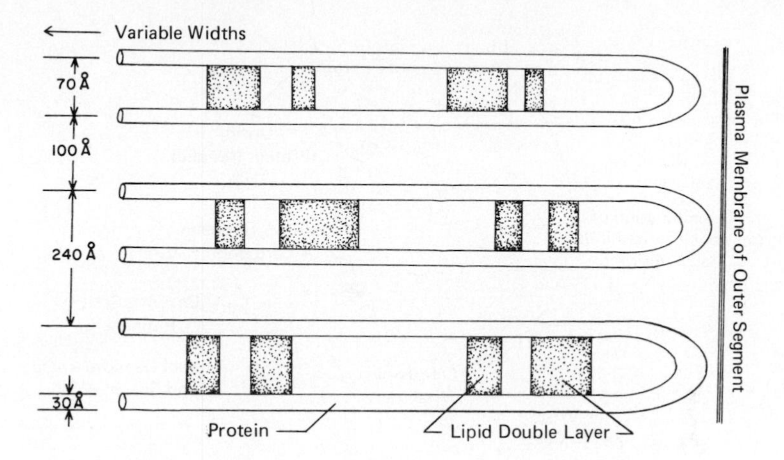

FIGURE 7-23. The detailed structure of the rod sacs.

PHOTOCHEMISTRY OF VISION

The light-sensitive pigment in the rod sacs of the retina is *rhodopsin*, sometimes called *visual purple*. Figure 7-24 shows the absorption spectrum of rhodopsin (i.e., the relative absorption of light by rhodopsin as a function of wavelength) compared to the sensitivity curve of the human eye. This comparison leaves little doubt that rhodopsin is the photosensitive absorber in the eye.

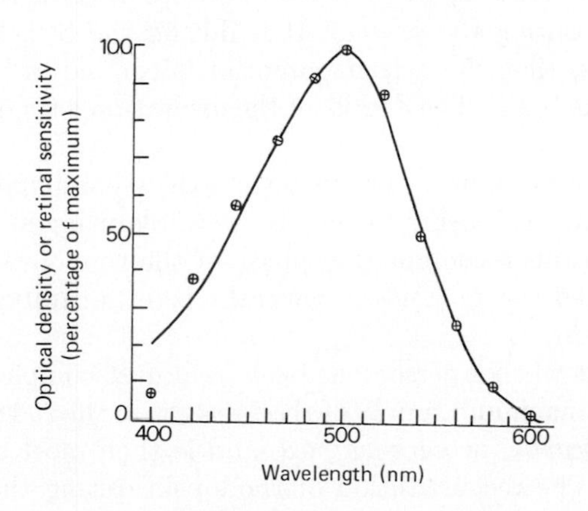

FIGURE 7-24. Comparison between the absorption spectrum of human rhodopsin (continuous curve) and the sensitivity of the dark-adapted human eye (circles). From C. H. Graham (1965). Copyright 1965 John Wiley and Sons. Reproduced by permission.

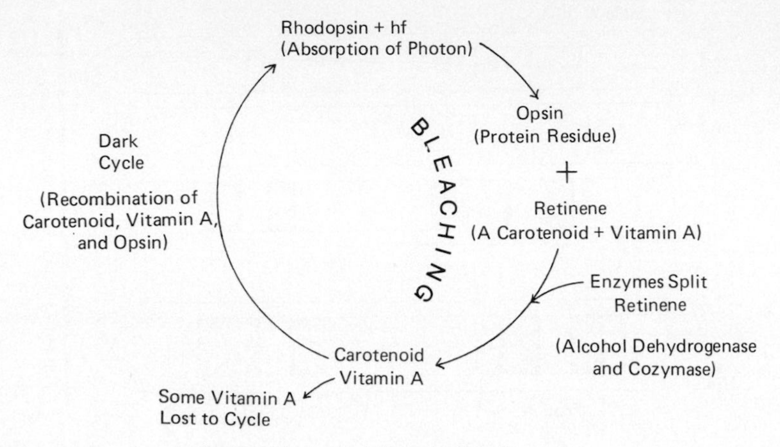

FIGURE 7-25. The rhodopsin cycle.

Upon exposure to light, rhodopsin begins a long cycle of chemical reactions that finally reconstitutes rhodopsin again. The more important steps in the rhodopsin cycle are summarized in Figure 7-25. Upon absorption of a photon, rhodopsin breaks apart into a purely protein residue called *opsin*, and a *carotenoid* (a chemical similar to carotene, the yellow pigment we see in many autumn leaves) called *retinene*. Chemically, retinene is an aldehyde of vitamin A, and the enzymes *alcohol dehydrogenase* and *cozymase* convert retinene to vitamin A. The process to this point is called the *bleaching* of rhodopsin because the visual purple turns to a pale yellow during the process. It is during the bleaching,which is a very fast process, that the action potential is excited on the rod and is transmitted to the brain. The details of the excitation are not known at the present time.

The reversion of vitamin A to retinene and the recombination of retinene with opsin to form rhodopsin is a slow process, requiring about 30 minutes in humans. Since this recombination phase of the rhodopsin cycle requires no light, it is called the *dark phase*. Several enzymes catalyze the reconstitution of rhodopsin.

It is found that when a person has been in almost complete darkness for 20–30 minutes, a maximum supply of rhodopsin is in the rod sacs. The eye is then said to be *scoptic*, or *dark-adapted*, and is then most sensitive to detection of light. The reconstitution of rhodopsin during the dark cycle is somewhat inefficient with respect to vitamin A, i.e., some fraction of vitamin A is lost during the cycle, and for this reason a constant supply of vitamin A is necessary to the eye. This is the reason for the necessity of a continual intake of vitamin A in the diet. Animals deprived of vitamin A develop *nyctalopia*, or *night-blindness*, a relative insensitivity of the eye due to a deficiency of rhodopsin.

The photochemistry of the cones is not as well understood as that of the rods. The photosensitive material in human retinal cones has neither been isolated nor identified. Nevertheless, a number of facts are clear. Only the *photopic*, or *light-adapted*, eye can perceive color; the cones and never the rods, are involved in color vision.

FUNCTIONAL AND PHYSICAL ASPECTS OF VISION

Scotopic rod vision is more sensitive than cone vision. Since the fovea contains only cones, the eye perceives color clearly when focused directly on an object (so that the image is directly on the fovea) but is most sensitive when an image is formed away from the fovea, i.e., when it is viewed by the peripheral vision, or not looked at directly. This phenomenon is clearly observed by allowing the eye to become dark-adapted and looking at the binary stars that form the bend in the handle of the big dipper. The brighter star, Mizar, can be seen by focusing directly on it, but the fainter one, Alcor, cannot. Yet if one looks about 20° away from these stars, the presence of the fainter one becomes obvious. Thus, although the *resolution* of the eye, i.e., its ability to separate two images, is greater for light impinging on the fovea, the sensitivity of the eye is greater in peripheral vision, using rods. Because the cones in the fovea have a one-to-one association with neurons in the optic nerve, cone vision is "fine-grained," or sharp, whereas rod vision is "coarse-grained," or diffuse. This explains why it is sometimes said that "in the dim light (where only rod vision is used), all cats are gray (because rods have no color capability)." Indeed, because of the coarse-grained nature of rod vision, one may not even be sure that they are cats.

The ability of the eye to resolve two points on an object is called *visual acuity*. The angle subtended by an image on the retina from the inner nodal point of the schematic eye is the *visual angle*. Visual acuity is defined as the reciprocal of the minimum resolvable visual angle, measured in minutes of arc. In the normal eye, the minimum resolvable visual angle is about one minute. "Eye charts" used to measure visual acuity in humans use letters designed so that the separations that distinguish between letters subtend a visual angle of one minute when viewed from a distance of 20 feet. The measured visual acuity is usually expressed as the ratio of the distance at which the subject's eye resolves two points to the distance at which the "average" eye resolves the same points. Thus, a person with 20–30 vision resolves at 20 feet what the average person can resolve at 30 feet.

Visual acuity is affected seriously by the illumination of the object being viewed. This is because in bright light the iris constricts the pupil and allows only paraxial rays to enter the eye. This in turn reduces both spher-

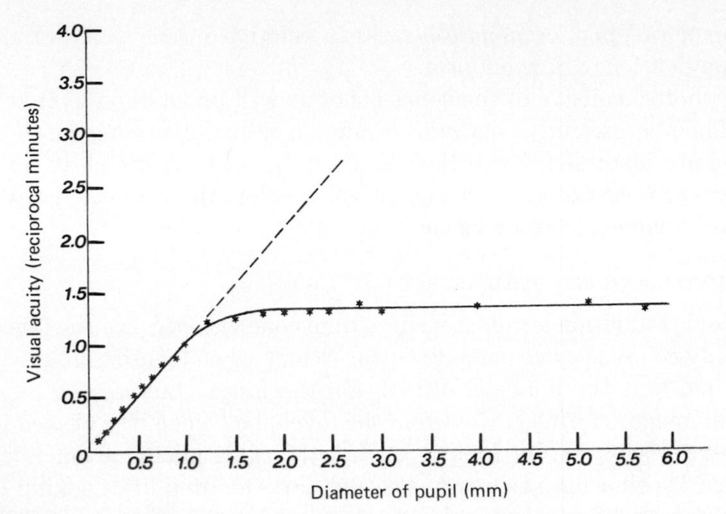

FIGURE 7-26. Experimental data relating visual acuity and pupillary diameter. The straight line defines the Rayleigh limit attributable to diffraction. Adapted from C. H. Graham (1965). Copyright 1965 by John Wiley and Sons. Reproduced by permission.

ical and chromatic aberrations of the eye, sharpening the image on the retina. These facts explain why resolution is noticeably improved when objects are observed in a strong lighting situation.

An interesting aspect of visual acuity can be examined by a few simple calculations. Within the fovea, the cones are closely packed, and each cone is about 0.003 mm in diameter. Why, in the course of its evolution, has the eye not developed thinner cones to provide us with sharper vision? Perhaps the answer lies in the limitations imposed by the physics of the optical system of the eye. Consider some of the physically significant dimensions of the optical system. Experiment shows (see Figure 7-26) that maximum acuity occurs when the pupil diameter is about 2 mm and remains about the same until the diameter of the pupil reaches 6 mm. Then, using the aperture of the eye's lens to be 2 mm, we find, using Equation (7-7), the angular radius α that can be resolved within diffraction limitations of the Rayleigh criterion is

$$\alpha = 1.22\,\frac{\lambda}{d} = 1.22\,\frac{\lambda}{2\,\text{mm}} \tag{7-14}$$

Since the cones are most sensitive to yellow light, we may assume $\lambda = 5000$ Å $= 5 \times 10^{-4}$ mm; α then becomes

$$\alpha = \frac{1.22 \times 5 \times 10^{-4}}{2} = 3.05 \times 10^{-4}\ \text{rad} \tag{7-15}$$

Now 1 rad = 3.43×10^4 minutes of arc, so α becomes

$$\alpha = 3.05 \times 10^{-4} \text{ rad} \times \frac{3.43 \times 10^4 \text{ minutes}}{\text{rad}} = 1.05 \text{ minutes} \qquad (7\text{-}16)$$

This corresponds to our earlier assertion of one minute of arc as the minimum resolvable angle for the human eye. Now the inner nodal point of the eye is about 17 mm from the surface of the retina, and the angle $\alpha = 1.05'$ subtends an arc length s on the retina given by

$$s = r\alpha \qquad (7\text{-}17)$$

$$s = 17 \text{ mm} \times 3.05 \times 10^{-4} \text{ rad} = 0.0052 \text{ mm} \qquad (7\text{-}18)$$

This means that because of diffraction the eye cannot resolve images separated on the retina by less than about 0.0052 mm. The diameter of cones is already less than this value.

It would seem, then, that the size of the cones in the eye have evolved to the optimal diameter. One may wonder why the acuity of the eye does not increase as the pupil diameter increases from 2 to 6 mm. The limitations of diffraction by the eye's aperture indeed improve the resolution as the diameter of the pupil is increased; on the other hand, the confusion of the image by aberrations of the lens increases as the pupil opens to permit entry of nonparaxial rays. These opposing effects essentially cancel each other for pupil diameters from 2 to 6 mm. The eye is therefore capable of adjusting its aperture over this range to compensate for varying brightness conditions without sacrificing any visual acuity.

It is perhaps interesting to consider at this point the factors in adaptation that have permitted some species to improve their visual acuity. Many predatory birds are known to possess sharper eyesight than humans. In general, predatory birds have eyes whose lenses are quite large relative to their overall size. This permits efficient gathering of light and a large aperture. A more important effect is the particular shape of the fovea in these birds. Figure 7-27 shows a comparison of the shapes of a bird fovea and a mammalian fovea. The sharp indentation in the bird fovea spreads an image over a large number of receptor cells, thereby effectively magnifying the image. Hawks, eagles, and a few other birds who hunt on the wing even have a second fovea in each eye. Since their eyes are on the sides of their heads, they have foveae on the medial position of their eyeballs. A second fovea in each eyeball is in a posterior position, an arrangement that allows the bird to see clearly in front of him and utilize binocular vision to provide him with depth perception.

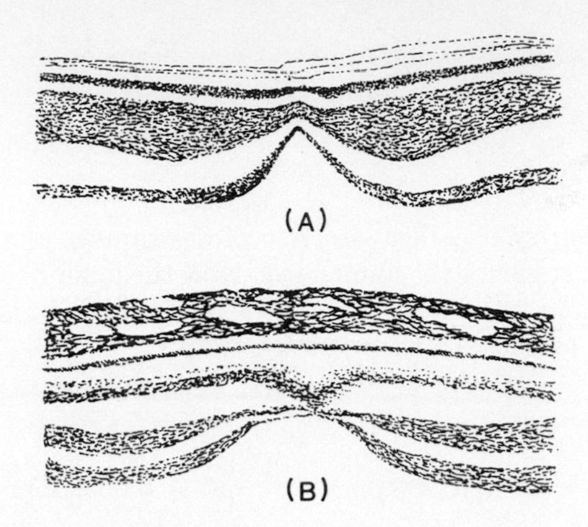

FIGURE 7-27. Cross sections of foveae of (A) a bird and (B) a mammal.

Theory and Experiment of the Visual Process

In this section we shall consider two problems associated with the visual process, both of which have been mentioned earlier. Our considerations do not represent complete, nor even particularly modern, examples of biophysical research. It is hoped that they typify the approaches, both theoretical and experimental, to problems of the visual process that have been the subject of study for many decades. The first of these areas is color vision, which has puzzled physicists, chemists, psychologists, and physiologists for so long—and continues to defy definitive understanding. The other area we shall consider is an example of a classic biophysical experiment, the measurement of the minimal threshold of vision, i.e., an experiment to determine how few photons the human eye can perceive.

Color Vision

The facts of visual perception garnered from experiments with humans are perhaps too often interpreted in anthropomorphic terms, i.e., the visual characteristics of the entire animal kingdom are often interpreted in terms of human visual experience—and even human visual limitations. In particular, the physical factors that produce sensations we recognize as color perception may be quite different in other species. Further, the physiological mechanisms for processing and interpreting the effects of these physical factors may also be different from those in humans.

Three of the physical parameters that we can associate with light stimuli are *spectral quality* (wavelength content), *intensity,* and *duration of expos-*

ure. In humans, the intensity and duration of exposure affect our sense of color only in minor ways. This may not be the case in other animals. Even the response to spectral quality may occur by either differential sensitivity of receptors to wavelength (as would be the case if several photopigments were involved) or by interpretation by the brain of various neural processes that may be wavelength dependent (like time lags or other temporal modifications in the nerve pulses to the brain).

Realizing that we may not impute the limitations and capabilities of human vision to other animals, we may describe color vision in terms of the sensations with which we are familiar. The human eye recognizes three *primary colors*, so called because combinations of these can be found to give rise to all the rest. The accepted primary colors are *red, green,* and *violet*. Not only can all colors be produced by these, but also *white, black,* and *gray* can be obtained.

Three attributes of color are used to specify the physical qualities we associate with color (see Figure 7-28):

1. *Brilliance*, called "value" among artists, is that attribute of any color by which it may be classed as equivalent to some member of a series of grays that range continuously from black to white. Black is said to have zero brilliance. "Glare" is a state of brightness that causes discomfort. Physically, brilliance is a measure of intensity.

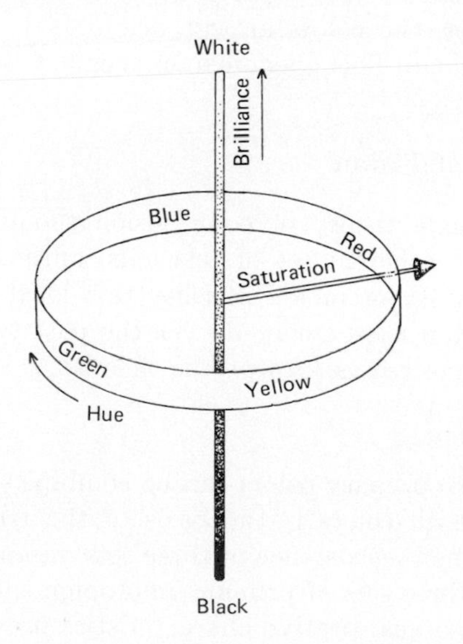

FIGURE 7-28. An illustration of the characteristics of color: brilliance, hue, and saturation.

2. *Hue* is determined by wavelength; it is that attribute of a color by which it differs characteristically from the gray of the same brilliance. It is in respect to hue that colors fall into classes designated as reddish, yellowish, greenish, bluish, etc. Hue is what most people mean by "color" in a non-technical sense.

3. *Saturation* is an attribute of all colors that determines their degree of difference from a gray of the same brilliance. Saturation may be thought of as "purity of hue."

For any given color there is a *complementary color* that combines with it to produce white. Complementary colors are 180° apart on the hue circle of Figure 7-28. If the visible spectrum is considered as a linear sequence of colors from red to violet, it is found that colors closer together than complementary colors give on fusion some intermediate color. Thus, red and yellow, when fused, produce orange. Colors separated in wavelength more than complementary colors produce some shade of purple upon fusion.

Two further color effects are observed in human vision. *After-images* are perceived for a short time after a person stops looking at a color. He may continue to perceive the same color (*positive* after-image) for a brief time, or he may see the complementary color (*negative* after-image) for a more extended time. Under certain conditions, e.g., when a piece of paper of a given color is laid upon a paper of another color that is close to its complementary color, the colors appear heightened, or more saturated, than when viewed alone. This phenomenon is called *color contrast*.

Research into Color Vision

Ideally, an adequate theory of color vision should describe all color phenomena within a coherent set of relations, optimally in quantitative form. No extant theory or model satisfies this ideal. More than eighty theories of color vision have emerged over the past two centuries. Many of these can be categorized as *trichromatic* theories.

TRICHROMATIC THEORIES

The fact that three primary colors can be combined in appropriate proportions to produce all colors is the basis of the trichromatic theories. These theories assume the existence of three sets of sensory mechanisms in the retina, that is, three sets of primary photopigments, or perhaps three types of cones, whose quantitative characteristics provide a basis for discriminating between colors. The trichromatic theories also assume specific types of nerve fibers that correspond, respectively, to the red, green, and violet photochemical substances. (Actually there are many combinations of three wavelengths with which the spectrum can be matched.) In these theories, equal excitation of the three cone types of three photopigments results in a sensation of white. All other colors are compounded by stimulation of the three receptor materials in different proportions.

The phenomenon of negative after-images is explained convincingly by the trichromatic theories as follows: If one stares at a red object, the corresponding photopigment is acted upon. When all the cones are subsequently exposed to white light, the green and violet photopigments, which have been used less, respond in greater proportion to the white light, and the after-image sensed takes on a blue-green (complement of red) color.

Some of the trichromatic theories hypothesize specific zones of the cerebral cortex of the brain, each of which is associated with one of the primary colors. In doing so, these theories relegate to the brain many of the interpretive functions necessary to assimilate the three channels into meaningful color information. This, in itself, is not a particularly damning defect of the theories, for we have seen (in Chapter V) that the subtle and complex brain functions are indeed a vital part of other sense perceptions.

Several objections have been raised to the trichromatic theories. They fail to explain certain aspects of color blindness. The most recent anatomic and physiological studies suggest that the neural aspects of these theories are oversimplified. Finally, it is somewhat unsatisfying that the photosensitive pigments that form the basis of these theories have neither been isolated nor identified.

THE LAND EXPERIMENTS

In 1959 Land reported a series of experiments describing some striking color effects that were quite unexpected on the basis of earlier color theories. He produced natural scenes in full color by mixtures of *two* colors. The effect was produced by combining in precise register two superimposed projected photographic positives of the same scene. The two uncolored positives were produced by taking two photographs of the same scene from exactly the same viewpoint. One photograph was taken through a filter transmitting light in the long wavelength third of the visible spectrum, the other through a filter transmitting the middle third. When the long record was projected in red light and the short record in the "white" light of an incandescent lamp, the full gamut of colors appeared. In some experiments, yellow light was used to project the short record. The colors perceived in the experiments were largely independent of the intensities of the two projection lamps.

Land concluded that the classical laws of color mixing probably conceal broader basic laws of color vision. The apparent inadequacy of present color theories suggests that he might have been correct.

Experimental Determination of the Threshold of Vision

The minimum amount of light that must reach the eye in order to produce a visual sensation is called the *absolute visual threshold.* Determination of the absolute visual threshold is understandably one of the more interesting biophysical problems of the visual process. In 1942, Hecht, Schlear, and Pirenne made experimental measurements to determine quantitatively the minimum number of photons to which the human eye can respond. This is one of the classic experiments that exemplify the value of techniques of the physical sciences to the life sciences.

The experimenters utilized every known physiological fact about the eye to ensure that the measurements provide information that represents the maximum sensitivity of the eye. They provided for the following conditions:

1. The subjects of the experiment were placed in darkness for about 30 minutes before testing to assure dark-adaption.
2. Since the rods are densest on the retina about 20° from the axis of the eyeball, the light used to test the eye was arranged to fall on this region of the subjects' retinae.
3. An optimally sized test spot of light was chosen in accordance with earlier experiments that had shown that the scotopic eye was most sensitive when a circular test spot subtends an arc of 10 minutes at the cornea.
4. The optimum duration of a light flash was chosen. The product of duration and intensity produces constant response up to about 0.01 second; longer durations are affected by resynthesis of rhodopsin. Therefore, light flashes with durations of 0.001 second were chosen.
5. The scoptic eye is most sensitive to light at a wavelength of 510 nm. In these experiments, the light used was restricted to a wavelength band 10 nm wide centered at 510 nm.

In the experimental apparatus, light from a carbon arc at constant current was filtered, then passed through a double monochrometer (a device, described in Chapter XI, for isolating narrow wavelength bands of light), and shuttered by an arrangement that accurately and reproducibly provided single flashes of 1 millisecond duration. The light intensity was measured by a thermopile, a device that absorbs the light, converts it into heat, and in this case, uses the heat to provide a theomoelectric potential that can be measured extremely accurately. Such an arrangement can be calibrated against a standard lamp of known intensity.

The experiment was conducted under the conditions described above by having the experimenter produce a flash of light, after which a subject reported whether or not he saw it. As might be expected, there were some variations in individual subjects; indeed there were statistical variations among the data of each subject.

The minimum energies perceived by the subjects of these experiments varied beteeen 2.1 and 5.7 $\times$ 10^{-10} erg of light energy incident on the cornea per flash. For light at 510 nm, we may calculate, using Equation (7-3), that each photon has an energy of 3.84 $\times$ 10^{-12} erg. Then the number of photons incident on the cornea that caused visual sensation lies in the range from 54 to 148.

It is even more interesting to consider the sensitivity of the retina, or even a rod, for although the sensitivity of the eye as a whole is significant in terms of human vision, it reveals almost no information that furthers our understanding of the photochemical processes in the retinal pigments. To proceed, then, we must know the degree to which light is attenuated, either by reflection or absorption, in traversing the eye from cornea to rod sac.

Under the conditions of the experiment, it was found that

1. 4% of light incident on the cornea is reflected away from it.
2. 50% of light entering the cornea is absorbed by the refracting media of the eye along its path to the retina.
3. 80% of the light incident on the inner surface of the retina is absorbed in passing through the several layers of the retina before reaching the outer segment of the rods.

Therefore, of the 54 to 148 photons incident on the eye, we can conclude that the actual number reaching the rod sacs is in the range from 5 to 14. This was considered an upper limit by these experimenters. Some workers, with similar experiments, have found the minimum number of visible photons to be 2. Recalling that the spot on the subjects' eyes subtended an arc of 10 minutes, we can suppose that the likelihood of more than one photon in each flash entering a single rod was quite small. Therefore, it seems reasonable that even though 2 to 14 rods must be simultaneously activated to produce a visual sensation, a single photon is sufficient to activate a single rod.

We have shown that the energy in a single photon at 510 nm is 3.84 $\times$ 10^{-12} erg. A mole (6.02 $\times$ 10^{23}) of photons therefore has an energy equal to

$$E = 3.82 \times 10^{-12} \text{ ergs/photon} \times 6.02 \times 10^{23} \text{ photons/mole}$$

$$= 2.30 \times 10^{12} \text{ ergs/mole}$$

$$E = 2.30 \times 10^{5} \text{ joules/mole}$$

and since a joule is equal to 2.39 $\times$ 10^{-4} kcal, the energy of a mole of these photons can be expressed as

$$E = 2.3 \times 10^5 \text{ joule/mole} \times 2.39 \times 10^{-4} \text{ kcal/joule} = 55 \text{ kcal/mole}$$

In Chapter II we saw that energies of this magnitude are required to break a single covalent bond. Indeed, this result implies that a single photon of visible light, upon absorption by rhodopsin in the retina, could at most break only a few bonds of any variety. It is impressive that our eyes are so constructed that the energy of a single photon can initiate an action potential on neurons that can be interpreted as a sensation of vision.

References and Suggested Reading

Graham, C. H., ed. (1965). "Vision and Visual Perception." Wiley, New York.

Hecht, S., Shlear, S., and Pirenne, M. H. (1942). Energy, quanta, and vision. *J. Gen. Physiol.* **25,** 819.

Hubel, D. H. (1963). The visual cortex of the brain. *Sci. Amer.* **209,** 54 (Offprint 168).

Land, E. H. (1959). Experiments in color vision. *Sci. Amer.* **200,** 84 (Offprint 223).

Lipson, S. G., and Lipson, H. (1969). "Optical Physics." Cambridge Univ. Press, London and New York.

MacNichol, E. F. (1964). Three-pigment color vision. *Sci. Amer.* **211,** 48 (Offprint 197).

Michael, C. R. (1969). Retinal processing of visual images. *Sci. Amer.* **220,** 104 (Offprint 1143).

Pirenne, M. H. (1952). Quantum physics of vision. *Progr. Biophys.* **2,** 319.

Pirenne, M. H. (1967). "Vision and the Eye." Chapman & Hall, London.

Ruston, W. A. H. (1962). Visual pigments in man. *Sci. Amer.* **207,** 120 (Offprint 139).

Sommerfeld, A. (1954). "Lectures on Theoretical Physics," Vol. 4. Academic Press, New York.

Tansley, K. (1965). "Vision in Vertebrates." Chapman & Hall, London.

Wald, G. (1956). Eye and camera. *Sci. Amer.* **183,** 32 (Offprint 46).

CHAPTER VIII

The Circulatory System

Humans and other higher forms of life enjoy the freedom of living under a wide variety of conditions—considerable temperature variations, many levels and types of activity, and great differences in chemical composition of ingested foods. We are afforded so much choice by virtue of a complex mechanism that maintains for our cells an environment that is always within a relatively narrow physiological range of constancy. The *circulatory system* conducts fluids throughout the body to provide the necessary cellular environment, and it transports materials required for the metabolism, nutrition, and protection of the cells.

The circulatory system consists of a *blood vascular system* and a *lymphatic system*; the former delivers the intracellular fluids to the cells, and the latter helps to collect the fluids to be recirculated. In this chapter we shall consider the fluid tissue that circulates, the *blood*; the distribution system, the *blood vessels*; and the pump that provides the hydrodynamic energy required for continuous circulation, the *heart*. We shall focus our attention on some of the physical aspects of the circulatory system. In particular, we shall examine the energetics and the velocity and pressure relationships throughout the system. Some of the electrical aspects of the heart are particularly informative, and these will be examined in some detail.

Blood and its Distribution

Blood serves a number of functions in maintaining the constant cellular environment necessary in advanced organisms. The *respiratory function* of blood is accomplished by conveying oxygen from the lungs to the cells and carbon dioxide from the cells to the lungs. The *nutritive function* consists of carrying glucose, amino acids, fats, and vitamins from the digestive tract to the cells. Blood provides an *excretory function* in transporting nitrogenous wastes of metabolism to the organs of excretion and a *protective function* in containing agents that defend the cells against foreign bodies and injurious microorganisms. The *regulatory functions* of blood include distribution of hormones from ductless glands to the cells which they affect, the conduct of water to excretory organs to maintain a relatively constant volume of water in the body, and the equalization of body temperature by giving off heat from superficial blood vessels.

Characteristics of Blood

Blood is a tissue composed of a fluid within which are many kinds of solids, cells, and solutes. In humans, blood makes up about one-thirteenth of the total mass of the body; the average adult male, whose mass is between 65 and 75 kg, has about 5 to 6 liters of blood. Blood has a specific gravity of about 1.05, and its viscosity is about 5 times that of water. Blood is slightly alkaline, with a pH very near 7.4 (pH is a measure of hydrogen ion concentration, and therefore a measure of the acidity or alkalinity of a solution. A pH of 7.0 is neutral; greater than 7.0 is alkaline, and less than 7.0 is acidic.)

If a specimen of blood is centrifuged, it will separate into two portions. The *formed elements*, or the cells, of the blood are slightly more dense than the plasma, a straw-colored liquid in which the cells are normally immersed. The plasma constitutes about half the volume of the blood. The formed elements are *red blood cells* (or *erythrocytes*), *white blood cells* (or *leukocytes*), and *platelets*. Blood cells are often called *corpuscles*.

BLOOD CELLS

Erythrocytes, shown in Figure 8-1, are disks that are concave on both sides and look somewhat like a dumbbell when viewed edgewise (the cells are rather transparent). The disks have a tendency to stack like coins in the bloodstream. Each disk is about 8 microns (μm) in diameter. In the human embryo, erythrocytes are formed in bone marrow and in the liver; but in adults, they are formed only in red bone marrow, which is found in the sternum (breast bone), ribs, vertebrae, and the proximal epiphyses (knobs at the ends) of the femur and humerus (thigh bone and upper arm bone). During their immature stages the red blood cells have nuclei but no *hemoglobin*. After mitosis is complete, the nucleus is eliminated from the cell, and the cell is transformed into an erythrocyte, mostly composed of hemoglobin enclosed in the supporting framework called the stroma. Hemoglobin is a porphyrin–protein complex, in which a heme group containing iron is

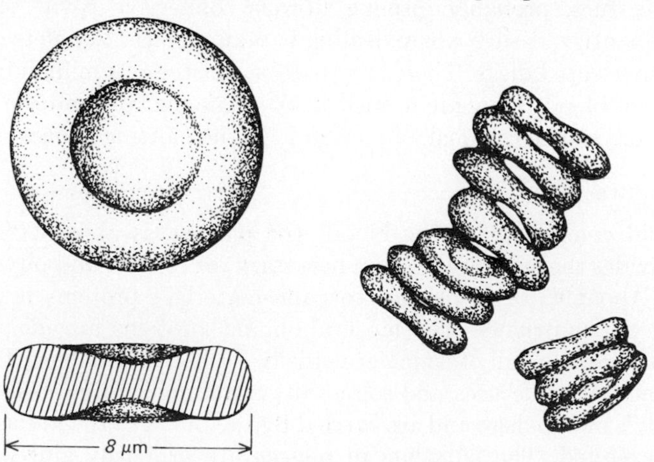

FIGURE 8-1. Red blood cells (erythrocytes).

attached to the protein globin, which is capable of joining onto oxygen. The resulting oxyhemoglobin is the vehicle of oxygen transport throughout the body.

There are about 5 million erythrocytes in a cubic millimeter of blood at any given time, though considerable transient variations may occur. The ability of the body to maintain a relatively constant number of red blood cells is called *hemopoiesis,* and although the mechanism is not well understood, it is known that production of erythrocytes is stimulated by a decrease in blood oxygen.

The life span of erythrocytes is not known exactly, but they probably survive between 10 and 120 days. When their usefulness is over, they are destroyed in the liver and in the spleen.

Leukocytes are nucleated cells that occur in a number of varieties that can be classified as *granular* or *agranular,* depending on whether or not they contain granules in their cytoplasm. The normal white blood cell count is far less than the erythrocyte count. There are between 5000 and 10,000 leukocytes in a cubic millimeter of blood. All white blood cells are capable of ameboid motion. The granular corpuscles protect the body from bacterial invasion. They are capable of passing out of small blood vessels by flattening themselves. Then they migrate through tissues infected by bacteria and act as *phagocytes,* cells that engulf and digest bacteria and other foreign matter. Like the red blood cells, the white corpuscles are formed in red bone marrow; they are destroyed either in the liver or by migrating into the intestinal tract. It is thought that leukocytes have a life span of only 1 or 2 days in the circulatory system.

The agranular leukocytes are believed to play a role in the repair of damaged tissues, probably forming fibrous connective tissue. They are found in quantity at sites where healing is taking place. Platelets are small clear non-nucleated disks. There are about a quarter of a million in a cubic millimeter of blood. They are formed in the bone marrow and are replaced every few days. Their primary function is in the clotting process.

BLOOD PLASMA

The fluid component of the blood, the plasma, is about 90% water, which provides the aqueous medium necessary for cellular and physiological processes. About 9% of the plasma is organic material—proteins, nitrogenous materials, and nutrients. The most abundant proteins are *albumin, globulin,* and *fibrinogen,* all of which are chiefly formed in the liver. The nitrogenous materials, like urea and some salts of ammonia, are waste products of the body's metabolism and are carried by the blood to the kidneys where they are excreted. The nutrients in plasma are primarily glucose (called *blood sugar*) fatty acids, and cholesterol.

The blood gases are dissolved in the plasma. Oxygen, nitrogen, and carbon dioxide are only slightly soluble in water and are therefore carried in small amounts by the plasma. Varying quantities of special substances like vitamins, hormones, antibodies, and enzymes are also transported by the blood and are found in the plasma.

Coagulation of Blood

The ability of the blood to *coagulate*, or clot, is an adaptive mechanism of the body that is necessary for survival. Blood clots occur as semisolid plugs at the ends of ruptured blood vessels and prevent excessive loss of blood. Obviously, without this mechanism, the most trivial accidental damage to vessels would result in death. Even when no damage is visibly apparent, minute hemorrhages constantly occur throughout the body; as a result, blood clots are constantly being formed throughout the body.

If freshly shed blood is examined with a microscope, it is found that it quickly forms a gelatinous mass that is a semisolid fresh clot. The mass is seen to be blood cells that are enmeshed in an interlaced network of fine threads. There is in the fresh clot considerable fluid trapped within the mesh. The threads that form the clot are a contractile, insoluble material called *fibrin*.

In a short time, the fibrin contracts, squeezing the fresh clot into a more compact solidified mass and forcing out the fluid component. The completed clot is then surrounded by a straw-colored fluid called *serum*. Since the clot includes the formed elements of blood and fibrin, serum is the blood plasma less the fibrin.

The details of the coagulation process are still under investigation, but all the current theories agree on certain processes that are part of coagulation. We shall consider the established processes in the reverse order of their actual occurrence, beginning with the formation of fibrin that establishes the clot.

Fibrin is, as we have noted, insoluble in water. It is formed by transformation of a soluble protein component of blood, *fibrinogen*. The transformation is mediated by the protein *thrombin*, which is not present in normally circulating blood (otherwise clots would form in the bloodstream).

$$\underset{\text{(soluble)}}{\text{Fibrinogen}} + \text{thrombin} \longrightarrow \underset{\text{(insoluble)}}{\text{fibrin}}$$

One of the proteins in the normal blood stream is a form of globulin called *prothrombin*. In turn, thrombin is formed in shed blood by the action of calcium ions on a phospholipid called *thromboplastin*.

$$\text{Prothrombin} + \text{thromboplastin} + Ca^{2+} \rightarrow \text{thrombin}$$

The thromboplastic substance is particularly available where injuries to tissues have occurred. Although blood drawn from a blood vessel without having come into contact with injured tissue will coagulate, it is a much slower process. Normally, *bleeding time,* the time required for the blood resulting from a pin prick to clot, is about two and a half minutes for humans. *Coagulation time,* the time required for drawn blood to clot in a test tube, ranges from three to ten minutes in blood from normal humans.

Blood Vessels and Circulation

A vast, complex network of blood vessels transports the blood throughout the body. The heart provides the pressure necessary to maintain the flow through the vascular circuit. The high-pressure distributing system is composed of *arteries,* which repeatedly branch. The smaller branches, called *arterioles,* in turn branch into a system of minute vessels, the *capillaries,* through which substances are exchanged between the blood and tissues of the body. The pressure of the blood is very reduced in the capillary system, and from the capillaries blood is collected into a low-pressure network. From the capillaries, blood flows into small *venules,* which join together to form *veins.* The veins return the blood to the heart.

ARTERIES

In an adult the largest artery, the *aorta,* is about an inch in diameter. The aorta conducts oxygenated blood from the heart and begins branching into arterioles that go to all parts of the body. An artery is elastic in a normal person and contains a muscle layer between two layers of elastic tissue. Figure 8-2 shows a cross section of a typical artery.

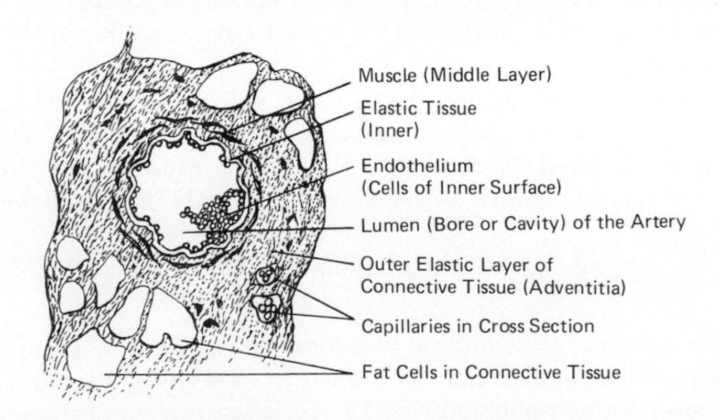

FIGURE 8-2. Cross section of a small artery.

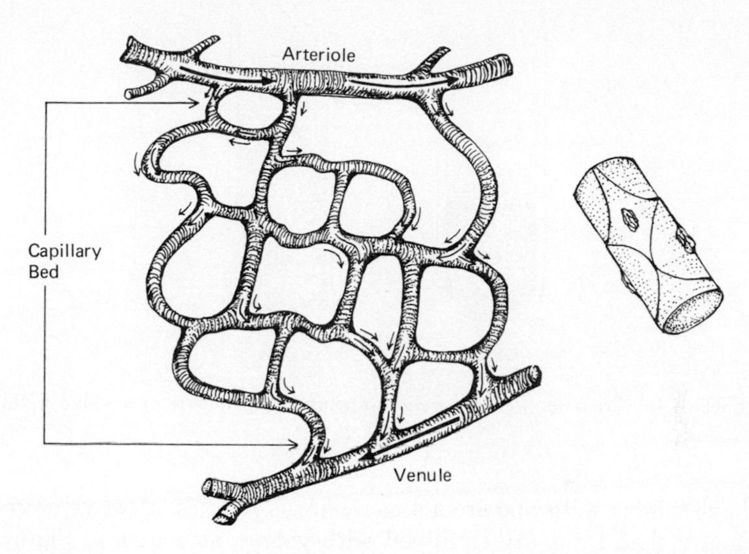

FIGURE 8-3. A capillary bed. The inset shows the cellular structure of a capillary.

THE CAPILLARIES

The distal regions of the arterial tree end in systems of capillaries called *capillary beds*. A representation of a capillary bed is shown in Figure 8-3. The blood flow, which is quite swift in the larger vessels, is quite leisurely in the capillaries, and this slow movement provides the time necessary for exchanges between blood and tissues. The capillaries are tubes formed of single layers of cells held together by a gluey substance, and leukocytes can ooze out through the cell boundaries to migrate through tissues. Each capillary is about 8 μm in diameter, barely large enough for red blood cells to pass through; under the microscope, erythrocytes can be seen passing through capillaries in single file, causing the walls to bulge somewhat as they pass. The capillaries empty into venules.

In some areas of the body, blood, in going from arterioles to venules, passes not through capillary beds but directly through *arteriovenous shunts*. These bypass routes permit relatively large volumes of blood to flow rapidly from the arterial to the venous system. The shunts occur particularly in areas like the skin, where the blood serves to regulate body temperature against environmental temperature changes.

VEINS

The veins form a system of vessels that collects blood from the capillaries and arteriovenous shunts and returns it to the heart. The structure of veins is similar to that of the arteries; each usually has three layers. Veins, how-

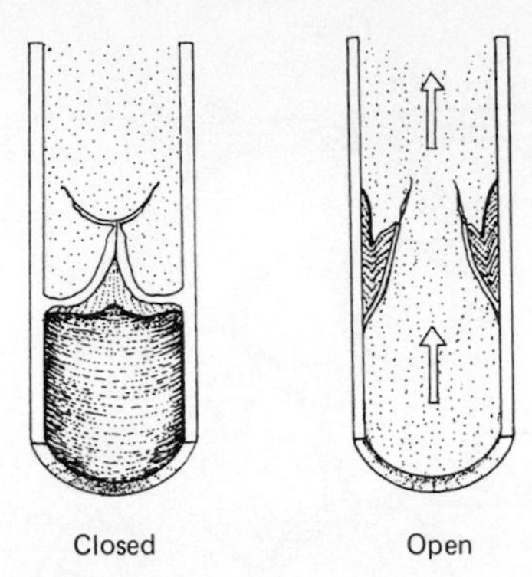

FIGURE 8-4. Cross section of a vein, showing the structure of a valve within the vein.

ever, have thinner walls and are not as elastic as arteries. Most veins greater in diameter than 1 mm are equipped with valves, as shown in Figure 8-4. The valves are folds on the inner layer of the vein that have free edges lying centrally and in the direction of blood flow. This arrangement allows blood to flow freely toward the heart but is effective in preventing a reversal of flow. Valves are most numerous in the legs, and they are absent in the veins of the intestine.

Blood Pressure and Velocity

A direct measurement of blood pressure in arteries, veins, or capillaries requires some surgical procedure. A completely severed small artery can be connected directly to a column of mercury or water. The height of the column supported by the blood is a measure of arterial pressure. The arterial pressure varies cyclically with the phase of the heartbeat; the maximum pressure during the cycle is the *systolic blood pressure*, and the minimum is the *diastolic blood pressure*. In healthy young adults, the systolic and diastolic pressures are typically 120 mm Hg and 80 mm Hg, respectively. Such measurements are customarily reported as systolic pressure/diastolic pressure, or 120/80. The arithmetic difference between the systolic and diastolic pressures is called the *pulse pressure*.

Indirect measurements of arterial pressure can be easily and fairly accurately obtained by using a *sphygmomanometer*, the familiar cuff device used on the arm in the most routine physical examination. This device consists of a rubber bag covered with an unyielding material, a rubber hand bulb for inflating the bag, an escape valve to release the air from the bag at a convenient rate, and a *manometer* (a calibrated mercury column for measuring pressure) connected to the bag by a length of tubing. The cuff is usually placed on the upper arm, and the brachial artery is the site of measurement. By placing a stethoscope over the artery at a point distal to the cuff, an observer can hear characteristic sounds of blood flow. By inflating the bag and slowly releasing the air until the pressure in the bag is matched by the pressure in the artery, the observer can note the manometer reading when blood begins to flow through the artery. This reading represents the systolic pressure. The reading at which blood begins to flow continuously throughout the heart cycle corresponds to the diastolic pressure.

As the heart pumps blood into the vascular system, the volume of the system does not remain constant but changes because of the elastic stretching of the arteries. Blood is forced into the arteries faster than it can escape into the capillaries and veins, so energy is stored in the stretched arteries and is expended in accelerating the blood before the next heartbeat. The result is a pulsating pressure in the arteries but a continuous pressure throughout the capillaries and veins. Figure 8-5 shows the *gauge pressure*, i.e., the the difference in pressure from atmospheric pressure, of the vascular system. Notice that the arterial pressure pulsations are apparent until the blood reaches the small arterioles. The pressure in the large veins reaches a value slightly below atmospheric pressure before entering the heart.

Blood flows rapidly in the arterial system, slightly faster than 20 cm/second in the aorta, but slows to less than 0.1 cm/second in the capillaries. The flow speed increases again in the veins to about 10 cm/second just before entering the heart. Of course, we expect blood to flow faster in a small

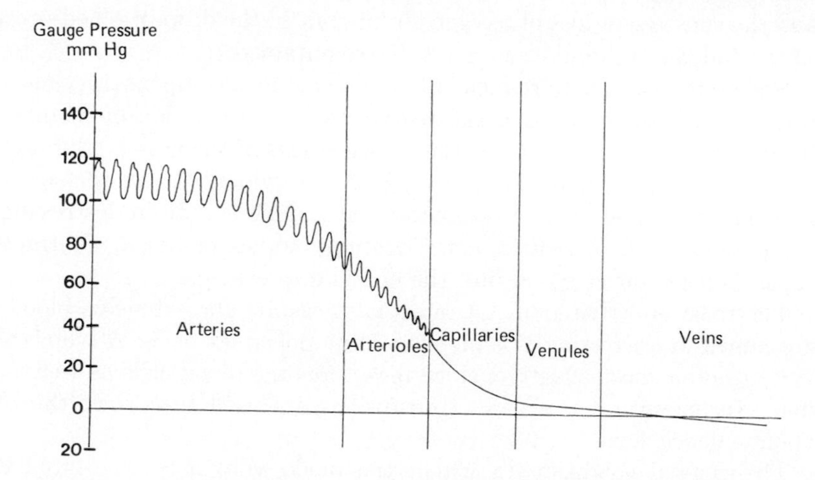

FIGURE 8-5. Blood pressure in various regions of the vascular system.

caliber region of a closed system, but the slower velocities in the capillaries occur because the *total* cross-sectional area of all the capillaries far exceeds that of the great vessels.

An interesting phenomenon associated with blood flow that may be easily observed is the *arterial pulse.* When we place a fingertip over a superficial artery such as that in the wrist, we feel a surge that closely coincides with the heartbeat. The pulse that is felt is the result of the sudden pressure variation that occurs when the heart suddenly forces blood into the already filled arterial system. The elastic expansion and recoil of the artery sets up a wave in the arterial wall that propagates along the artery toward the peripheral system. These pulses travel much faster than the column of blood in the artery. The pulse velocity is nearly 10 m/second, cf. blood velocity of 20 cm/second. The pulses provide a convenient means for measuring the rate of the heartbeat. Sensing the pulse with the fingertips is technically called *palpation of the pulse.*

Further Physical Aspects of Circulatory Function

BLOOD VOLUME

The total volume of blood in the circulatory system is related to the blood pressure throughout that system. Since the circulatory system is closed and elastic, the greater the volume of fluid it contains, the greater will be the pressure within the fluid. Thus, in any situation in which the body experiences extensive hemorrhaging, there is the danger of a serious diminution in blood volume with a concomitant fall in arterial pressure. The immediate danger to life in such a case lies not in the loss of erythrocytes and the consequent loss of oxygen supply but in the drop in blood pressure so that adequate circulation cannot be maintained.

The body responds to reduced blood volume by a compensatory mechanism in which there is a general *vasoconstriction* (reduction of diameter of blood vessels by contraction of the muscle layers in the vessels) throughout the circulatory system. There is also some evidence that the *spleen,* a large organ located near the stomach that functions both in destroying old erythrocytes and in storing considerable volumes of blood, contracts to expel its reservoir of blood into the circulatory system.

The most effective artificial remedial measures for extensive blood loss are aimed at correcting the blood volume deficit at once. A whole-blood transfusion is most effective, since it restores the blood cells as well as the fluid. In emergencies, plasma transfusions are used to correct the blood volume deficiency.

The general condition in which the blood volume in the circulatory system is greatly reduced (for whatever reason), causing the reflex vasoconstriction throughout the system, is called *shock*. To complete the picture of blood volume control, we may note that any increase in blood volume, by whatever means, is reflexively counteracted by *vasodilation* (enlargement of the diameter of blood vessels by relaxation of the muscles in the walls of the vessels).

EFFECTS OF GRAVITY

Gravity exerts a downward force on all the blood and lymph in the circulatory system. A hydrostatic effect therefore causes the blood pressure in the lower parts of the body to be greater than in higher parts. We may observe this effect in the veins of the hand which are distended when the hand is held downward but which are nearly collapsed when the hand is held higher than the heart.

When a person suddenly assumes an upright posture, the gravitational pull on the blood may reduce the pressure in the arteries supplying the head to such an extent that fainting results. Fainting is probably an adaptive mechanism, because the result of fainting is that the long axis of the body is automatically restored to the horizontal position, allowing an increased flow of the blood to the brain. The obvious first aid procedure for fainting is therefore to permit the subject to remain lying down or even to be placed so that the head is lower than the rest of the body.

The effects of gravity on the circulatory system can be profound in the extreme cases presented by modern air and space travel. The gravitational pull may be increased manyfold on pilots pulling out of a dive or otherwise being subjected to large accelerations. The pull is likewise decreased to zero in the weightless conditions encountered by astronauts. These are current biophysical problems that are being studied intensively.

The Lymphatic System

In order that the blood vascular system deliver nutrients to body tissues and collect metabolic wastes, it is necessary that nutrients leave the capillaries. Similarly, the fluid that bathes the cells in tissues must leave the blood vascular system. As tissues use up oxygen and other nutrients and produce wastes, the concentrations across capillary walls will favor migration of oxygen and nutrients to the tissues and of wastes to the blood stream. The fluid, having left the blood vascular circuit, is called *lymph*. A complex system of *lymphatic capillaries* and *lymphatic veins* receives the lymph and returns it to the venous sytem at two points just before the great veins join the heart, where the gauge pressure is zero or slightly negative.

The lymph enters the lymphatic system through the walls of the lymphatic capillaries, which are very similar in structure to blood capillaries except that they begin blindly in small cul-de-sacs. The lymph capillaries lead to the larger lymph vessels, which are similar to veins in that both have valves to restrain a reversal of flow. The lymph vessels have thinner walls than veins, and the valves in lymph vessels are more numerous than in veins. *Lymph nodes* are bean-shaped bodies located at frequent intervals along the lymphatic system through which the lymph passes. The nodes are the sites of the germination of the agranular leukocytes called *lymphocytes*. There are also cells called *histocytes* within the lymph nodes that ingest bacteria and debris from the lymph. The nodes therefore serve to filter the lymph before it reenters the bloodstream.

The flow of lymph through the lymphatic system is primarily due to the motions of skeletal muscles, which on contraction squeeze the lymph vessels throughout the body. Since the lymph vessels are equipped with valves, the repeated squeezing results in a net flow toward the larger vessels that empty into the venous system.

The Heart

The heart is a muscular mechanical pump that provides the power to maintain the circulation of the blood. It functions by cyclically receiving and expelling the blood with sufflcient pressure to drive it through the circulatory system. In a resting adult male, the heart has an output of about 85 ml of blood per cycle (or beat). A healthy adult at rests has a heart rate of about 60 beats/minute so that about 5 liters/minute pass through the resting heart. During periods of maximum exertion, the *cardiac output* may exceed 35 liters/minute. (The Latin word for "heart" is *cors,* and the prefixes "cardi" and "cardio" refer to the heart; *cardiac* is the adjective form.)

Anatomy of the Heart

The heart, a muscular organ about the size of a person's fist, is located between the lungs and behind the breastbone. About two-thirds of the heart lies to the left of the medial plane. Figure 8-6 shows the location of the heart. It rests on the diaphragm, a muscular partition that divides the chest cavity and the abdominal cavity. The *base* of the heart is upward and to the left; the *apex,* the point of the heart, is located downward and to the left. The entire heart is enclosed by a double-walled sac, the *pericardium.*

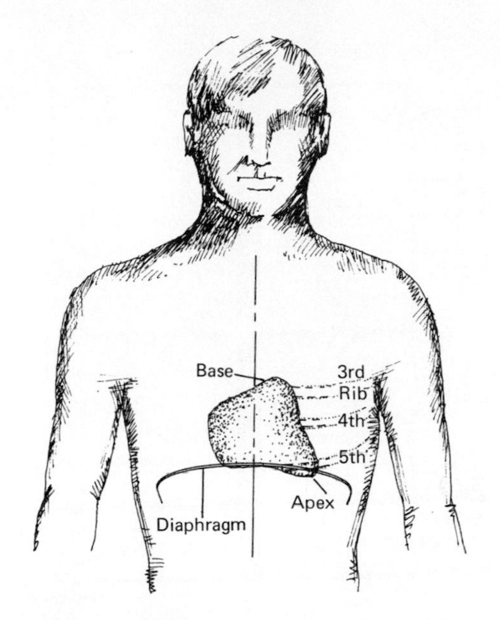

FIGURE 8-6. The position of the heart in the chest.

The heart contains four chambers, two on each side. A *septum* completely separates the heart into a right and left half. In terms of the chambers, the heart is constructed like two adjacent, but noncommunicating, townhouse apartments. The chambers communicate vertically but not horizontally. Figure 8-7 illustrates schematically how the heart chambers are arranged. The heart actually functions as two pumps, the right side pumping blood through the *pulmonary* (referring to the lungs) circulation system for exchange of gases with the lungs, the left side pumping oxygenated blood through the *systematic* circulatory system. Figure 8-8 is a frontal section of the heart showing the actual placement of the chambers. The upper chambers are the right and left *atria* (singular, *atrium*). The lower chambers are the right and left *ventricles*. The path of blood flow through the heart is summarized in Figure 8-9. The blood leaves the great veins and empties into three cavities adjacent to the heart: the *superior vena cava* that collects the blood from the head and upper chest; the *inferior vena cava*, which receives blood from the lower portion of the venous system; and the *coronary sinus*, a wide channel that collects blood from the coronary veins on the surface of the heart. In the course of the cardiac cycle, this systemic venous blood enters the right atrium of the heart, passes to the right ventricle, and is pumped to the lungs through the *pulmonary artery*. After being oxygenated by the lungs, the blood is returned to the heart via the right

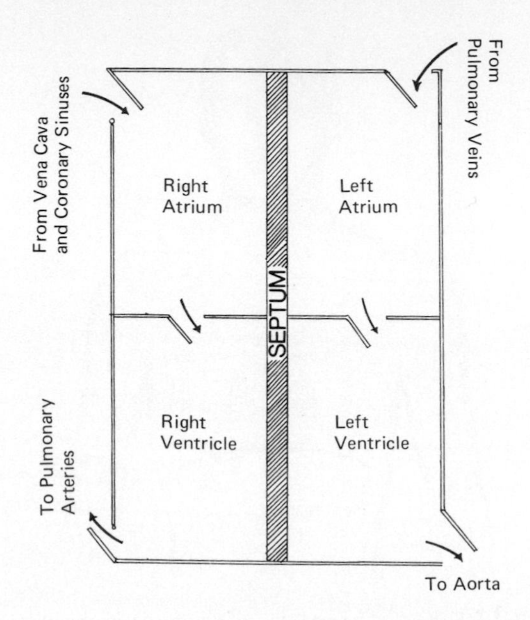

FIGURE 8-7. A schematic representation of the chambers of the heart, the valves of the heart, and the direction of blood flow through the heart.

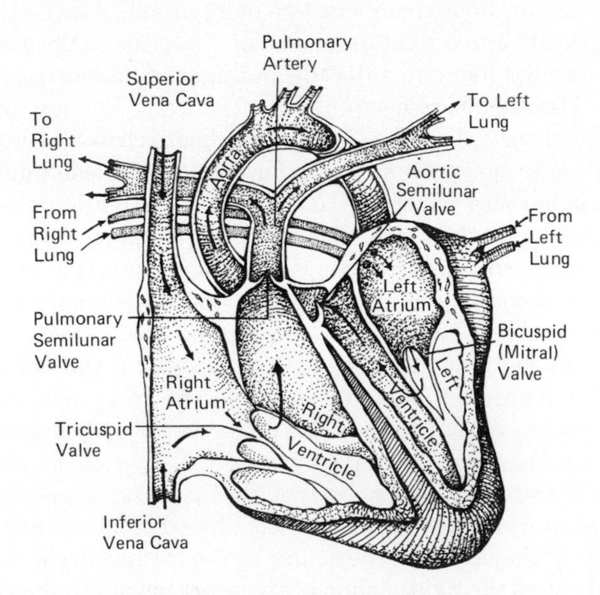

FIGURE 8-8. A section of the human heart showing the placement of the chambers and valves, the great vessels, and the directions of blood flow.

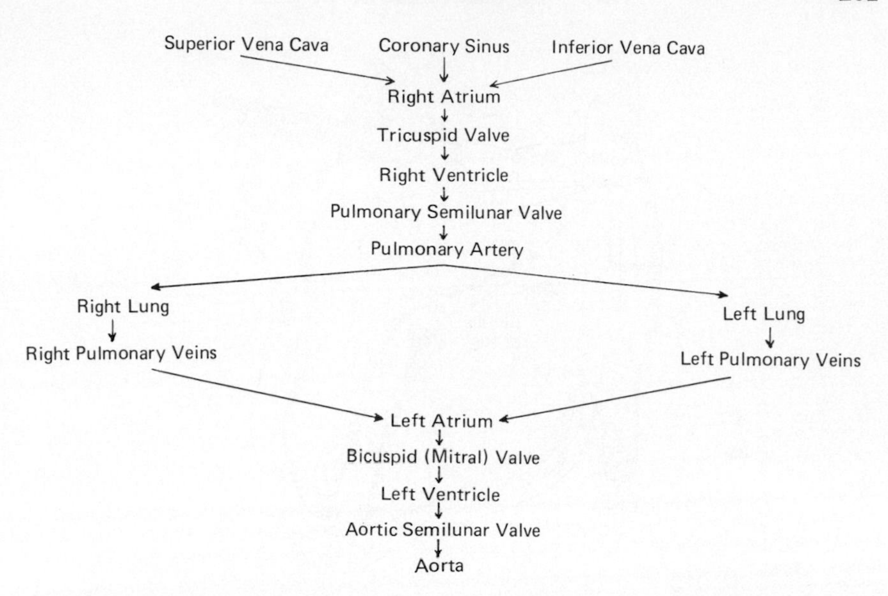

FIGURE 8-9. A schematic flow chart showing the course of blood through the heart and lungs.

and left *pulmonary veins*. Oxygenated blood reenters the heart at the left atrium and passes to the left ventricle. The contraction of the left ventricle forces the blood out of the heart into the aorta where it proceeds through the systemic arterial system. The wall of the left ventricle is about three times the thickness of the wall of the right ventricle. This difference is necessary because the muscle tissue of the left ventricular heart wall must eject the blood from the left ventricle against the systemic pressure.

The four valves of the heart are utilized to keep the blood from flowing in the wrong direction during the cardiac cycle. Both valves leading out of the ventricles have leaves shaped like half-moons and are called pulmonary and aortic *semilunar valves*. The other two valves, each leading from an atrium to a ventricule, are named for the number of cusps, or leaves, that comprise the valve. The blood passes between the right chambers through the *tricuspid valve* and between the left chambers through the *bicuspid valve* (sometimes called the *mitral valve*).

Origin and Propagation of the Heartbeat

Cardiac muscle is different from all other muscle in that it has inherent rhythmicity. This means that, even when completely severed from all nervous connections, the heart muscle will continue to beat. Indeed, if a

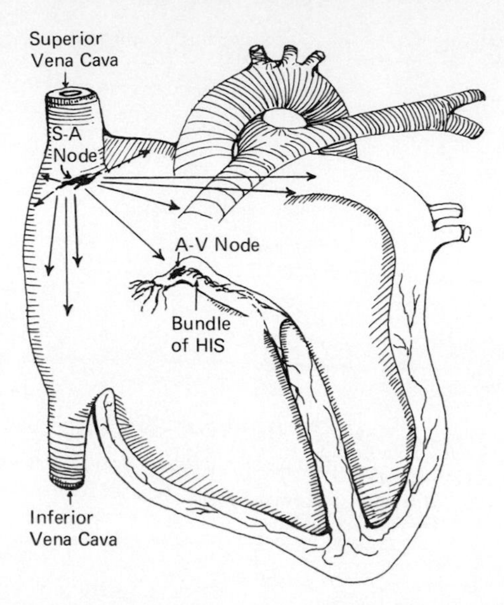

FIGURE 8-10. Diagram illustrating the spread of the contraction wave through the heart from the S-A node.

chunk of the heart muscle is excised and placed under a microscope, it continues to contract and relax periodically.

The signal to start the heartbeat begins in the medulla of the brain and is transmitted along the *vagus nerve* to a mass of specialized tissue on the surface of the right atrium near the superior vena cava. This tissue called the *sinoatrial node* (or S-A node) is the starting point of the heart's contraction. Figure 8-10 shows how the heartbeat proceeds. The contraction that starts at the S-A node radiates outward from the node as a contraction wave through both atria. On the lower part of the wall of the right atrium is another mass of specialized tissue, the *atrioventricular node* (A-V node), which relays the wave signal from the S-A node downward through the septum to the apex of the heart and up the outer ventricular walls on strands of specialized cardiac muscle called the *bundle of His.* As the signal reaches each region, it causes contraction of the cardiac muscle in that region of the heart. Thus, the S-A node is the pacemaker of the heart, and the beat of the heart is an orderly progression of a contraction wave that causes the proper chamber to be squeezed at the proper time.

The rate at which the normal heart beats is under the control of the central nervous system. The rate is determined by a combination of the chemical environment and the temperature of the S-A node. If the S-A node is inactive, the A-V node may function as the pacemaker. The normal 60 beats/minute is reduced to about 50 beats/minute when the A-V node controls the pace. When no nodes are functional, the heart rate drops to about 30 beats/minute. In any case, the rate is determined by the fastest pacemaker.

The Cardiac Cycle

The sequence of events that occurs during one heartbeat constitutes the cardiac cycle. The term *systole* refers to the contraction phase of the cardiac cycle, and *diastole* refers to the relaxation phase of the cycle. Thus, "atrial systole" refers to the state of activity of the two atria. The sequence of events is most easily described graphically. Figure 8-11 shows a number of parameters simultaneously as a function of time during the cardiac cycle. The period of the cycle is about 0.8 seconds. The bars at the top of the graph show the states of the atria and ventricles during the cycle. Notice that the entire heart is resting for about half the cycle while there is an overlap in atrial and ventricular diastole. The curves show the blood pressure in the aorta and ventricle, the relative volume of the ventricles, and a representation of heart sounds as they might have been recorded from the output of a microphone placed on the chest.

Energetics of the Heart Function

The heart is a transducer, converting chemical energy into mechanical energy of contraction and finally transmitting that energy to the blood in the form of *hydrodynamic* and *hydrostatic* energy. Hydrodynamic energy refers to the kinetic energy of moving fluids, and hydrostatic energy is the energy that a fluid possesses by virtue of its position, i.e., its potential energy. The heart moves the blood, providing kinetic energy; and if we describe the kinetic energy per unit volume T_{vol} for a fluid whose density is ρ kg/liter and whose velocity is v m/second, we have

$$T_{\text{vol}} = \tfrac{1}{2}\rho v^2 \tag{8-1}$$

The potential energy of a fluid depends on its height above some arbitrary level (arbitrary, because potential energy, or any other kind of energy, cannot be determined absolutely. Only differences in energy can be measured; therefore differences in energy can be measured using any level as the zero of energy). Since we shall be considering the energy and power de-

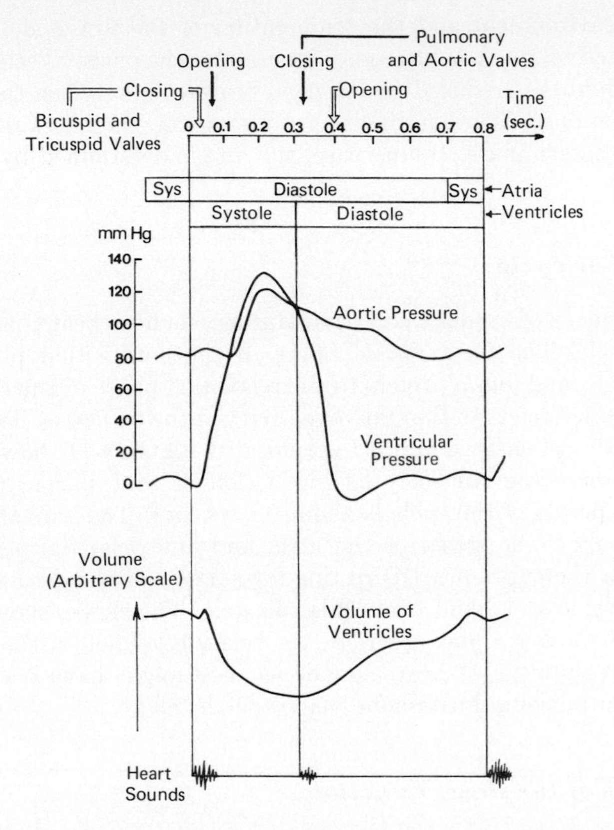

FIGURE 8-11. Several parameters associated with cardiac activity shown simultaneously as a function of time.

livered by the heart to the circulatory system at the aorta, i.e., at the same level as the heart, the potential energy due to gravity is zero. We must, however, consider the pressure against which the heart works—and we shall consider the pressure as a form of potential energy. When a fluid is under pressure it has the capacity to do work by converting the pressure into kinetic energy. As an example, when a volume of liquid is pumped upward to a height h into a tank connected to the liquid at ground (zero) level through a small standpipe, the pressure at ground level is

$$p = \rho g h \tag{8-2}$$

where ρ is the density of the fluid, and g is the acceleration of gravity. Now

ρ is the mass per unit volume, so Equation (8-2) can be written as

$$p = \frac{mgh}{\text{volume}} = \frac{\text{potential energy}}{\text{volume}} = V_{\text{vol}} \tag{8-3}$$

We may therefore express the potential energy per unit volume as pressure p, and the heart's total mechanical energy per unit volume can be expressed as the sum of the kinetic and potential energies per unit volume:

$$H_{\text{vol}} = T_{\text{vol}} + V_{\text{vol}} \tag{8-4}$$

$$H_{\text{vol}} = \tfrac{1}{2}\rho v^2 + p \tag{8-5}$$

Here p is the blood pressure in the aorta.

We may assume for our purposes that blood is an *incompressible fluid*. This means that for the range of pressures encountered in the bloodstream, the volume of the blood itself will not change appreciably. When an incompressible fluid is contained within a closed system, either the volume flow rate K (the volume passing a point in the system per unit time) must be constant or the total volume of the system must change. Although arteries are elastic to some extent, it is a very good approximation to assume that the volume of the circulatory system is a constant. There are certain conditions under which capillaries are opened (as when the blood flow to muscles increases during strenuous activity or erectile tissue is flooded with blood). We shall consider steady-state conditions of the circulatory system and may therefore consider the blood volume and the blood's volume flow rate to be constants.

The power P (work or energy per unit time) developed at any instant by the heart can be obtained by multiplying H_{vol} by the volume flow rate K:

$$P = H_{\text{vol}}K \tag{8-6}$$

Of course, v and p vary considerably during the cardiac cycle, and the velocities and pressures in the aorta and pulmonary artery are not the same. Therefore, we may express the average power developed by the heart in terms of the *average* velocities and pressures in the great arteries by

$$P = \tfrac{1}{2}\rho\overline{v_A{}^2}K + \tfrac{1}{2}\rho\overline{v_P{}^2}K + \bar{p}_A K + \bar{p}_P K \tag{8-7}$$

in which the subscripts A and P refer to the aorta and pulmonary artery and the bars represent average values.

We must now invoke experimental information in order to proceed. It has been determined that the blood velocities of the aorta and pulmonary artery are essentially the same, whereas the pressure inside the aorta is six times greater than that inside the pulmonary artery. Incorporating these facts into Equation (8-7) gives

$$P = \rho \overline{v_A{}^2} K + \frac{7\bar{p}_A K}{6} \tag{8-8}$$

We can relate the average linear velocity of the blood $\bar{v}$ to the volume flow rate K and the cross-sectional area A of the aorta by

$$\bar{v} = \frac{K}{A} \tag{8-9}$$

The mean square velocity that appears in Equation (8-8) is very different from the square of the mean (or average) velocity, because the blood is flowing from the heart only a fraction of the cardiac cycle. Again, it has been found experimentally that $\overline{v^2}$ and $(\bar{v})^2$ are related by

$$\overline{v^2} = 3.5\ (\bar{v})^2 \tag{8-10}$$

Substituting Equations (8-9) and (8-10) into Equation (8-8) gives

$$P = \frac{3.5\rho K^3}{A^2} + \frac{7\bar{p}_A K}{6} \tag{8-11}$$

Equation (8-11) represents the average power developed by the heart for a steady-state condition, i.e., for a condition of rest, moderate exertion, or extreme activity. The first term on the right of Equation (8-11) represents that part of the power developed by the heart that is used in moving the blood, that is to say, the kinetic portion of the power. The second term is the hydrostatic term, representing that portion of the power devoted to maintaining the pressure in the circulatory system.

We have already introduced all of the numerical values necessary to evaluate Equation (8-11). The specific gravity of blood is very near unity, so we may set $\rho = 1$ kg/liter $= 10^3$ kg/m^3. The aorta is about 3 cm in diameter at the aortic semilunar valve. In Equation (8-9) we find the mean linear velocity in terms of the mean flow rate and the cross-sectional area A; therefore, we will use a mean area for A. In Figure 8-11 we may note that the aortic valve is open about one-third of the cardiac cycle, so we use an average diameter of 1 cm; A then becomes $\pi(0.5)^2$ cm$^2 = 0.79 \times 10^{-4}$ m^2. The arterial blood pressure of a normal adult is 120/80, so the average pressure $\bar{p}_A$ is 100 mm Hg $= 13.3 \times 10^4$ newtons/m^2. [One mm Hg is the pressure at the bottom of a vertical column of mercury 1 mm high; therefore, 1 mm Hg $= \rho_{Hg} gh =$ (13.6 gm/cm^3) (980 cm/sec^2) (0.1 cm) $=$ 1.33×10^3 dynes/cm$^2 = 1.33 \times 10^2$ newtons/m^2.] For a person at rest, the average flow rate K is about 5 liters/minute = 0.0833 liter/second = 8.33×10^{-5} m^3/second. If we substitute these mks values into Equation (8-11), we obtain

$$P_{\text{rest}} = \frac{3.5\,(10^3\ \text{kg/m}^3)\,(8.33 \times 10^{-5}\ \text{m}^3/\text{second})^3}{(0.79 \times 10^{-4}\ \text{m}^2)^2}$$

$$+ \frac{7\,(1.33 \times 10^4\ \text{kg/m sec}^2)\,(8.33 \times 10^{-5}\ \text{m}^3/\text{second})}{6}$$

$$P_{\text{rest}} = 0.32\ \frac{\text{kg m}^2}{\text{sec}^3} + 1.30\ \frac{\text{kg m}^2}{\text{sec}^3} \qquad \text{(8-12)}$$

$$P_{\text{rest}} = \underset{\text{(kinetic)}}{0.32\ \text{W}} + \underset{\text{(hydrostatic)}}{1.30\ \text{W}} = 1.62\ \text{W}$$

In the case of a person strenuously exerting himself, all the values above are the same except the flow rate, which becomes K = 35 liters/minute = 5.82×10^{-4} m^3/second. Substituting this value into Equation (8-11) gives

$$P_{\text{active}} = \frac{3.5\,(10^3)\,(2 \times 10^{-10})\ \text{joule}}{0.62 \times 10^{-8}\ \text{second}}$$

$$+ \frac{7\,(1.33 \times 10^4)\,(5.82 \times 10^{-4})}{6} \times \frac{\text{joule}}{\text{second}}$$

$$P_{\text{active}} = \underset{\text{(kinetic)}}{113\ \text{W}} + \underset{\text{(hydrostatic)}}{9.1\ \text{W}} = 122\ \text{W} \qquad \text{(8-13)}$$

In comparing Equations (8-12) and (8-13), it is seen that in the resting state, a person's heart delivers less than 2 W, and most of the power is used in maintaining the pressure; in the active state, however, while developing over 120 W, the heart is expending most of its energy in providing flow of the blood.

Electrocardiography

Every time the heart beats, changes in electrical potential occur on the surface of the heart and within the myocardium. These potential changes spread throughout the body to the surface of the skin. Electrodes placed on any pair of points on the body will show differences in potential that are related to the heartbeat. A record of such potential changes is called an electrocardiogram (EKG or ECG). Standardized procedures for placement of the sensing electrodes, amplification of the potential differences observed, and speed and method of recording have been adopted so that EKG is now understood to refer to specific types of recordings. In this section we shall consider the origin and manifestations of the electrical changes in the heart that accompany the heartbeat. We shall become acquainted with some of the standardized procedures that are useful in analyzing the electrical activity of the heart and shall relate the EKG to some physiological functions of the heart.

Electrical Events in Cardiac Cells

The cells of the myocardium are electrically very similar to neurons in that there exists a *dipole layer* across the plasma membrane, i.e., the outer surface of the membrane is positively charged with respect to the inner surface. The electrically polarized membrane can be depolarized locally, after which the depolarization spreads along the cell as a depolarization wave. This is exactly what occurs on the neuron as an action potential propagates (see Chapter V). The action potential in cardiac muscle is of much longer duration than in neurons, lasting up to 300 milliseconds, compared to 1 millisecond for neurons. Cardiac tissue also requires a much longer repolarization period than nerve tissue.

The cells of the heart are joined together by structures called *intercalated disks*, once thought to be special anatomical units of the heart tissue but now considered anatomically indistinguishable from the cellular membrane. Electrically, these disks cause conduction (or depolarization wave propagation) to be *syncytial*, i.e., large masses of myocardium behave as a single network. As a consequence, when a wave of depolarization is begun at any point on the myocardium, it proceeds throughout the network.

Some Physics of Depolarization Waves

Figure 8-12 shows an electrical *dipole* and a dipole layer. When two equal and opposite charges, each of whose magnitude is q, are separated by a small distance δ, they constitute a dipole whose *dipole moment* has a magnitude $p = q\delta$. A dipole layer is formed by separating positive and negative charges across a layer of thickness δ; each region of the layer contains equal numbers of positive and negative charges, so the net charge of any region of the layer is zero. Figure 8-13 shows the geometrical arrangements used in

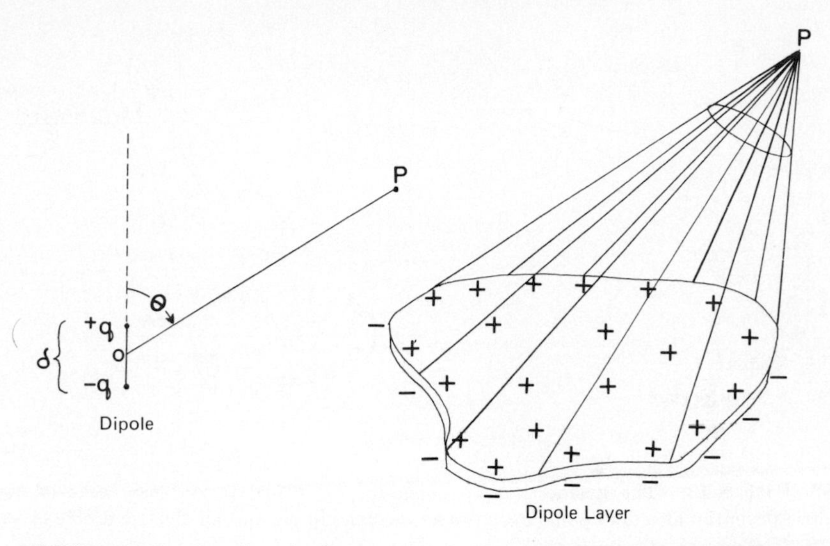

FIGURE 8-12. An electric dipole and an electric dipole layer. The field point P is the point at which the electrical effects of either configuration are being considered.

computing the potential V at a point P some distance (large compared to δ) from a dipole and a dipole layer. For a dipole, the potential at P due to the charges $-q$ and $+q$ at distances r_1 and r_2 from P, respectively, is given by

$$V = \frac{q}{r_2} - \frac{q}{r_1} \tag{8-14}$$

In terms of r (the distance from the center of the dipole to P) and θ (the angle between the lines from $-q$ to $+q$ and from the center of the dipole to P), r_1 and r_2 are approximately

$$r_1 = r + \frac{\delta}{2} \cos \theta \tag{8-15}$$

$$r_2 = r - \frac{\delta}{2} \cos \theta \tag{8-16}$$

Substitution of Equations (8-15) and (8-16) into (8-14) gives

$$V = q\left(\frac{1}{r - (\delta/2) \cos \theta} - \frac{1}{r + (\delta/2) \cos \theta}\right) = \frac{q\, \delta \cos \theta}{r^2 - (\delta^2/4) \cos^2 \theta} \tag{8-17}$$

In this result we may note that since $\delta \ll r$, $\delta^2/4$ is negligible compared

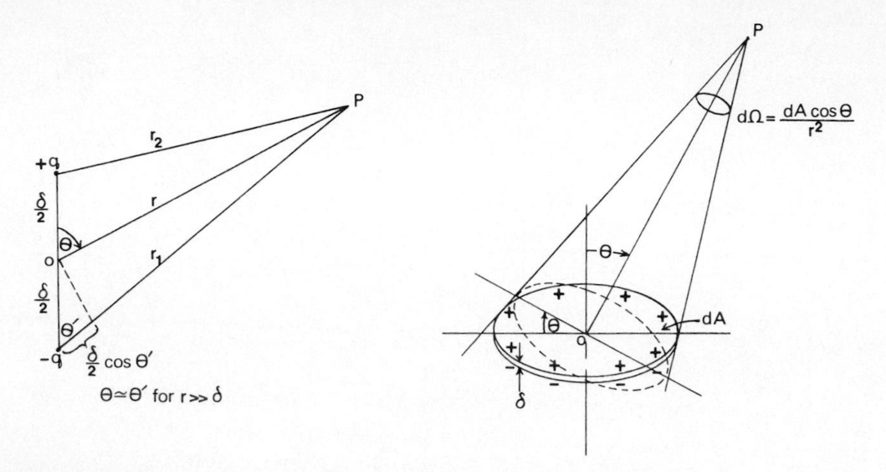

FIGURE 8-13. The geometrical arrangement for computation (see text) of the electric potential at a field point P due to an electric dipole and an electric dipole layer.

with r^2. In this approximation the potential at P due to the dipole is

$$V = \frac{p \cos \theta}{r^2} \tag{8-18}$$

where we have used $p = q\delta$.

Similar arguments apply to the dipole layer. A small area of the layer dA, which has surface charge densities (charge per unit area) $+\sigma$ and $-\sigma$ and a thickness δ, produces a contribution to the potential at P given by

$$dV = \frac{\sigma\, dA\, \delta \cos \theta}{r^2} = \frac{p_A\, dA \cos \theta}{r^2} \tag{8-19}$$

in which p_A is the dipole moment per unit area, defined as $p_A = \sigma\, \delta$. Now the solid angle subtended by the area dA at P is

$$d\Omega = \frac{dA \cos \theta}{r^2} \tag{8-20}$$

so Equation (8-19) may be written

$$dV = p_A\, d\Omega \tag{8-21}$$

Integration over the entire surface S of the dipole layer gives

$$V = \int_S p_A\, d\Omega \tag{8-22}$$

and if the dipole moment per unit area is constant over the surface, Equation (8-22) becomes simply

$$V = p_A \Omega \tag{8-23}$$

where Ω is the solid angle of the surface seen from P. Equation (8-23) means that the potential depends only on the apparent size (projection) of the layer from P and is independent of the detailed shape of the layer.

In a resting heart cell, when the surface dipole layer is not changing, the net potential at an external point is zero. This is the case because the point outside the cell is faced by two dipole layers that subtend the same solid angle; as depicted in Figure 8-14, one layer has the positive charges closer to P while the other layer has the negative charges closer to P. Note that at a point P inside the cell, the potential is given by Equation (8-23) with the solid angle set equal to 4π, and the potential at any point inside the cell is $-4\pi p_A$. Since moving an electrode from outside the cell to the inside changes the potential from zero to $V_m = -4\pi p_A$, the transmembrane potential, we may express p_A in terms of the potential across the membrane as

$$p_A = \frac{V_m}{4\pi} \tag{8-24}$$

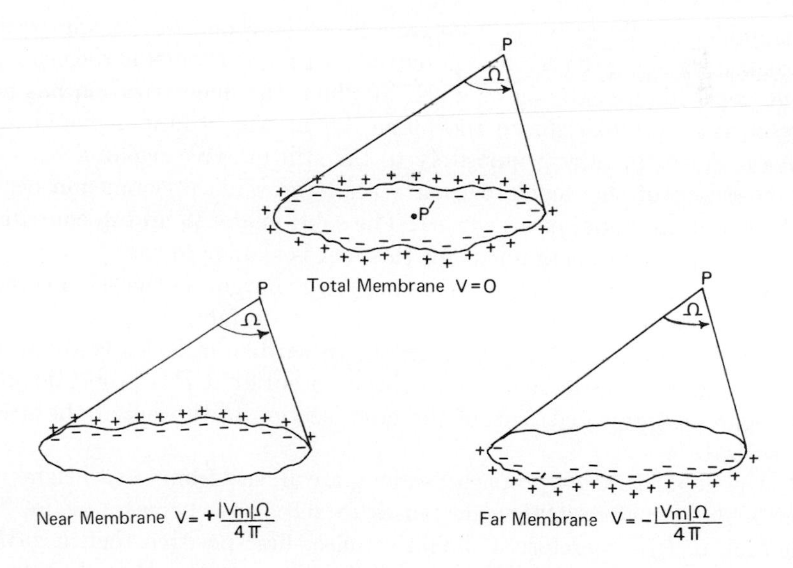

FIGURE 8-14. The electric potential at a point outside of a closed cell that has a dipole layer at its plasma membrane.

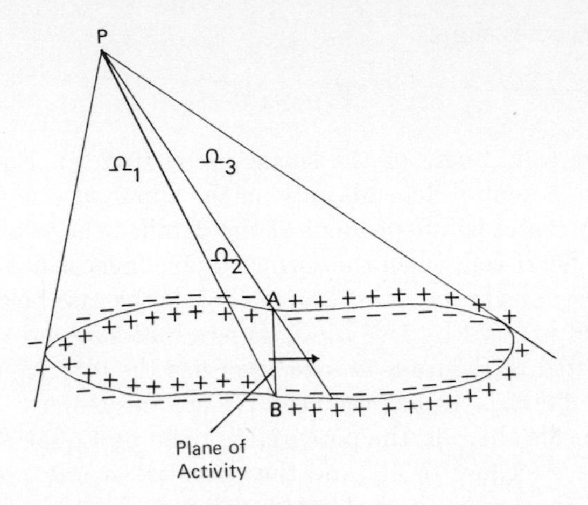

FIGURE 8-15. Diagram illustrating the electric potential at a point P outside of a cell that is undergoing depolarization activity.

and Equation (8-23) becomes

$$V = \frac{V_m \Omega}{4\pi} \tag{8-25}$$

In an electrically active cell, i.e., one in which a plane of depolarization is progressing along the cell, the potential at a point P outside the cell is no longer zero. Figure 8-15 shows a cell in which the depolarization has progressed from left to right to the plane AB at the instant depicted. The active region is polarized oppositely to the still inactive region because of the overshoot of the depolarization (recall the same phenomenon occurs in the action potential of a neuron). The solid angles Ω_1 and Ω_3 contribute no potential at P; the argument for this fact is similar to that for a resting cell. The membranes in Ω_2 both have negative charges on the side nearer P and both contribute a negative potential to P. In fact, the potential at P is, from Equation (8-25), $-(V_m/4\pi)\Omega_2$. In cardiac cells V_m is about 120 mV. In Figure 8-15, the magnitude of the potential at P is proportional to the apparent (projected) size of the cross section of the cell at the site of depolarization.

As the depolarization propagates along a cell, say from left to right, the solid angle Ω_2 increases, then decreases to zero, and increases again. The potential at P is therefore a diphasic pulse, first positive then negative. This effect is illustrated in Figure 8-16. An electrode placed at B records a diphasic pulse, but an electrode at A records only a negative pulse and one at C only a positive pulse as the wave of activity proceeds. The arrow in

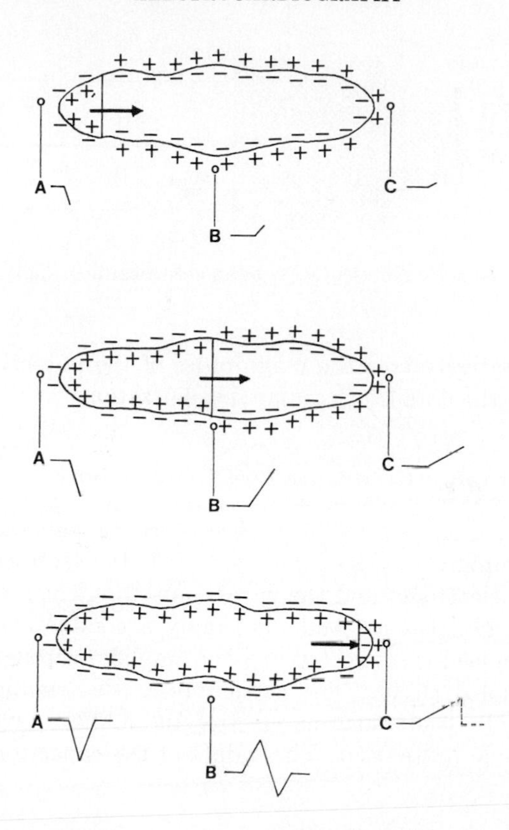

FIGURE 8-16. Electric potentials recorded by electrodes at various locations near a cell as a depolarization wave moves along the cell from left to right.

the figure points in the direction of the depolarization wave motion; we shall refer to such an arrow as a *depolarization vector*. As a depolarization wave sweeps down the heart from the S-A node, we shall refer to the *cardiac vector* (sometimes called the *axis of the heart*), which designates the net direction of the polarization wave (though no magnitude is associated with the cardiac vector).

Electrodes placed on the surface of the skin detect the changes in potential at their points of contact produced by the depolarization wave that moves down the heart. Of course, the potential at the skin surface is diminished from the values described so far because of the polarizability of the tissues between the heart and the skin. When a quantitative model of electrocardiography is developed, the tissue polarizability must be taken into account. Meanwhile, considerable information can be obtained without

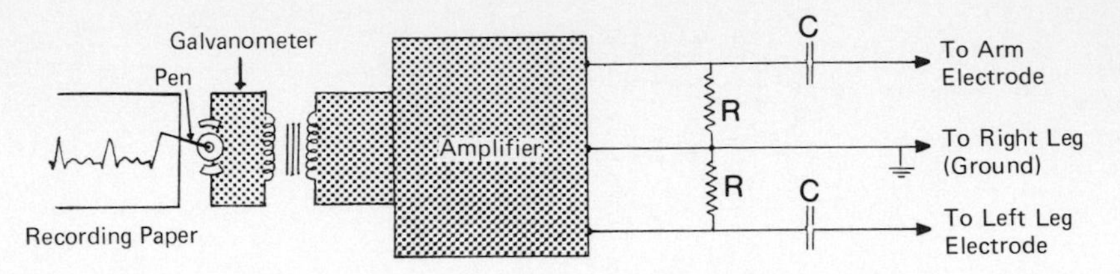

FIGURE 8-17. Schematic representation of an electrocardiographic recording device.

knowing quantitatively how the magnitudes of the potentials at the skin surface relate to the details of cardiac depolarization.

Electrocardiographic Recording and Conventions

A modern *electrocardiograph* is a direct-writing galvanometer on the output of an amplifier, whose input is a *lead*. In electrocardiography, a lead is a *pair* of electrodes and the wires connecting them to the amplifier (in most technical areas a "lead" is simply a conducting wire). Figure 8-17 shows how a lead is connected into the amplifier through a "balanced" capacitively coupled circuit. The capacitors in the circuit eliminate dc components on the lead (such as electrolytic voltages produced by perspiration, electrode paste, etc). The values of the capacitors and resistors

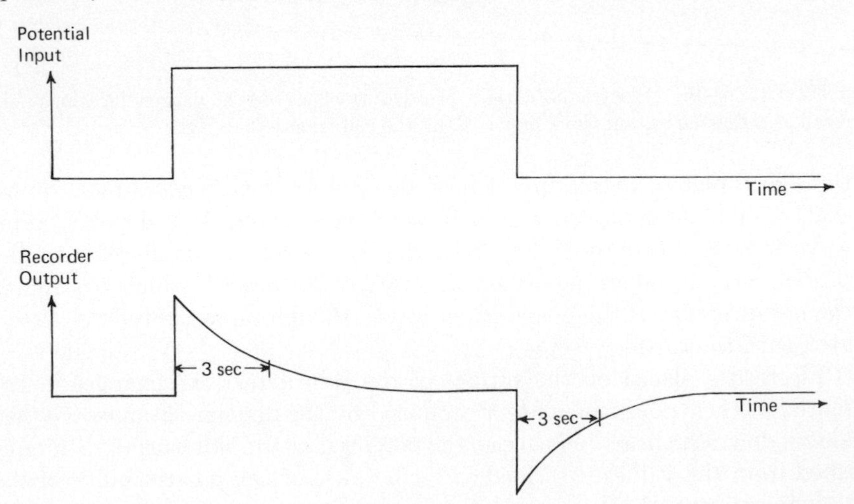

FIGURE 8-18. The differentiating response of an EKG recorder to a square-wave input.

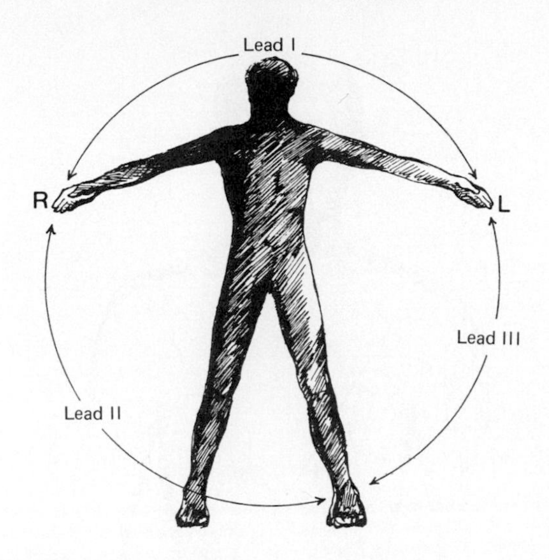

FIGURE 8-19. The three basic leads of an electrocardiograph.

of the input are adjusted so that the circuit has a time constant of 3 seconds, i.e., a voltage applied across the lead will produce a galvanometer pen deflection that, upon removal of the applied voltage, will return about two-thirds of the way to its zero position in 3 seconds. This time constant is sufficiently long that the rapid changes in potential across a lead during a cardiac cycle are faithfully recorded. The capacitive coupling also causes the response of an electrocardiograph to "differentiate" the input. Figure 8-18 shows the response of a recording pen on the output of the device when a square wave input is applied at the lead.

Convention specifies that 1 mV input into the electrocardiograph will produce a recording pen deflection of 1 cm. The recording paper moves at a speed of 25 mm/second. The three basic leads, designated I, II, and III, are shown in Figure 8-19. The leads are between electrodes on the arms (usually at the wrists) and left leg. An electrode on the right leg is grounded. The extremities actually function as connecting wires, and the lead connections might as well have been placed inside the body at points shown in Figure 8-20 as far as their electrical directional properties are concerned. The polarity convention of each lead is indicated in the figure.

The three leads I, II, and III are usually connected into a switch box similar to that shown in Figure 8-21, by which any lead can be made the input to the amplifier. The other leads shown in the figure will be explained subsequently. It will be seen that the leads measure differences

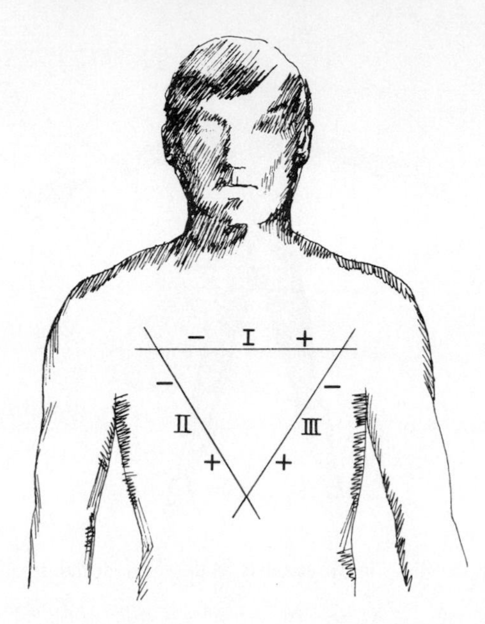

FIGURE 8-20. The equivalent electrode points in the body, shown as intersections of the sides of a triangle. Each side of the triangle is a lead axis, identified by a Roman numeral. The conventional polarity of each lead is shown by plus and minus signs.

in potentials according to

$$\begin{aligned} \text{Lead I:} \quad & V_{\mathrm{I}} = V_{\mathrm{L}} - V_{\mathrm{R}} \\ \text{Lead II:} \quad & V_{\mathrm{II}} = V_{\mathrm{F}} - V_{\mathrm{R}} \\ \text{Lead III:} \quad & V_{\mathrm{III}} = V_{\mathrm{F}} - V_{\mathrm{L}} \end{aligned} \tag{8-26}$$

where the subscripts L, R, and F refer to the left arm, right arm, and foot, respectively. This arrangement of leads was chosen by Willem Einthoven (1860–1927), the original developer of electrocardiography, who assumed that the electrode positions are electrically equidistant from the heart. Obviously, the Equations (8-26) are related algebraically by the relation

$$V_{\mathrm{II}} = V_{\mathrm{I}} + V_{\mathrm{III}} \tag{8-27}$$

which has been entitled *Einthoven's law* by physiologists (though Einthoven himself must have surely realized the trivial nature of the relation).

Lead Representations of Cardiac Electrical Activity

Einthoven's assumption that the three electrodes, L, R, and F, are equidistant from the heart suggests that the voltage differences appearing on each lead may be used as vector components of a resultant vector related to the direction of depolarization of the heart, i.e., to the cardiac vector. Figure 8-22 shows the relationship between the resultant vector **V** and the lead voltages (and their associated directions, which permits us to call them vectors) $\mathbf{V}_{\mathrm{I}}$, $\mathbf{V}_{\mathrm{II}}$, and $\mathbf{V}_{\mathrm{III}}$. The angle θ measures the orientation of **V** with respect to a reference axis that is chosen to be parallel to the line joining R and L. From the geometry of the figure, we see that the equilateral "Einthoven triangle" relates the magnitudes of the voltages according to

$$\begin{aligned} V_{\mathrm{I}} &= V\cos\theta \\ V_{\mathrm{II}} &= V\cos(\pi/3-\theta) = \tfrac{1}{2}V\cos\theta + (\sqrt{3}/2)\,V\sin\theta \qquad (8\text{-}28) \\ V_{\mathrm{III}} &= V\cos(4\pi/3-\theta) = -\tfrac{1}{2}V\cos\theta + (\sqrt{3}/2)\,V\sin\theta \end{aligned}$$

from which it is clear that $V_{\mathrm{II}} = V_{\mathrm{I}} + V_{\mathrm{III}}$, in agreement with Equation (8-27).

We may now associate each lead output of the EKG with the cardiac vector. Figure 8-23 illustrates the relative sizes and polarity of the responses of a particular lead of an EKG to a cardiac vector having various orienta-

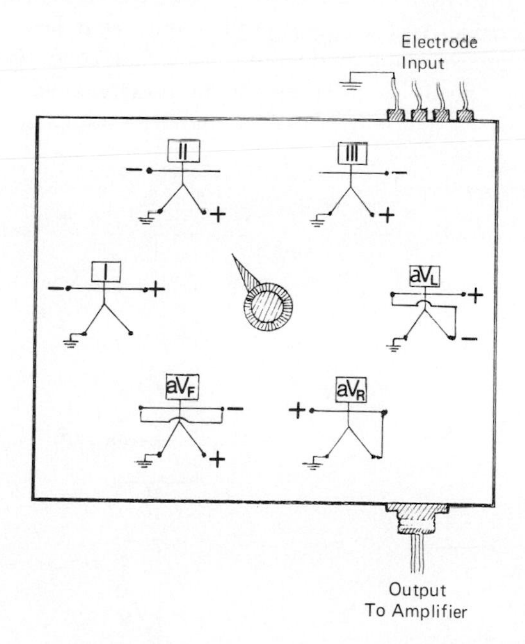

FIGURE 8-21. A switchbox for an electrocardiograph. This device permits the selection of any one of the six standard leads when the subject has four electrodes placed one on each arm and leg.

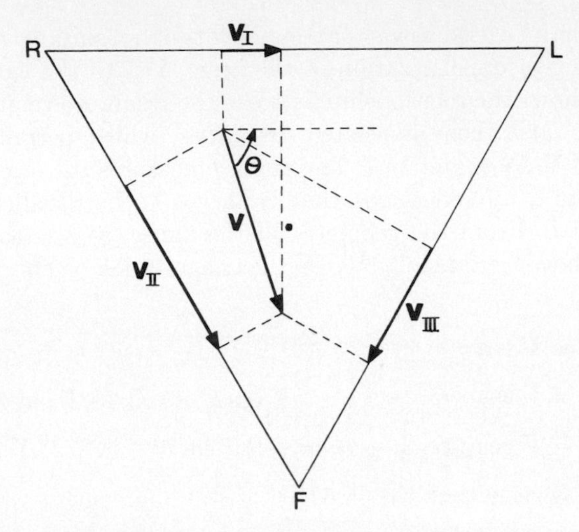

FIGURE 8-22. The Einthoven triangle. The cardiac vector **V** has components $\mathbf{V}_{\mathrm{I}}$, $\mathbf{V}_{\mathrm{II}}$, and $\mathbf{V}_{\mathrm{III}}$, along the three lead axes.

tions relative to the lead axis. It should be clear that a lead will produce no response if the cardiac vector is perpendicular to that lead's axis. A maximum positive response is obtained when the cardiac vector is parallel to the lead axis and pointing in the direction of the positive end of the lead axis;

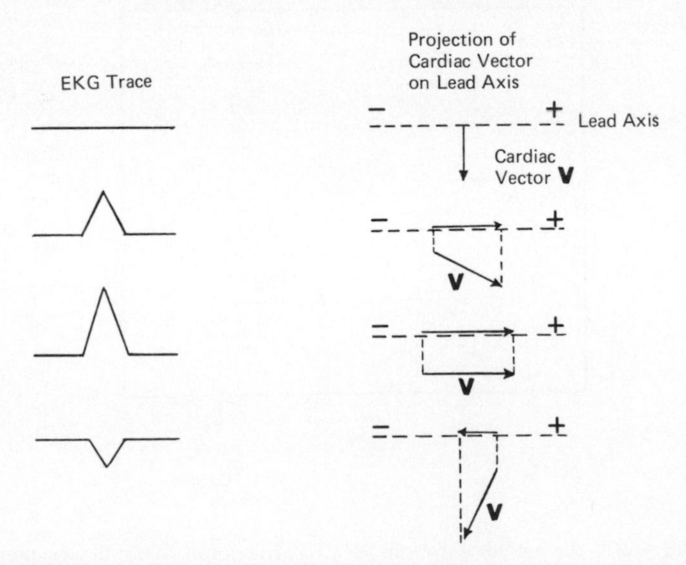

FIGURE 8-23. The effect of orientation of the cardiac vector **V** with respect to a lead axis on the EKG trace corresponding to that lead. This figure illustrates the relative size and polarity of the EKG trace for various orientations of **V** relative to the lead axis.

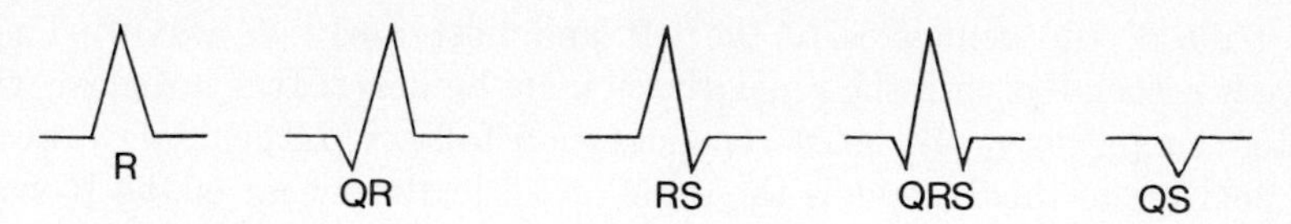

FIGURE 8-24. The various types of EKG waves that occur after ventricular activity has begun.

maximum negative response occurs when the cardiac vector is antiparallel to the lead axis.

EKG WAVES

The patterns recorded by an EKG consist of deviations from a zero level. Each significant deviation is called a wave, which is named with a letter. Each wave can be associated with electrical activity in a specific region of the heart.

Figure 8-24 may assist in identifying the characteristics associated with the terminology. An R wave is the first positive deflection in lead I that occurs after ventricular activity has begun. If the R wave is preceded by a negative deflection, the negative wave is termed a Q wave. A negative deflection following an *R* wave is called an S wave. A totally negative wave is a QS wave. Two waves occur in diastole—P and T. The P wave precedes the QRS complex, and the T wave follows it. A typical EKG is shown in Figure 8-25. Beginning with the P wave, the waves occur in alphabetical order in those leads in which all the waves are present.

The P wave is associated with atrial depolarization, which proceeds from right superior to left inferior. Thus, the P wave is a positive deflection on leads I, II, and III. The QRS complex is the result of ventricular depolarization, which proceeds first by depolarization in the septum from left to right and anteriorly (Q wave), then by depolarization in the region of the apex to the left and inferiorly (R wave), and finally by depolarization of

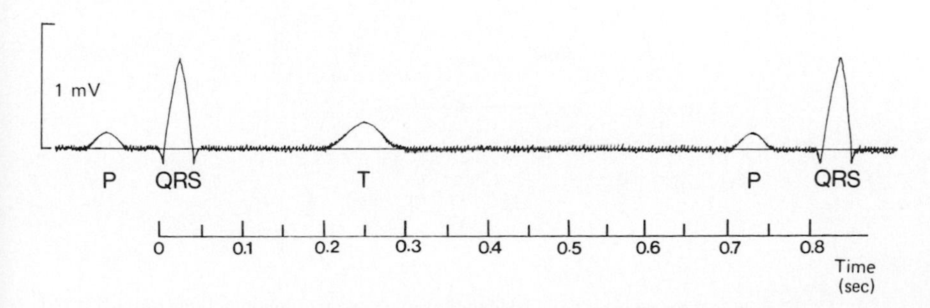

FIGURE 8-25. A typical EKG trace from a single lead. The letters under each formation are conventionally used to identify the various formations.

the walls of the ventricles to the left and posteriorly (S wave). The T wave is associated with the repolarization of the heart. It is not clear what pattern or pathway the heart repolarization follows. In general, however, the polarity of the T wave is the same on all leads as that of the R wave.

OTHER EKG LEADS

Linear combinations of the potentials at the electrodes L, R, and F can provide additional axes (other than the axes of leads I, II, and III) in the frontal plane of the body. The standard set of additional leads are chosen so that the axes bisect the angles between leads I, II, and III. These additional leads are called *augmented leads*. They are shown along with their conventional polarities in Figure 8-26 labeled aV_L, aV_R, and aV_F. Each augmented lead is said to be *unipolar*, referring to the fact that the axis of each augmented lead points directly from the heart toward the electrode for which it is named.

The standard augmented leads are achieved by wiring the three electrodes as shown on the switchbox in Figure 8-21. It will be seen from Figure 8-26 that the positive direction of aV_L is $(\mathbf{I} - \mathbf{II})/2$. Similarly, the directions of aV_R and aV_F are $-(\mathbf{I} + \mathbf{II})/2$ and $(\mathbf{II} + \mathbf{III})/2$, respectively. Here the positive directions of leads I, II, and III have been treated as unit vectors.

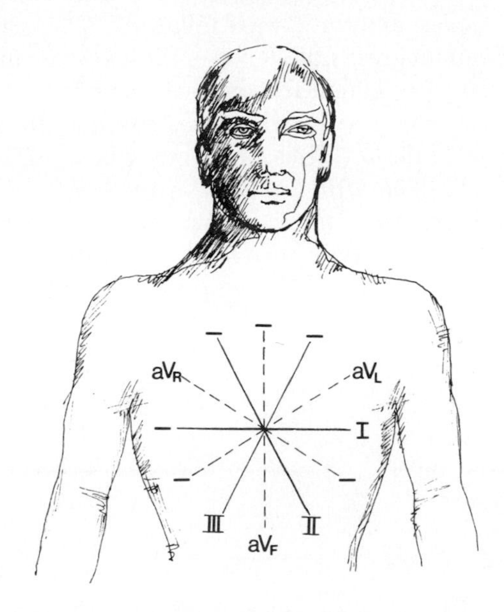

FIGURE 8-26. The axes and polarities of the six frontal leads conventionally used in electrocardiography.

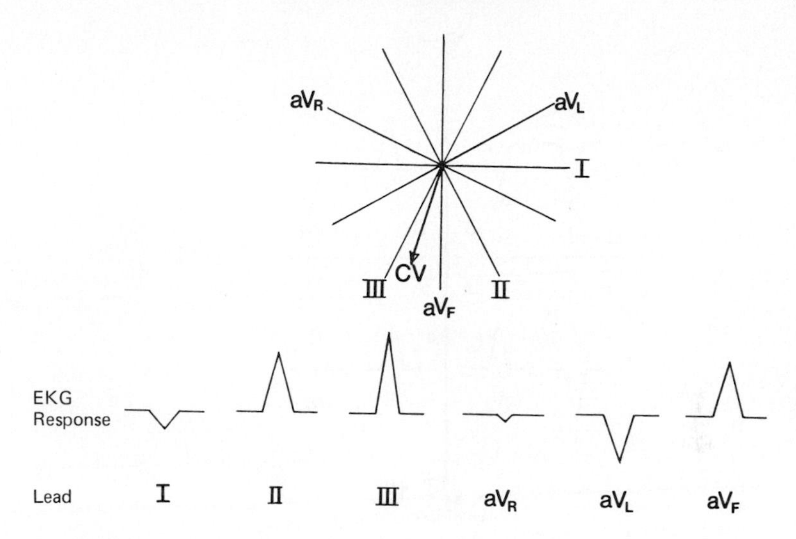

FIGURE 8-27. The response of each of the frontal leads of an EKG to a particular cardiac vector **CV**.

The augmented leads provide no new information over that available from leads I, II, and III; they are used so that the components of the cardiac vector can be more carefully plotted in the frontal plane of the body. Figure 8-27 illustrates how the components of the cardiac vector **V** appear on each of the six leads in the frontal plane.

The quadrant in which the cardiac vector lies can be quickly determined by examining one of the depolarization waves (P or R) on leads I and aV_F. Referring to Figure 8-28, we see that a vector lying in quadrant A produces positive waves on both the leads; a vector lying in quadrant D gives a positive wave on aV_F but a negative wave on lead I. The deviation of the cardiac vector from its normal position (in quadrant A of Figure 8-28) is a powerful diagnostic indicator of heart defects. Right deviation, i.e., deviation of the cardiac vector clockwise from the normal position, is seen in pulmonary disease (like emphysema) and in hypertrophy (enlargement) of the right ventricle. Left deviation is seen in coronary artery disease and aortic valvular disease.

To this point we have considered leads that can provide information only in the frontal plane of the body. Six other leads are used to provide information in the horizontal plane. The placement of the electrodes forming these leads is shown in Figure 8-29. Since these leads are placed "in front" of the heart, they are called *precordial leads*. Each precordial lead is unipolar, using the sum of the limb electrode potentials V_L, V_R, and V_F (which is the potential at the center of the equilateral triangle formed

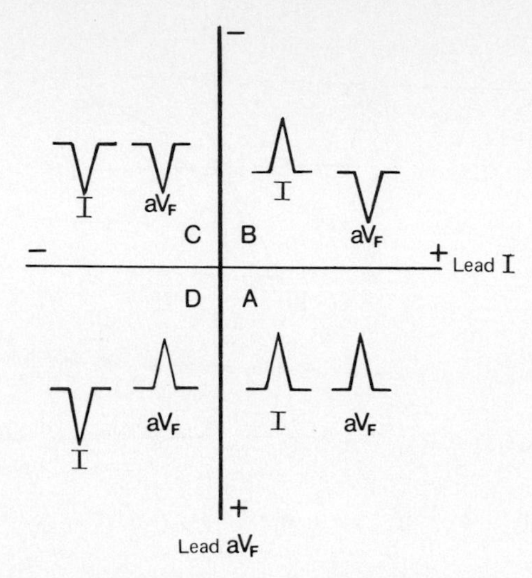

FIGURE 8-28. The determination of the quadrant in which the cardiac vector lies by examination of the polarities of a wave trace on leads I and aV_F.

by leads I, II, and III) as the reference potential. The precordial leads are labeled V_1 through V_6. A complete twelve-lead EKG taken from a normal heart is presented in Figure 8-30. A careful plot of the voltages on each of the twelve leads provides a good orientation of the cardiac vector in three dimensions.

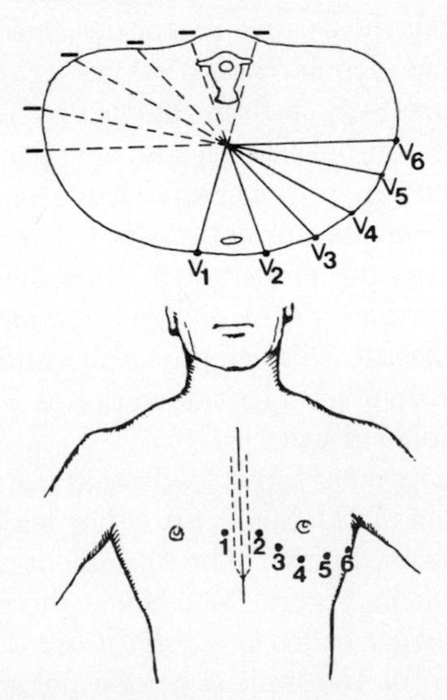

FIGURE 8-29. The placement and polarities of the six conventional precordial leads.

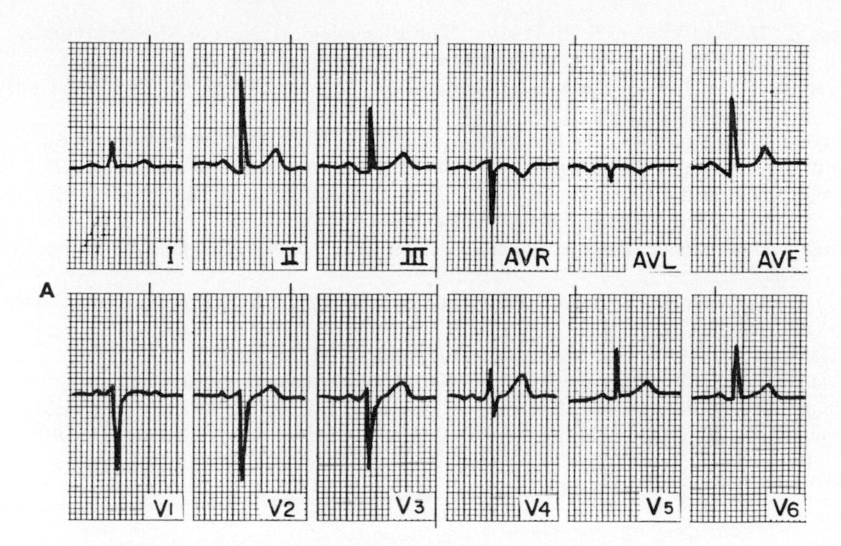

FIGURE 8-30. A normal twelve-lead EKG. From Andreoli *et al.* (1971). "Comprehensive Cardiac Care," 2nd ed. The C. V. Mosley Co., St. Louis. Reproduced by permission.

CLINICAL USE OF THE EKG

We have already mentioned the diagnostic significance of deviations of the cardiac vector from a normal position. There are a few other clinical uses of the EKG that can be understood from a theoretical basis. As an example, the location and identification of *infarcts* (regions of dead myocardial tissue resulting from lack of oxygen) that occur in many kinds of heart attack are common uses of the EKG. In general, the depolarization waves in the myocardium are reflected from the dead tissue or scar tissue.

Cardiologists, medical specialists in diseases of the heart, are often able to obtain from EKG's a wide range of information and in some cases surprisingly specific indications of cardiac defects. It should be pointed out, however, that most of the medical usefulness of electrocardiography depends not on theory but on rules that are the result of extensive experience and careful correlation of EKG's and cardiac defects. Because of the very complex parameters and geometries that affect the surface potentials resulting from the heartbeat, present theoretical models are inadequate to interpret the details of cardiac activity from surface potential information.

References and Suggested Reading

Andreoli, K. G., Hunn, V. A., Zipes, D. P., and Wallace, A. G. (1971). "Comprehensive Cardiac Care," 2nd ed. C. V. Mosby, St. Louis, Missouri.

Best, C. H., and Taylor, N. B. (1961). "The Physiological Basis of Medical Practice," 7th ed. Williams & Wilkins, Baltimore, Maryland.

Johnson, F. D. (1944). Electrocardiography. *In* "Medical Physics" (O. Glaser, ed.), p. 352. Yearbook Publ., Chicago, Illinois.

Mayerson, H. S. (1963). The lymphatic system. *Sci. Amer.* **208,** 80 (Offprint 158).

Perutz, M. F. (1964). The hemoglobin molecule. *Sci. Amer.* **211,** 34 (Offprint 196).

Ruch, T. C., and Patton, H. D., eds. (1965). "Physiology and Biophysics," 19th ed. Saunders, Philadelphia, Pennsylvania.

Waugh, D. F. (1959). Blood coagulation—A study in homeostasis. *Rev. Mod. Phys.* **31,** 557.

Whitelock, O. V. S., ed. (1957). The electrophysiology of the heart. *Ann. N.Y. Acad. Sci.* **65,** 653.

Wiggers, C. J. (1957). The heart. *Sci. Amer.* **196,** 74 (Offprint 62).

Wood, J. E. (1968). The venous system. *Sci. Amer.* **218,** 86 (Offprint 1093).

Wood, W. B. (1951). White blood cells vs. bacteria. *Sci. Amer.* **184,** 39 (Offprint 51).

Zweifach, B. W. (1959). The microcirculation of the blood. *Sci. Amer.* **200,** 54 (Offprint 64).

CHAPTER IX Muscle and Bone

The skeleton is the body's framework of bones, hard materials that serve as the principal support of the body and as the passive instruments of locomotion in the body. The active organs of motion are the muscles. Motion in this sense includes not only movements of the entire body but those of breathing; beating of the heart; and the movements of esophagus, stomach, and other *viscera* (internal organs of the body).

Muscles, which comprise 40 to 50% of the body's mass, are tissues that have developed to a high degree one of the basic properties of living material—*contractility*, the capability of a cell to reduce its length in opposition to a force. The voluntary motions of the body are accomplished by contractions of *skeletal muscles,* which move the bones. In skeletal motions the bones serve as levers that *articulate* (meet or join together) at movable joints. Indeed, all physiological activities of the body are closely related to motion brought about by muscular contraction.

In this chapter we shall examine some of the general anatomical features of muscle and bone and become acquainted with some of the important terminology associated with these tissues. Some of the physiological aspects of bone and muscle will be described. The interrelationship between the muscular activity of the body and the central nervous system are pointed out. Finally, the most obvious purpose of muscle and bone—their mechanical properties and capabilities—will be considered in terms of simple physical principles and mechanical models.

Anatomy and Terminology

Bone

Bone is one of a larger class of tissues called *connective tissues,* which are characterized by having relatively few cells interspersed in a matrix of intercellular material. In general, connective tissue serves to connect and support other tissues of the body. In the case of bone, the intercellular substance is very hard as a result of impregnation with mineral salts, chiefly calcium compounds. The inorganic component of bone constitutes about two-thirds of its mass. The organic component consists of cells, blood vessels, and *cartilage* (which is itself a connective tissue that is tough and flexible, yet inextensible, i.e., it bends easily but does not stretch or break easily). When the inorganic material is removed from bone, the remaining material is called *decalcified bone,* which is very similar to cartilage in its physical properties. Bone that is free of organic material is extremely brittle and can be crushed between the fingers.

Bones are classified as *long* (such as the *femur,* or thigh bone), *short* (like the bones in the hand or foot), *flat* (like the bones of the skull), and *irregular* (like the vertebrae or ossicles of the ear). A typical long bone, shown in Figure 9-1, will serve to introduce the important features of *osseous* (bony) tissue. The long thin body of the bone is the *shaft,* whose strength derives from *compact bone* surrounding a central *medullary cavity* filled with bone *marrow,* a connective tissue supporting blood vessels and cells. The compact bone is surrounded by a membrane called the periosteum. The articulating ends of the bone, called *epiphyses,* are covered with cartilage to reduce friction in the joints. The interior of the epiphyses is *cancellous* (spongy) *bone,* having a lattice work of cavities that contain *red bone marrow,* the source of red blood cells.

Bones are well supplied with blood. Arteries enter long bones near the middle of the shaft through a *nutrient foramen* (a foramen is a hole through or opening into a bone). Nerves accompany the arteries into and throughout the bone. The arteries, once inside the compact tissue, lie within the center of a complex structure called a *Haversian system*, a large number of which lie within compact bone tissue. Figure 9-2 shows the details of a Haversian system in cross-section. The Haversian system includes a central canal containing arterioles, venules, nerves, and lymph vessels. The central canal is surrounded by concentric rings of fibers called *lamellae* (singular, lamella). Aligned around the lamellae are *lacunae* (singular, lacuna, a little lake), the almond-shaped spaces within the hard matrix that contain the *osteocytes* (bone cells). The osteocytes are responsible for having formed the bony material in which they are imprisoned. Very small canals, *canaliculi*, extend from one lacuna to another and to the surfaces of the bone, where capillaries supply nutrients to the osteocytes. The connecting canaliculi contain processes from the osteocytes that reach out to each other, forming a nutritional lifeline throughout the bone.

There are 206 bones in the adult skeleton. A few, whose names are relatively commonplace and should be familiar to the student of biophysics, are labeled in Figure 9-3.

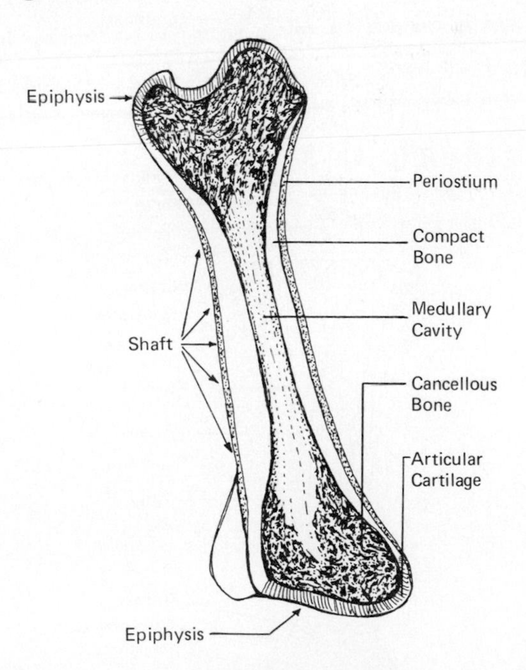

FIGURE 9-1. A typical long bone, shown in cross section.

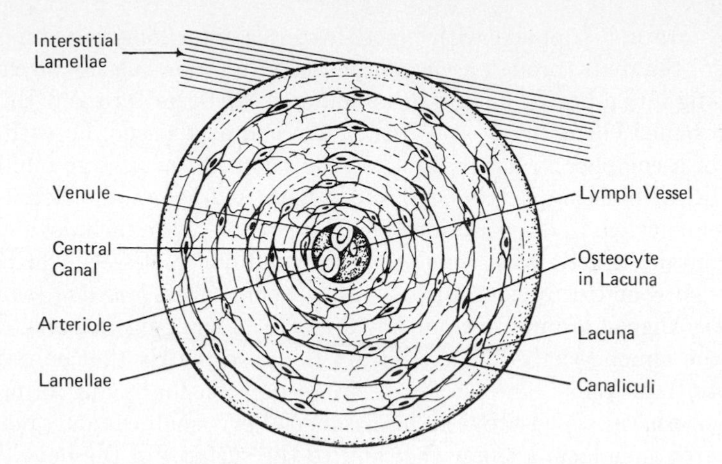

FIGURE 9-2. A Haversian system in compact bone.

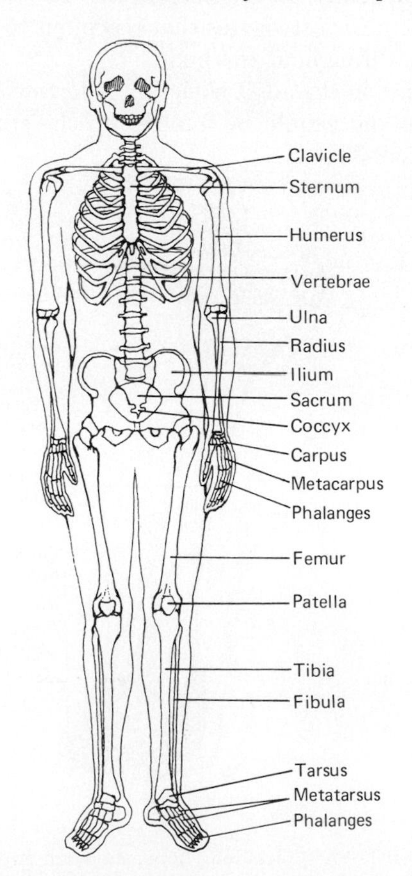

FIGURE 9-3. The principal bones of the human skeleton.

Muscle

There are more than 500 muscles in the human body. They vary in length from a small fraction of a centimeter to about two feet. They cover the skeleton almost entirely and make up nearly all the fleshy parts of the body. Muscles may be classified in a number of ways: by function, by innervation (how they are supplied with and controlled by nerves), by embryological development, or by histology (microscopic tissue structure). The least ambiguous classification, and the one we shall adopt here, is the histological, which distinguishes between three types of muscle: *striated, smooth,* and *cardiac.* We shall consider each type briefly.

STRIATED MUSCLE

The anatomy of striated muscle may be approached by considering the appearance of a sirloin steak or a slice of ham. The red, fleshy, lean portion of the steak is striated skeletal muscle, which is separated into large bundles by thin white sheets of connective tissue called *fascia.* The large bundles are sections of individual muscles. One can often see small holes in the

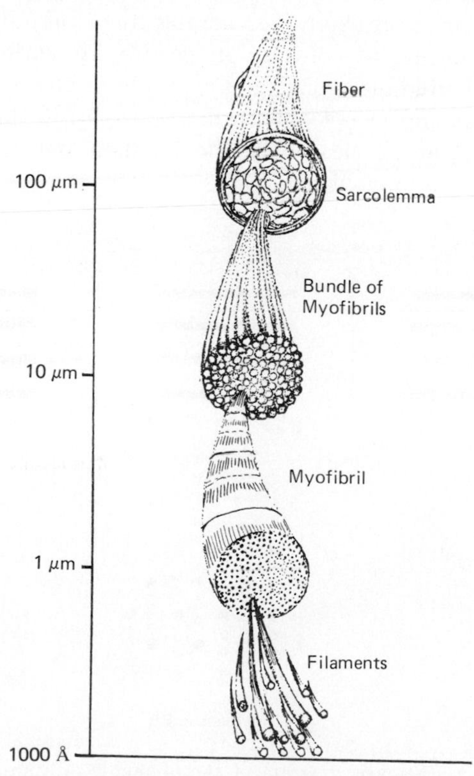

FIGURE 9-4. A logarithmic diagram of the components of a muscle fiber.

section where blood vessels and nerves pass between bundles. The bundles are surrounded and often infiltrated by fat.

A single muscle is composed of a pyramided system of strands made up of strands and so on to the molecular level. Figure 9-4 depicts the substructures, their names, and typical dimensions associated with each component. The gross muscle is made up of *fibers*, each of which is from 1 to 40 mm in length and from 10 to 100 microns (μm) in diameter. Each fiber is bounded by its *sarcolemma*, the membrane of a cell with many nuclei. Inside of each fiber are a large number of bundles about 5 μm in diameter, each of which in turn contains *myofibrils*, strands about 0.5 μm in diameter. The myofibrils are the smallest units of muscle tissue that have been stimulated to contract. Myofibrils contain strands of *filaments* that occur in two sizes (diameters). The filaments are of the order of 10 nm thick. The filaments contain at least two protein chains, *actin* and *myosin*; the former is found in the thin filament, the latter in the thick filament.

Electron microscopy has provided evidence of some interesting relationships between muscle components. The thick and thin filaments are arranged as shown schematically in Figure 9-5. Notice that the ends of each kind of filament are in register. The arrangement permits the filaments to slide past one another as contraction occurs. In various cross sections, hexagonal and double hexagonal arrays appear. The arrangement causes the striations that appear not only in filaments but in the larger myofibrils (though it is not at all clear why the filament patterns are in register throughout the entire myofibril).

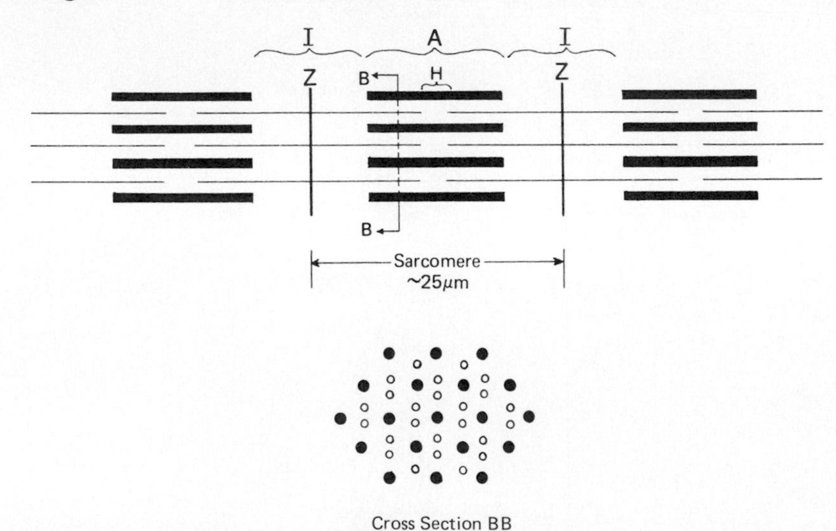

FIGURE 9-5. A schematic diagram of the filaments in a muscle fiber. The region between the Z-lines is called a sarcomere, the basic repetitive unit of a muscle fiber.

The striations of muscle have been studied for many years, and a traditional terminology has come to be associated with the various regions. The lettered designations of the various regions are shown in Figures 9-5 and 9-6. The *A-band* corresponds to the length of the thick filaments, and the A-band strains darkly. The *I-band* corresponds to the space between the ends of the thick filaments and stains lightly. The *H-band* is the region

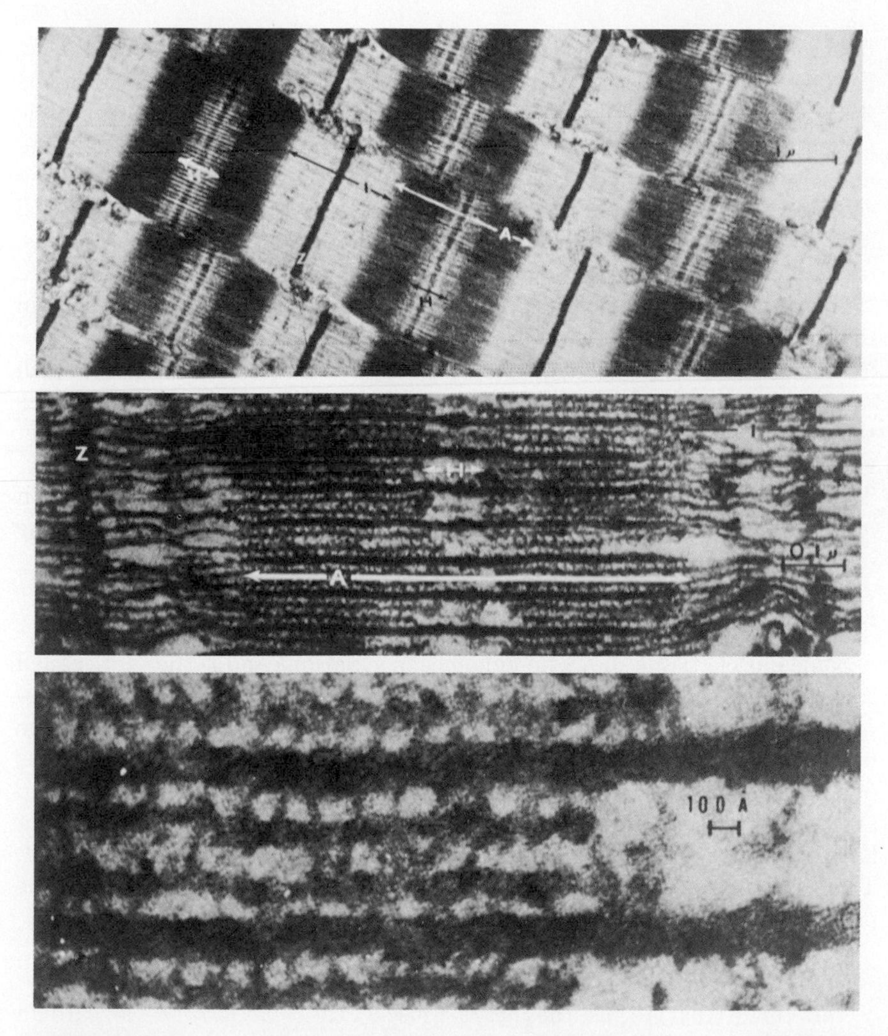

FIGURE 9-6. An electron micrograph of striated muscle. (From Brachet and Mirsky, 1962).

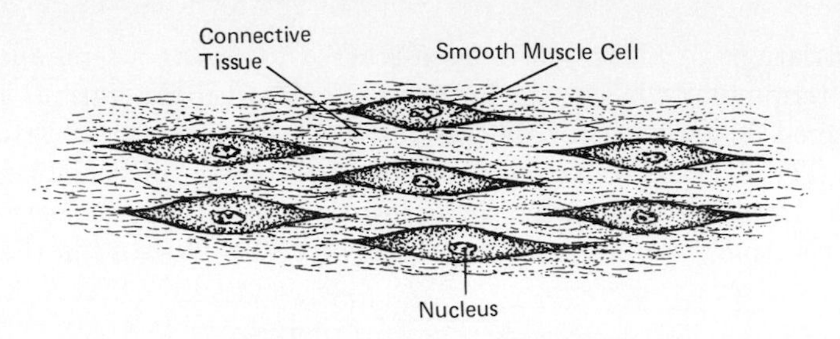

FIGURE 9-7. Smooth muscle.

between the ends of the thin filaments; H lies within A and A stains darkly. In the center of I-band is a very dark line, the *Z-line*. The region from one Z-line to the next is called a *sarcomere*, the basic repetitive unit of muscle structure. In an uncontracted muscle, the sarcomere is about 25 μm long and shortens as contraction takes place. Notice that the sarcomere is in no sense a cell. Since myofibrils are multinucleated, containing many nuclei as well as mitochondria, a sarcomere may contain no nuclei, one nucleus, or even occasionally more than one nucleus.

Near the end of a skeletal muscle, the connective tissue increases in relative content until, as the muscle fibers disappear, only a dense white *tendon* remains. The tendon, inextensible yet flexible, serves to attach muscle to bone. Tendons that are flattened are called *aponeuroses*. Muscles that attach to movable bones usually pass over a joint, and one bone is in general more movable than the other in the ordinary activity of the body. The points of attachment of the muscle to two bones are called either an *origin* or an *insertion*. The origin is attached to the less movable bone; the insertion is attached to the more movable bone.

SMOOTH MUSCLE

Some muscle tissue shows no evidence of striations and is therefore called smooth muscle. Tissues of smooth muscle are said to be *visceral*, as opposed to skeletal, because they form the walls of the internal organs of the body, including blood vessels, intestines, esophagus, and bladder. The cells of smooth muscle are spindle-shaped (see Figure 9-7), have lengths up to about 0.5 mm and have diameters up to about 0.02 mm. Each cell has a large nucleus. Tissues of smooth muscle are not under voluntary control. They are nourished by blood vessels that generally run among the cells parallel to the lengths of the cells.

CARDIAC MUSCLE

The tissue that forms the heart muscle, the *myocardium*, is cardiac muscle. The myofibrils of the heart have faint striations, much less distinct than in skeletal muscle. Cells are smaller and mitochondria are more

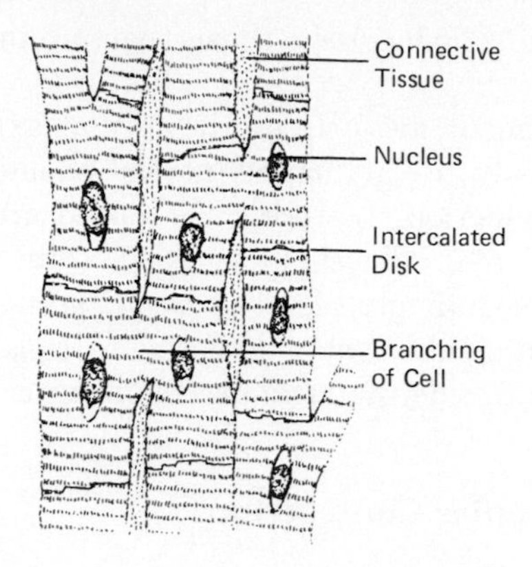

FIGURE 9-8. Cardiac muscle tissue.

abundant in cardiac muscle than in striated muscle. Each cell has a large nucleus placed centrally. Figure 9-8 shows two unusual features of cardiac tissue: (a) The cells often branch between each other. (b) There are dark-staining bands, the *intercalated disks,* across the muscle at the junction of cells. The cells of the myocardium are surrounded by connective tissue containing fine fibers of *collagen,* a strong elastic material, along with blood and lymph capillaries. The heart muscle is nourished by blood from the *coronary artery* system.

Cardiac muscle tissue has essentially no capacity to reproduce itself after injury. Healing of the myocardium after the death of cells is accomplished by the formation of tissues that include connective tissue. Considerable scar tissue may be incorporated into the myocardium as the result of a heart attack, a situation in which a significant volume of the myocardium is deprived of oxygen by a reduction or stoppage of blood supply from the coronary arterial system.

Terminology

In describing the movements of muscle–bone combinations, we use a number of technical, yet pointedly descriptive terms. It is worthwhile to be acquainted with a few of these terms.

Flexion is bending. When the angle between adjacent parts of the body is decreased, as when the calf of the leg is bent toward the thigh, we say that flexion has occurred at the knee joint. *Extension* is the reverse of flexion; the angle between adjacent parts is increased in extension. Thus the straightening of a bent leg causes extension at the knee. Notice that flexion and extension are events referring to a joint. Muscles that effect flexion at a joint are called *flexors*; those that produce extension are *extensors*.

Abduction is movement away from the medial axis of the body, e.g., raising the arm constitutes abduction at the shoulder joint. *Adduction* is the opposite of abduction. Lowering the raised arm to the side of the body effects adduction in the shoulder joint. Muscles producing these movements are correspondingly called *abductors* and *adductors*.

Rotation is the turning about an axis without displacement, as when the head turns from side to side. *Rotators* are the muscles that effect rotation.

The Events of Muscular Contraction

Muscle is irritable tissue, i.e., it responds to stimulation, and its response is contraction. The stimulus is initiated in the central nervous system, from which an action potential on a motor neuron transmits the signal to the muscle. A group of muscle fibers contract after the stimulus arrives at the muscle.

We shall consider these events in the order in which they take place.

Nervous Control of Muscle

The brain controls motor function. Considering our earlier encounters with brain function, it should not be surprising that the brain's control of muscular function is very complex and subtle. Nor should it be surprising that we understand very little about how the brain accomplishes this control. Although we have characterized skeletal muscle as voluntary, it is clear from experience that we cannot voluntarily cause an individual muscle, independent of all others, to contract. We can move a finger or perhaps execute a complex gymnastic movement, but we can voluntarily control only the gross movement; the brain handles the details by issuing commands to the muscles. The timing and finesse of muscular control is, of course, commonplace, yet a moment's reflection on a simple act like tossing a ball to a given spot can inspire awe. Hundreds of muscles are involved, each of which must exert precise tensions in very precise time sequences.

Not only are contractions involved in all our motions, but relaxations of muscles are equally important. In Figure 9-9 we can see the interrelationships between an extensor and a flexor in a simple bending motion of the arm. When the flexor contracts to bend the arm, the extensor must simultaneously relax, and when straightening the arm, the flexor must relax as the extensor contracts. These facts suggest that muscles are in a

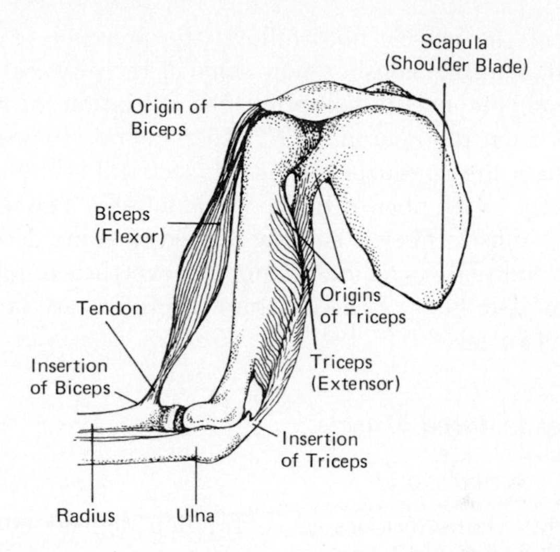

FIGURE 9-9. The bones and muscles of the upper arm. This figure illustrates the antagonistic arrangement of the flexor and the extensor muscles.

continuous state of controlled tension. All skeletal muscles are normally under slight tension, a state referred to as *tonus*, or muscle tone. Normal support of the body against the pull of gravity, for example, requires that all the skeletal muscles supply the necessary forces to maintain posture. Therefore, the brain is continuously monitoring and directing skeletal muscles without any conscious thought or act of will on our part.

Neuromuscular Transmission

Motor neurons end on surfaces of muscle fibers. The junction between the neuron's axon terminal fibers and the sarcolemma covering the muscle fiber has a transmembrane potential of about 70 mV in the resting (uncontracted) state. The *end-plate* mechanism (as this junction is called) functions almost exactly like a synapse between two neurons (see Chapter V). When an action potential on the motor neuron reaches the end-plate, acetylcholine (ACh) is released from the axon terminal fiber. The ACh causes a change in the permeability of the sarcolemma to Na^+ and K^+ ions, initiating a depolarization wave along the muscle fiber. Thus, an action potential (with an overshoot of about 10 mV) is actuated on the muscle fiber that is very similar to that on the neuron. Just as in the case of synaptic conduction between neurons, the transmission capability of an end-plate is restored by the enzyme acetylcholinesterase, which destroys the ACh.

Contraction of the muscle fiber follows the presence of the muscular action potential. We shall consider now some of the physical and chemical characteristics of the events following the initiation of a stimulating electrical signal from the neuron. Most experimental studies that provide us with these data are conducted using an electrical stimulus, applied by electrodes directly to the fiber or to the whole muscle. The resulting action potential on the muscle fiber is slightly altered in some details from that produced by an end-plate in an intact animal; nevertheless, for our purposes we may assume basic facts of the contraction process are not significantly different in the two cases.

Contraction of Isolated Muscle

CONDITIONS OF CONTRACTION

The observable characteristics of contraction depend not only on the means of stimulation (to which they are relatively independent), but on the physical and chemical conditions that exist at the time of stimulation (to which they are quite sensitive). When muscles contract, they may either produce movement and thereby do work, or they may produce tension, i.e., force, without motion. Measurements of muscular contraction are therefore made under one of two physical constraints, illustrated in Figure 9-10.

1. The muscle is fixed at both ends, so that no shortening of the muscle can take place upon contraction. Obviously, no work is done in this case. This type of contraction is termed *isometric* (equal length).

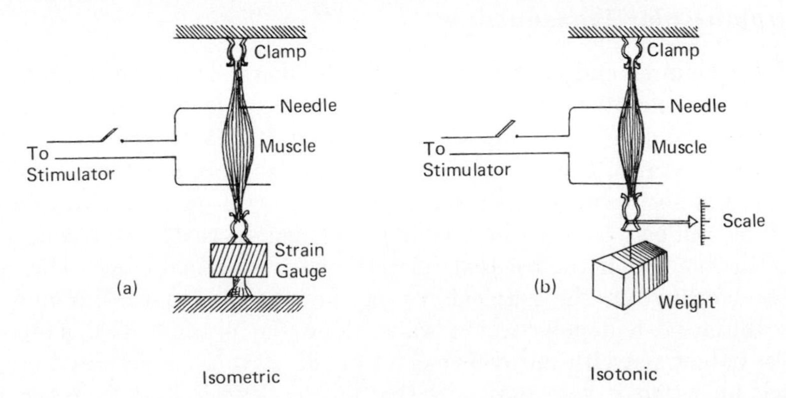

FIGURE 9-10. Arrangements for recording the response of a muscle to stimulation. (a) Isometric arrangement, in which the muscle's length is constrained to remain fixed. (b) Isotonic arrangement, in which the change in length of the muscle may be measured under various loading conditions.

2. The muscle is fixed at one end while the other end is loaded with a fixed mass m. In this case the muscle, upon contraction, becomes shorter as it exerts a constant force mg. The work done in moving the muscle through a vertical distance x is mgx. This type of contraction—shortening under constant load—is called *isotonic* (equal force).

The chemical environment of a muscle is important to the function of contraction. Measurements of contraction characteristics should specify the chemicals used, if any, to bathe or periodically wash the muscle. As metabolic products of contraction accumulate in and around the muscle, significant changes in contraction characteristics can take place.

PHYSICAL CHARACTERISTICS OF CONTRACTION

When a muscle is subjected to a short electrical stimulus above some minimal threshold value of potential, the muscle responds with a single contraction called descriptively a *twitch*. The amplitude of the twitch is proportional to the amplitude of the stimulus; this fact does not violate the all-or-nothing character of response of a single fiber to stimulation, but is explained by the fact that more fibers in the muscle are fired as the stimulus voltage is increased; the fibers are separate units and somewhat insulated from each other by connective tissue. The minimal stimulus is one that evokes firing of a single fiber, and the maximal stimulus is one that causes contraction of every fiber present. An isometric twitch contraction resulting from maximal stimulation of a typical skeletal muscle is shown graphically in Figure 9-11. The abscissa is in arbitrary units of force, the values de-

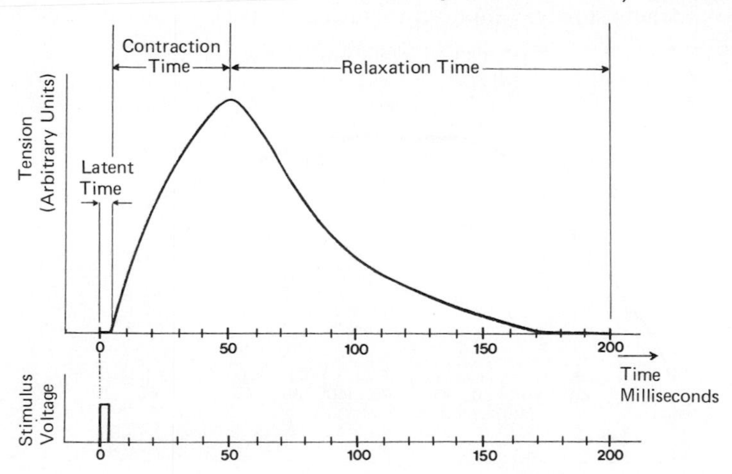

FIGURE 9-11. The twitch contraction. This diagram shows the response of a muscle (measured under isometric conditions) to electrical stimulation of a short duration.

pending on the particular muscle being observed. A single fiber gives essentially the same characteristics as a whole muscle. The time between the application of the stimulus pulse and the initiation of contraction is called the *latent time*. The time interval from initiation of the contraction to the time of maximum tension is called the *contraction time*. The time interval between the time of maximum tension and the time at which the resting state is restored is the *relaxation time*. During a very short period of the early part of the contraction time, the muscle cannot be stimulated to fire by any stimulus, however strong. This "dead time" interval is the *absolutely refractory period*.

If a muscle receives a second stimulus during a time interval near the time of maximum tension, an increased tension is observed whose amplitude may be even greater than twice that of a single twitch. This effect is called *summation of contraction*. When stimuli are repeatedly applied to a muscle in rapid succession before the fibers have relaxed from the one precedign, the muscle assumes a state in which no relaxation is apparent. This effect is shown in Figure 9-12. The state of contraction in which successive applications of stimuli do not increase the tension in the muscle, but do maintain the contraction, is called *tetanus*. A sustained state of tetanus results in a gradual diminution of the tension; this weakening of contraction is termed *fatigue*.

In skeletal muscles that move limbs, contraction times are of the order of 100 milliseconds in mammals. The fastest acting muscle in mammals (about 7 milliseconds) is one that moves the eyeball in its socket. Smooth muscle contracts slowly compared to skeletal muscle. The contraction time

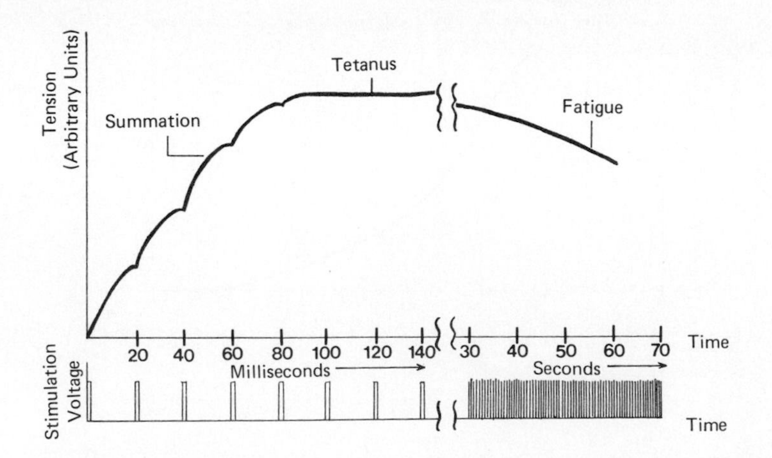

FIGURE 9-12. The response of muscle to repeated stimulation. Note the change of time scale in the figure.

of muscles in the uterus and urinary bladder is several seconds. Latent times are typically about 10 milliseconds.

CHEMISTRY OF MUSCLE CONTRACTION

During muscular contraction, chemical energy is converted to mechanical energy and heat. If muscles are analyzed chemically, it is found that they are about 80% water, with most of the remaining mass being the proteins myosin and actin. The remainder of the muscle is fat, glycogen, and two phosphorus compounds, *creatine phosphate* (CP) and adenosine triphosphate (ATP; see Chapter II). Neither of the proteins is capable of contracting alone, but when mixed together in a test tube with potassium and ATP, contraction occurs. It has been suggested that the contraction of the protein occurs by folding of the amino acid "links" of the protein chain by expulsion of water molecules from around the surface of the proteins.

If a chemical analysis is made of the substances in and around muscle before and after contraction, it is found that glycogen, oxygen, CP, and ATP are used up during contraction, and CO_2, *lactic acid,* and inorganic phosphates are produced. The fact that oxygen is consumed in contraction suggests that oxidation furnishes the energy of contraction. Experiments show, however, that contraction can occur without oxygen present. Furthermore, although we breathe rapidly during muscular exertion, the rapid breathing continues for considerable time after the muscular activity has stopped. These facts suggest that oxidation is not directly involved in contraction but in the recovery processes following contraction.

ATP is the immediate agent of energy supply for contraction and relaxation. The activation or inhibition of an enzyme, ATPase, by Ca^{2+} ions determines whether the contractile proteins contract or relax in the presence of ATP. Then energy for contraction comes from the hydrolysis of ATP to ADP and a phosphate group, i.e., the breaking of the high-energy phosphate bond of ATP (this reaction was discussed in Chapter II). The anaerobic phase of contraction is completed by the following reactions:

$$\text{ADP} + \text{CP} \leftrightarrows \text{creatine} + \text{ATP} + \text{energy}$$

$$\text{Glycogen} \leftrightarrows \text{lactic acid} + \text{energy}$$

The energy liberated in the decomposition of glycogen is used in the resynthesis of CP and ATP.

The aerobic phase, which takes place as oxygen becomes available, consists of oxidation of the lactic acid to produce the final products, CO_2 and H_2O:

$$\text{Lactic acid} + O_2 \rightarrow CO_2 + H_2O + \text{energy}$$

The energy released in this case is used to convert some of the lactic acid into glycogen, which is then available to the anaerobic phase:

$$\text{Lactic acid} + \text{energy} \rightarrow \text{glycogen}$$

It is an important physiological fact of muscular activity that the contraction does not utilize oxygen directly yet ultimately depends on oxygen. When muscles perform work at a great rate, as in sprinting, oxygen cannot be supplied in sufficient quantities to keep lactic acid from accumulating in the muscles. In this circumstance, the muscles are said to have incurred an *oxygen debt,* which is repaid by rapid breathing and increased heart rate after the muscular activity has ceased. The increased breathing continues until the lactic acid has been converted to glycogen and CO_2 and until the blood sugar has restored the remaining glycogen deficit in the muscles.

In moderate exercise, as in a long but relatively slow race, the runner's heart rate and breathing increase until oxygen is supplied at a rate that equals the formation of lactic acid. If the rate of lactic acid formation is decreased as a result of continued rapid intake of oxygen, the runner gets a "second wind."

Repeated contraction of an excised muscle in the laboratory causes fatigue, but the muscles in an intact animal never reach the fatigue point. It can be shown that in an exhausted animal, i.e., an animal that can no longer contract a particular muscle that has been over-exercised, the muscle itself can be made to contract by direct electrical stimulation. Further, the motor nerves serving the muscle can be shown *still* to be able to transmit impulses. It follows then that the fatigue has occurred at the neuromuscular junction, the end-plate.

The Physics of Muscular Contraction

In this section we shall briefly examine one of the current theories of muscular contraction at the molecular level, then proceed to the macroscopic level of muscular activity as it is manifested in bodily motions.

The Sliding Filament Theory

Biochemical and electron microscopic data over the past few years have led to a theoretical picture of the physical mechanisms of muscular contraction. The most modern theory, referred to as the *sliding filament theory,* is an attempt to group the observed facts into a single model consistent with all the data.

How can the energy supplied by ATP to the contractile proteins produce a net translational force? If the bonds between ATP, actin, and myosin at two successive sites along the fiber exert equal forces in opposite directions, no net force is developed. In order to produce a net force, the bonding between ATP and the proteins must form in a preferential direction. In particular, it is assumed that the ATP bonds pull the thicker filaments of myosin from adjacent sarcomeres toward the Z-line between them (see Figure 9-5). The evidence from electron photomicrographs indicates that minute bridges can form between the thick and thin filaments. In this model, it is assumed that ATP is normally attached at only one side of the bridge site (the myosin side), preventing the combination of actin and myosin and thus permitting the filaments to slide freely past each other in the longitudinal direction. With the help of Ca^{2+} ions acting as linkages, ATP molecules at the bridge site form bonds between thick and thin filaments at an acute angle. The bonds shorten as ATP is hydrolyzed to ADP, thereby exerting a component of force parallel to the fibers. It is supposed that this happens at many bridge sites along the fibers, and many ATP molecules are used up by a single thick filament interacting with a number of thin filament bridge sites on the six neighboring thin filaments (see Figure 9-5) as they move by. The ATP is constantly being regenerated during this process by CP. In the case where tension is maintained, but no longitudinal contraction occurs in the muscle, i.e., in the isometric contraction, the force is produced by repetitive interaction and consequent turnover of ATP at fixed bridge sites. In this case, energy is consumed by the muscle without its performing work.

Relaxation in this model is accomplished by removal, or *sequestering* (hiding), of the Ca^{2+} ions, which removes a link in each of the bridges, thereby permitting the muscle to slide back to its uncontracted position. The concept of removing Ca^{2+} from the contractile elements and storing it in a separate compartment until the muscle is stimulated to contract again may seem somewhat farfetched. Nevertheless there is considerable evidence supporting this hypothesis. As an example, local contraction can be effected in an intact muscle by injection of Ca^{2+}. Considerable research is currently being directed toward the detailed study the effects of Ca^{2+} concentrations in muscles.

As a muscle begins to fatigue, it is not the contraction mechanism that fails but the relaxation mechanism that begins to become slower and more labored. When a muscle is completely fatigue, it is in a fixed, contracted state called *rigor*. This is the effect commonly called "cramps." When death occurs, the skeletal muscles can no longer relax, but become rigidly contracted; this is *rigor mortis* (stiffness of death), which sets in within 10 minutes to 7 hours, depending on the physical and chemical state of the muscles at the time of death. In the sliding filament model, rigor results from the formation of tight bridges between the actin and myosin filaments, i.e., the formation of *actomyosin*, so that the filaments lose their ability to slide past each other.

Heat Production in Muscle Contraction

A muscle, when considered as a machine for doing work, is about 25 to 30% efficient. This means that about three-fourths of the energy consumed by muscles is converted into heat. Although these values compare favorably with most man-made machines, it should not be assumed that the muscles are inefficient from a functional point of view. The heat geneated by the muscles of the body is important in the maintenance of a constant body temperature. When exposed to cold surroundings, a person may warm himself rather efficiently by vigorously exercising. A nonvoluntary response to the same conditions is shivering, in which the sketal muscles rapidly twitch on command of the central nervous system and generate significant heat near the surface of the body.

When a muscle twitch occurs, two distinct bursts of heat are generated in and on the muscle. The first is called *initial heat*, which is composed of heat produced during contraction (about 65% of the initial heat) and heat produced during relaxation (about 35%). The initial heat is generated in a very short time span, much less than a second. The second stage of heat production is called *recovery heat* (or sometimes delayed heat) because it is produced during the period following contraction when chemical changes take place that restore the muscle to the precontraction condition. Recovery heat is produced for several minutes, and the total heat generated in the recovery phase is about one and a quarter times that generated in the initial phase.

The Mechanics of Muscular Motion

Skeletal muscles move bones by *lever* action. A lever consists of a mechanical arrangement in which a force is applied to a rigid member that rotates under the influence of the force about a fixed pivot point, called the *fulcrum*, along the member. The rigid member is assumed to be loaded with a weight at some point along the member (we say point because a distributed load can be assumed to be concentrated at the center of gravity of the load). Levers are classified according to the relative locations of the force, fulcrum, and load. Figure 9-13 illustrates the three classes of mechanical levers and gives examples of the two classes found in the human body.

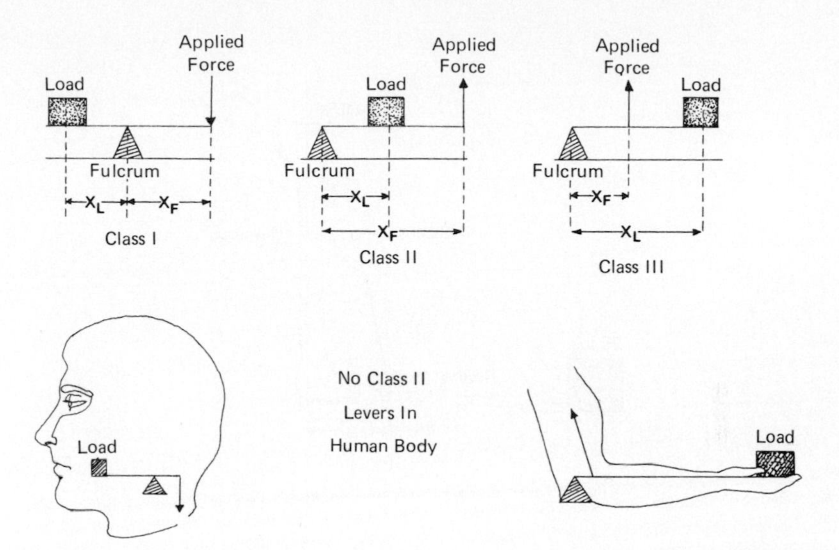

FIGURE 9-13. The three classes of lever systems. Examples of each class of lever system in the human body are shown, except class II, which is not represented in the human body.

Most lever systems in the body are class III. There are no class II levers in the human body. The theoretical mechanical advantage (TMA) of a lever system is defined as the ratio of the weight W of the load to the applied force F that will maintain the system in equilibrium. This mechanical advantage can also be represented in terms of the perpendicular distances from the fulcrum to the lines of action of the load and the applied force (in Figure 9-13, these distances are x_L and x_F, respectively):

$$\text{TMA} = \frac{W}{F} = \frac{x_F}{x_L} \tag{9-1}$$

Skeletal motion is rotational; bones rotate about axes through joints under the influence of forces produced by muscular contraction. A force is said to have a *lever arm* about a particular axis, where the lever arm is defined as the perpendicular distance from the axis of rotation to the line of action of the force. *Torque* about an axis is the product of a force and its lever arm about that axis.

The torques that rotate members of the human body about joints are produced by muscles that, because of the architecture of the body, are constrained to lie close to the bones. This means that the lever arms of bodily levers are usually very short. Since most lever systems in the body are class III, the forces required of individual muscles are often quite large.

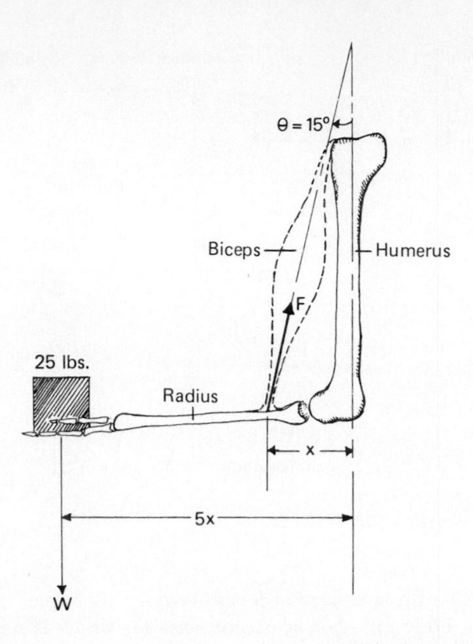

FIGURE 9-14. Arm supporting weight with forearm horizontal.

As an example, consider the case in which the upper arm is beside the body, the forearm is extended horizontally, and the hand is holding a 25-pound weight (see Figure 9-14). The *biceps*, the flexor muscle of the forearm, makes an angle θ of about 15° with the *humerus*, the upper arm bone. In equilibrium, the magnitude of the torque about the elbow produced by the weight (ignoring the weight of the forearm and hand) is equal to the magnitude of the torque due to the force F exerted by the biceps. The distance x from the insertion of the biceps on the bone of the forearm to the elbow is about one-fifth the distance from the hand to the elbow. In order to obtain the component of the force exerted by the biceps that is perpendicular to the lever (the radius), we multiply F by cos 15°. Then equating the magnitudes of the torques gives

$$
\begin{aligned}
25(5x) &= F(x \cos 15^\circ) \\
125x &= Fx(0.966) \\
F &= \frac{125}{0.966} = 129.5 \text{ pounds}
\end{aligned}
\tag{9-2}
$$

Thus the biceps develops a tension of 129.5 pounds in supporting the 25-pound weight.

If now the upper arm is raised while supporting the same 25-pound weight and at the same time keeping the forearm horizontal as in Figure 9-15, the angle φ between the tendon of the biceps and the bone of the forearm (the radius) to which it is attached is about 10°. The force F exerted by the biceps now has a lever arm of $x \sin 10°$, and equating the magnitude of torques about the elbow gives

$$25(5x) = F(x \sin 10°)$$

$$125x = Fx(0.174)$$

$$F = \frac{125}{0.174} = 720 \text{ pounds}$$

for the tension developed by the biceps. We would not expect the tension to increase indefinitely as the arm approaches the horizontal because the shape and positioning of the biceps does not permit the angle φ of Figure 9-15 to become much smaller than about 10°.

The *absolute muscle strength,* i.e., the maximum tension that can be developed by skeletal muscle, is about 140 pounds per square inch of cross-section through the largest diameter of the muscle. Some of the larger muscles in the leg are capable of developing forces in excess of 2000 pounds.

As a muscle shortens to develop tension, the tension is largest as the contraction begins and decreases approximately linearly with the shortening of the muscle. Figure 9-16 represents the smoothed data that might have been obtained by loading a muscle with a known weight, measuring the change in length ΔL of the muscle as it contracts isotonically, and then repeating the process for a series of weights. The magnitude of the weight is equal to the tension F developed by the muscle in moving the weight. The intercepts on the ΔL axis and the F axis are ΔL_{max} and ΔF_{max}, respectively. The slope of the straight line is $-(\Delta L_{max})/F_{max}$, and the

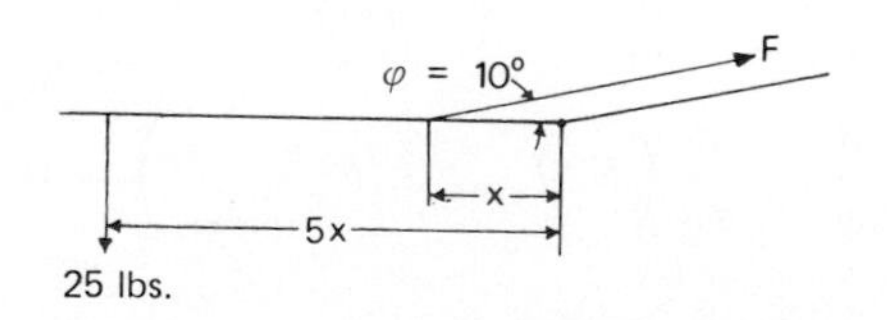

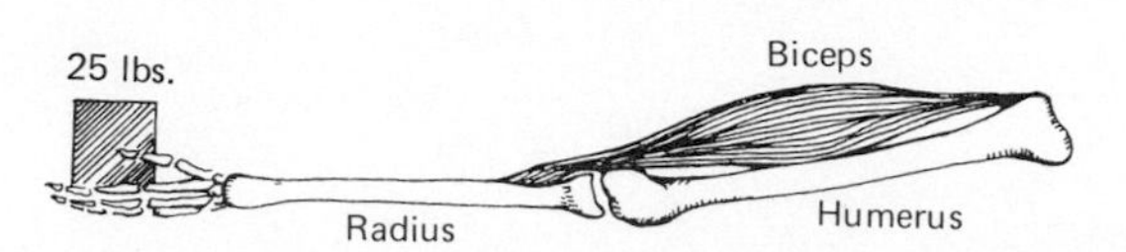

FIGURE 9-15. Arm supporting weight with entire arm nearly horizontal.

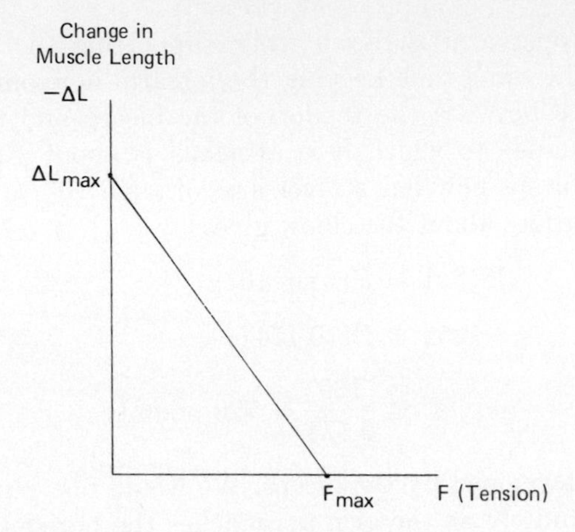

FIGURE 9-16. Shortening of isotonically contracting muscle as a function of tension in the muscle.

equation of the straight line can be written in the slope-intercept form, $y = mx + b$, as

$$\Delta L = -\frac{\Delta L_{\text{max}}}{F_{\text{max}}} F + \Delta L_{\text{max}} \tag{9-3}$$

$$\Delta L = \Delta L_{\text{max}}\left(1 - \frac{F}{F_{\text{max}}}\right) \tag{9-4}$$

The work done in moving the load through a distance ΔL is

$$W = F\Delta L = F\Delta L_{\text{max}}\left(1 - \frac{F}{F_{\text{max}}}\right) \tag{9-5}$$

The maximum work is accomplished when the derivative of W with respect to F is zero, i.e., when we have

$$\frac{dW}{dF} = \frac{d}{dF}\left(F\Delta L_{\text{max}} - F^2 \frac{\Delta L_{\text{max}}}{F_{\text{max}}}\right) = 0$$

$$\Delta L_{\text{max}} - 2F \frac{\Delta L_{\text{max}}}{F_{\text{max}}} = 0 \tag{9-6}$$

$$\frac{2F}{F_{\text{max}}} = 1 \tag{9-7}$$

$$F = \tfrac{1}{2}F_{\text{max}}$$

Equations (9-7) indicate that the muscle does the most work when the tension in the muscle is half its maximum possible tension.

References and Selected Readings

Bendall, J. R. (1969). "Muscles, Molecules, and Movement." Heinemann, London.

Bourne, G. H., ed. (1960). "Structure and Function of Muscle," 1st ed., Vol. 1. Academic Press, New York.

Brachet, J., and Mirsky, A. E., eds. (1962). "The Cell," Vol. IV, Chapter 7. Academic Press, New York.

Carlson, A. J., Johnson, V., and Cavert, H. M. (1961). "The Machinery of the Body," 5th ed. Univ. of Chicago Press, Chicago, Illinois.

Hoyle, G. (1970). How is the muscle turned on and off? *Sci. Amer.* **222,** 84 (Offprint 1175).

Huxley, H. E. (1965). The mechanism of muscular contraction. *Sci. Amer.* **213,** 18 (Offprint 1026).

Rosenbluth, J. (1965). Smooth muscle: An ultrastructural basis for the dynamics of its contraction. *Science* **148,** 1337.

Ruch, T. C., and Patton, H. D., eds. (1965). "Physiology and Biophysics," 19th ed. Saunders, Philadelphia, Pennsylvania.

Szent-Györgyi, A. (1953). "Chemical Physiology of Contraction in Body and Heart Muscle." Academic Press, New York.

Wilkie, D. R. (1968). "Muscle." St. Martin's Press, New York.

CHAPTER X Viruses

In Chapter I we considered those characteristics that distinguish between living and nonliving entities. *Viruses* are entities that occupy that obscure and fascinating region between living cells and molecules. They are far simpler, both in structure and function, than cells; yet under appropriate circumstances, they display characteristics and functions that are usually associated only with living systems. Viruses are a group of "organisms" that can be crystallized and used as a paperweight, showing no more signs of life than a stone; yet when placed in the presence of specific types of cells, they can reproduce themselves and transmit their genetic characteristics to progeny. Particles with such strange characteristics are inherently intriguing to a bioscientist. Viruses have traditionally been of interest to biophysicists for several reasons. Because of the extremely small size of viruses, biophysical methods such as ultracentrifugation, X-ray analysis, electron microscopy, and nuclear isotopic tagging techniques have played a primary role in all phases of virology since their discovery. The simplicity of viruses makes them attractive to all bioscientists for the same reasons that the hydrogen atom has been a prime target of physicists—the simplest systems seem to hold forth the greatest promise of being understood at a fundamental level.

In this chapter, we shall become aquainted with some of the properties of viruses. Subsequent chapters will consider some of the techniques that are necessary and useful in viral studies. To this end we shall describe the general nature of viruses—their possible origins and classification—and consider their structure. Viral structure is a classic example of simplicity and efficiency in a natural system. This same simplicity and efficiency is built into the genetic mechanisms of viruses, and we shall consider some of the genetic aspects of virology in describing the life cycle of a typical virus.

The Nature of Viruses

General Characteristics

Viruses are *parasites*. Biologists define parasites as organisms that live in or on the other living organisms from which they receive some advantage without compensation. Let us here concede that viruses may be called organisms. The organisms in or on which the parasites live are called *hosts*. When viruses are outside and away from an appropriate host (and, in general, viruses are quite specific for a particular host), they exhibit no characteristics associated with living things—not even *motility* (the capacity for spontaneous movement). At no time do viruses exhibit growth as we usually consider it. They do not metabolize, for they do not have metabolic enzymes. They do not respond to stimuli. In fact, the sole basis for considering the possibility of viruses being alive is their ability to reproduce, to maintain genetic continuity from generation to generation, and to alter themselves through mutation. The ability of viruses to reproduce is dependent on the fact that every type of virus contains nucleic acid—either DNA or RNA. However, no virus contains both DNA and RNA.

An infectious agent, i.e., an organism capable of infecting cells or producing disease in another organism, was at one time considered to be a virus if it was submicroscopic and *filterable* (small enough to pass through the pores of unglazed porcelain). It is for this reason that viruses are sometimes called "filterable viruses." The term no longer has any real significance, because we know that some bacteria are filterable and some true viruses are not.

Viruses have the ability to infect a specific host by attaching themselves to cells. By various means, depending on the type of virus involved, the viral nucleic acid invades the host cell. There, the viral genetic material usurps control of the host cell's metabolism, interferes with the cell's own functions, uses the cell's enzymes to construct replicas of the original infecting virus, and usually bursts the cell open, liberating many complete viruses that are each capable of infecting other host cells. Thus, the virus, in its life cycle, provides only the plan or pattern for its own welfare. The energy, building materials, and machinery for construction of viruses are obtained from host cells.

Since viruses are almost completely dependent on host cells, they cannot be *cultured* (grown or produced) on artificial media as bacteria can. Further, it should now be clear that viruses do not "grow" in the sense that there are no baby viruses, adolescent viruses, or mature viruses. Either they are in the process of being formed, or they are completely intact. The complete extracellular virus, capable of infecting a host, is called a *virion*.

Virus cultures can be obtained with sufficient purity that they can be crystallized. This fact is evidence that each virus of a "species" is morphologically identical, i.e., structurally identical. If they were not identical, it would be impossible for them to join together to form a crystal since crystals are composed of repeating units, each of which is structurally the same. The fact that viruses can be crystallized, along with the knowledge that these "crystals" can, in the presence of the proper host, reproduce themselves, tantalizes us with the question of whether or not viruses are living. In our further discussions here, let us dispense with this problem of semantics and simply deal with viruses. One of the great virologists, Lwoff, after considering this question eloquently and thoroughly concluded that "viruses should be considered as viruses because viruses are viruses."

Origins of Viruses

It is interesting to consider how viruses might have evolved. The simplicity of viruses immediately suggests that they might represent a very primitive form of life. They may have been the first aggregations of nucleic acids and proteins that formed the springboard to higher, more complex forms of life. Actually, this possibility does not seem likely, because viruses as we now know them place such restrictive demands on their environment, and it is difficult to imagine a feasible life cycle for such a simple system before cellular systems existed.

More probably viruses represent the evolutionary products of degenerate cells, which by mutating over long periods of time have lost one enzyme system after the other until they become completely dependent on the enzymes of a host. Different types of viruses may have evolved from different kinds of cells. On the other hand, it could be that only one cellular degeneration process took place, and all the currently present viruses may have evolved from a single ancestral virus. Although the origin of viruses is interesting to speculate upon, it is not likely that an experimental test can be devised that will provide the answer.

Some Medical History of Viruses

The general public more often than not thinks of microorganisms as "germs," the bad guys of the biological world. Bacteria are, for the most part, badly maligned in this respect. There is no more reason to consider bacteria in terms of a few *pathogenic* (disease-producing) types than to judge the human species in terms of its criminal element. Most bacteria are not only helpful but quite necessary to the other plant and animal life on earth. Viruses, on the other hand, provide no known benefits to the living forms they infect. They are either pathogenic or at best *benign* (of mild character).

Viruses have plagued mankind for thousands of years, though it is only recently that the culprits have been identified and labeled. There is evidence from Egyptian mummies that *poliomyelitis* (infantile paralysis) crippled people over 5000 years ago. *Smallpox* apparently afflicted Rameses, an Egyptian pharoah who died about 3000 years ago. Smallpox epidemics have killed untold numbers of people throughout the history of mankind. During the Renaissance, smallpox was an accepted part of life, much as the common cold is accepted today, and most people in Europe were disfigured to some extent by the disease. *Yellow fever* was prevalent for three centuries in tropical and temperate zones of the earth; its notoriety was underscored when it delayed construction of the Panama Canal. Today there are a large number of diseases that are recognized as viral infections: rabies, influenza, measles, mumps, chicken pox, shingles, warts, etc.

The control of viral diseases began with two major medical conquests by scientists who were not aware of the nature of the organisms involved. Jenner (in 1798) found that material obtained from lesions in cows having cowpox, a disease that produces no serious effects in humans, would confer immunity against smallpox upon humans if scratched into the skin. This process, called *vaccination*, has virtually eliminated smallpox from the earth.

In 1884, Louis Pasteur began studies of rabies, a disease almost always fatal in humans but rather common and mild in many mammals. He assumed that the microorganism responsible for rabies (which we now know to be a virus) was affecting the central nervous system. Therefore, he injected virulent (actively infective) viruses taken from a diseased dog into the brain of a rabbit. When the rabbit would die, he would take material from its brain, inject part of it into another rabbit and use part of it to infect another dog. When the next rabbit died, the procedure was repeated. With each subsequent passage of the virus, it caused more severe disease in the rabbits but less severe disease in the dogs. Thus, Pasteur was able to alter the virus, presumably by mutations, as it progressed through several generations. By injecting altered viruses into a human, Pasteur found that immunity to rabies was conferred upon that person. Similar procedures that produce "live" vaccines have since been prepared and used successfully in the treatment of poliomyelitis, measles, and yellow fever.

Modern medicine is still pitted against viral infections. Many aspects of the common cold are assumed to be due to viruses, and various strains of influenza viruses periodically strike down thousands of victims.

Classification and Nomenclature

The study of viruses is one of the few areas of biological study that have developed without an accepted taxonomy. Classifications in most areas of life science are based on morphology, function, physical properties, environmental associations, and so on. Many viruses are simply named for the disease that they cause, but in general the viruses have been named capriciously. The result is that no accepted system of classification or nomenclature has been established.

Some viruses have been named for their discoverer, as in the case of Rous tumor virus. Many are named for the geographical area of occurrence or discovery; Coxsackie, Fiji disease, and Sendai are examples. Some viruses have strange and exotic names: o'nyong-nyong, zika, spondweni, and—everybody's favorite—snotsieke, a virus causing disease in the gnu. At the other extreme, the viruses that attack bacteria have been the object of considerable research and have very businesslike names: T4, P22, K12(λ) and ϕX174.

A few "classes" of viruses have come to be accepted, at least within the jargon of virologists. We should be acquainted with them in order to maintain communication with the virologists.

Viruses are often categorized as *animal viruses* and *plant viruses*, referring, of course, to the hosts that are attacked. Animal viruses are the more clearly classified. We find among animal viruses the *arboviruses* (*ar*thropod *bo*rne viruses), which are transmitted by insects and include the viruses of yellow fever, dengue, and many forms of encephalitis. The *enteroviruses* mainly affect the gut and include many varieties of poliomyelitis viruses and the ECHO viruses. (One begins to be impressed by the propensity that virologists show for acronyms. ECHO, for example, stands for *e*nteric *c*ytopathic *h*uman *o*rphan. The "orphan" refers to the fact that, at the time of naming, the virus did not appear to be related to a known disease.) *Herpesviruses* are a variety of rather large viruses that produce cold sores. *Adenoviruses* got their name because they were first associated with diseases of the adenoids and include some of the types now known to produce cancer in mice. *Myxoviruses* (slime viruses) include the mumps and influenza viruses. They have in common an affinity for mucin (a protein associated with mucous membranes). *Poxviruses*, of which the smallpox virus is an example, are large viruses that form relatively large inclusion bodies (stainable masses) in the cells they infect. *Papovaviruses* are viruses that contain DNA and cause warts and wartlike bodies to appear on the skins of their host animals. *Reoviruses* (another acronym—*r*espiratory, *e*nteric, *o*rphan) are not clearly associated with any disease but are categorized by certain cytological effects they produce in tissue cultures in the laboratory.

One final category is a group of very small viruses containing RNA. The name of this group is a monument to the cuteness (maybe even to a form of sense of humor) among scientists. The *picornaviruses* include a number of viruses in the categories already mentioned. "Pico" is a prefix that suggests the viruses are small (pico = 10^{-12}). The letters in "picorna" are chosen to remind us that *p*olio viruses are in the group, that all members of the group are *i*nsensitive to ether, and that *C*oxsackie, *o*rphan, and *r*hino (rhino is Greek for nose, referring to those viruses causing the symptoms of the common cold) viruses are in this group. The last three letters in picorna point out that these viruses all contain RNA. Fortunately for the student, most names found in the life sciences do not contain so much information.

Plant and bacterial viruses have not been classified in any systematic way, though a few attempts have been made and ignored by the scientific community. We may hope that, as more facts are acquired in these areas, a systematic and universally acceptable organization will be forthcoming.

Structure of Viruses

We have already emphasized the exceedingly small size of viruses. The known viruses have diameters lying between 100 and 2500 Å. Not only is an electron microscope required to see most viruses, but very modern and refined ones are necessary to give any details of structure.

In 1956, before high-resolution electron microscopy was available, Crick and Watson, whom we have encountered (Chapter III) as the scientists responsible for clarifying the structure of DNA and elucidating its role in genetics, turned their attention to the question of what the structure of viruses must be like. They reasoned as follows:

1. Being very small, viruses can contain only limited genetic information.
2. Other than the genetic material, either DNA or RNA, the complete virus must be composed of very few different kinds of molecules, perhaps only one.
3. The limited variety of protein molecules available requires that the virion structure be formed from repeating units.
4. In order to minimize the energy of assembly, as well as the information necessary to direct construction, the repeated subunits must fall spontaneously into place to form a stable structure.
5. There are only a few possible ways in which repeating subunits can form roughly spherical shells.

The predictions of Crick and Watson (1956) are now known to be substantially correct for viruses that exhibit cubic symmetry, i.e., those whose geometric regularity (which may be far more complex than a cube) causes them to crystallize into a form that contains repeating cubic units. Other basic symmetries are found among viruses. Some viruses crystallize into *helical symmetry*, some have combinations of symmetries, and some display no symmetry at all; those in the latter group are called *complex*. We shall see that the rationale of Crick and Watson beautifully describes the structural facts of viruses.

The terminology of virus structure is fairly well established, and it may be helpful to define some of the parts of the virion at this point. The protein shell that encloses the nucleic acid in most virions is termed the *capsid*; if the nucleic acid is included, the combination is called the *nucleocapsid*. The structural subunits that form the capsid are called *capsomers* and are not necessarily made up of a single molecule. Some viruses, like those in the herpes group, are surrounded by a membranous envelope, called a *peplos*, whose subunits are *peplomers*.

Viruses with Cubic Symmetry

The crystallized viruses that show cubic symmetry when examined by X-ray diffraction usually appear individually as approximately spherical

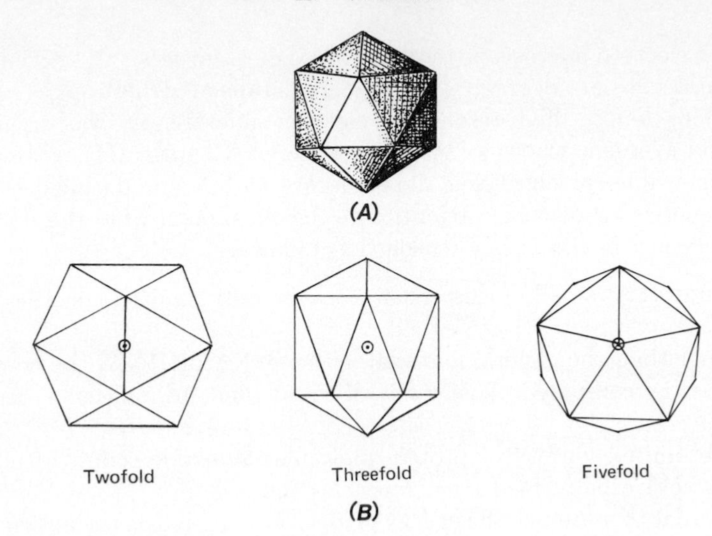

FIGURE 10-1. (A) A view of a regular icosahedron, a twenty-sided closed figure. (B) The three symmetry axes of a regular icosahedron.

shapes. If examined under very high magnification, however, each one is seen to be a *regular icosahedron*. A regular icosahedron, which is shown in Figure 10-1A is a closed solid figure with 20 faces of equal surface area, 12 vertices, and 30 edges. Each face is an equilateral triangle. The three *symmetry axes* of a regular icosahedron are illustrated in Figure 10-1B. A symmetry axis is an axis through a body such that rotation of the body through an angle θ (where $\theta \leq 180°$) produces a configuration equivalent to (indistinguishable form) the original configuration. An axis of symmetry of a body is said to have n-fold symmetry if rotation about that axis by $360°/n$ results in an equivalent arrangement. The icosahedra of Figure 10-1B are viewed along axes of twofold, threefold, and fivefold symmetry.

The icosahedral shape of viruses with cubic symmetry was strikingly demonstrated by a clever use of the electron microscope by Smith and Williams (1958). Figure 10-2 shows two photographs, one of a *Tipula* iridescent virus that was shadowed (see Chapter XI) from two different directions; the second photograph shows a large icosahedron photographed with light from two different directions. The shadow pairs are identical in the two photographs, showing both a blunt and a pointed shadow. This evidence leaves no doubt that the shape of the virus is icosahedral.

The triangular faces of the icosahedral viruses are made up of subunits, which are the capsomers. The shapes of the capsomers are restricted by geometric considerations to be either hexagonal (six-sided) or pentagonal (five-sided). More accurately, it is not the actual shape of the capsomer

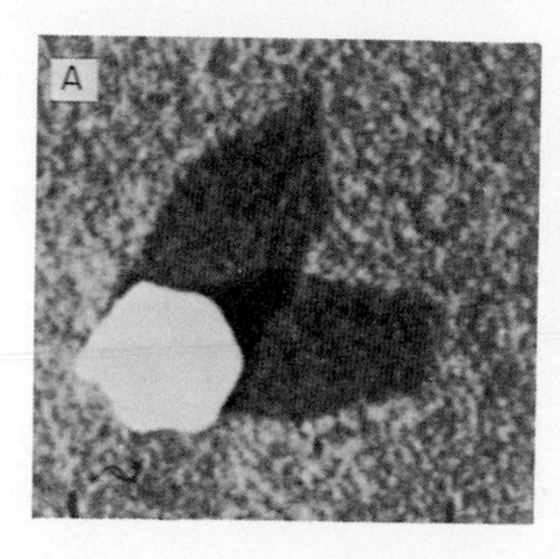

FIGURE 10-2. The icosahedral shape of the *Tipula* virus. (A) The virus preparation was photographed after having been shadowed by evaporated metal, rotated slightly, and shadowed again (see Chapter XI for shadowing technique). (B) A photograph of an icosahedron lighted from two directions. The shadows in each case are identical. From Smith and Williams (1958).

that is restricted but the number of neighboring capsomers it can be adjacent to. A capsomer itself may be spherical, but it may have six neighbors if it is on a face or edge or five if it is at a vertex. This can be seen in Figure 10-3, a model of an adenovirus that has the icosahedral shape. This particular virus has 252 capsomers of two different kinds—240 hexagonal and 12 pentagonal. The geometric plan of one of the 20 faces is shown in Figure 10-4.

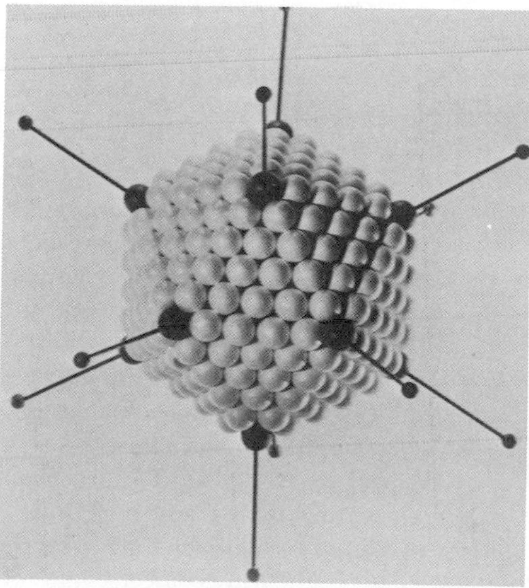

FIGURE 10-3. A model demonstrating the structure of the adenovirion. From Valentine and Pereira (1965).

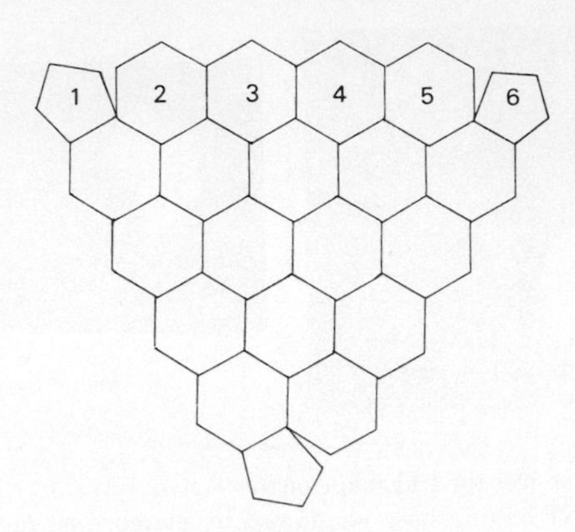

FIGURE 10-4. A capsomer of a virus. This diagram shows the arrangement of the hexagonal and pentagonal structural subunits in one capsomer. The complete capsid will contain 252 subunits ($x = 1$, $n = 6$).

A general prescription can be given for any possible configuration made up of hexagonal and pentagonal subunits. If n is the number of capsomers along an edge ($n = 6$ in Figure 10-4) and x is a positive integer, the total number of T of capsomers that can comprise a virion is given by

$$T = 10x(n - 1)^2 + 2 \tag{10-1}$$

In the example of Figure 10-4, n is 6, x is 1, and $T = 252$. There are many viruses with cubic symmetry, but all of them are bound by the restrictions of Equation (10-1). There are some known viruses that contain the minimum possible number of capsomers, 12 ($n = 2$, $x = 1$); others are known containing as many as 812 ($n = 10$, $x = 1$).

It now seems reasonable to assume that the viruses having cubic symmetry need only contain two kinds of protein, corresponding to the two types of positions (hexagonal or pentagonal) they assume on the surface of an icosahedron. Note that this fact is in agreement with the prediction of Crick and Watson.

In addition to adenoviruses, a number of other viral classifications display cubic symmetry. Herpesvirions are icosahedra. *Herpes simplex*, the one responsible for cold sores has 162 capsomers ($n = 5$, $x = 1$). Both adenoviruses and herpesviruses contain DNA within the capsid. The configuration of the DNA, whether folded, coiled, or interlaced among the capsomers is a topic of considerable research effort. The papovaviruses are DNA-containing viruses with icosahedral geometry, and some plant viruses that contain RNA, like turnip yellow mosaic virus (TYMV), have this geometry.

Viruses with Helical Symmetry

Many viruses, especially plant viruses, are rod-shaped bodies with helical symmetry. Some of the myxoviruses, such as the influenza virus and the mumps virus, are in this category.

The most intensively studied and best understood member of this group, however, is the tobacco mosaic virus (TMV), and the detail in which its structure is known is impressive. TMV has played a starring role throughout the history of virology. It was the object of the experiments in which the Russian botanist Ivanovski (in 1892) demonstrated the presence of a filterable microbe that was infectious. TMV was the first virus to be crystallized and the first virus studied by X-ray diffraction.

TMV is a hollow rod composed of a helical coil of RNA, to which are attached banana-shaped subunits of protein (see Figure 10-5). Each virion is 3000 Å long and 190 Å in diameter. The hole down the axis of the rod is about 40 Å in diameter, large enough for water and other small molecules to transit the tube. There are 2130 protein subunits, each of which is a single molecule of molecular weight 17,530. The complete amino acid sequence of the protein has been determined.

The helix of a TMV virion has a pitch of 23 Å, with $16\frac{1}{3}$ subunits per turn, or 49 subunits in three full turns. The single molecule of single-stranded RNA that is wound around the axial hole contains 6390 nucleotides, three for each protein subunit. The RNA chain in TMV has a molecular weight of 2.06×10^6.

Infective TMV rods can be dissociated chemically, i.e., the subunits can be stripped away from the RNA. If the RNA is then removed, the

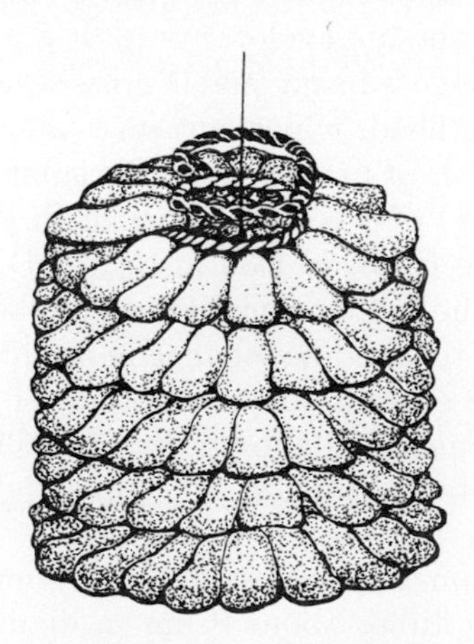

FIGURE 10-5. A schematic drawing of the tobacco mosaic virus (TMV) showing the internal coil of RNA and the protein units extending outward from the RNA.

protein subunits can be crystallized, just as could the virion. Upon crystallization, however, the protein subunits alone form disks instead of a helix, though the disks have a tendency to stack like Lifesavers so that a rodlike arrangement with a hole down the axis is sometimes observed. Each disk so formed has an integral number (16) of subunits, so that three disks contain 48 subunits instead of the 49 found in three turns of the helix of the virion. If these protein disks are dissociated and mixed again with the RNA that has been separated, the complete infective virion will reform.

Viruses with Complex Symmetry

Some viruses do not exhibit any of the symmetry properties we have discussed. Others have combinations of these symmetries, i.e., they have parts that are helical or parts that are icosahedral. These are complex viruses and are often given such descriptive terms as unsymmetrical, tadpole-shaped, or bullet-shaped.

The poxviruses are generally very large and without regular symmetry. Some (vaccinia) are about 2500 Å in diameter, the largest known viruses. They appear to have a core of nucleic acid wrapped in layers of protein.

A carefully studied group of viruses, known unromantically only as *T-even phages*, i.e., T2, T4 and T6, are among the more interesting complex viruses. "*Phage*" is an abbreviation of *bacteriophage*, a name applied to viruses that infect or "feed on" bacteria. The T-even phages are usually described as being tadpole-shaped, but to a nonzoologist they look almost exactly like a rocket vehicle. Figure 10-6 shows the configuration of T2 phages at two phases of the life cycle.

The head of the T2 phage is hexagonal in cross section and is capped by a hexagonal pyramid. The head, which measures about 950 Å by 650 Å, is made up of a large number of protein molecule subunits, each of which has a molecular weight (MW) of 8×10^4. The protein sheath of the head surrounds a single molecule (MW $= 1.30 \times 10^6$) of DNA, the configuration of which is not known. The protein sheath forming the head can be ruptured by subjecting the phage to a sudden change in osmotic pressure. The DNA will then be lost to the surrounding medium, and the remaining phage-minus-DNA is called a *ghost*. Phage ghosts are still capable of attaching themselves to host bacteria, though of course they are not capable of replicating.

The tail of a T-even phage is composed of a number of parts arranged about a central core, or tube, about 8 nm in diameter. The core is surrounded by a sheath, which, when extended as in Figure 10-6a, is composed of 24 rings. Each ring is made up of six protein subunits of one size and six

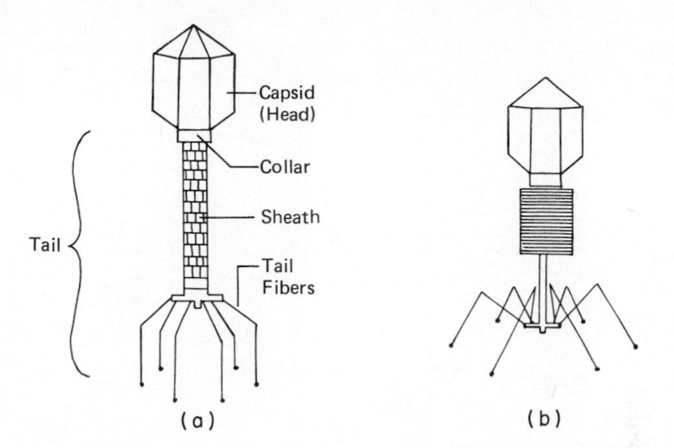

FIGURE 10-6. The T2 bacteriophage. (a) The intact phage with uncontracted sheath. (b) The phage after the sheath has been triggered.

of a larger size. This sheath is capable of contracting, the only example of purposeful internal motion in all of virology. Upon contraction, as in Figure 10-6b, the small subunits of the tail sheath merge into larger ones so that the contracted sheath has 12 rings of 12 subunits each. The volume of the sheath is conserved during contraction, i.e., the volume is the same before and after contraction. The energy for contraction is supplied by 144 ATP molecules carried in the phage—one ATP molcule for each subunit.

The six tail fibers are organs of attachment of the phage to the host. Once attached, the end of the core is lowered through the surface of the host cell, and the DNA from the head is extruded through the core as if through a syringe. The attachment process will be considered further in a subsequent section.

The Replication Cycle of Viruses

The details of the process of virus replication are available to us thanks to an experimental procedure developed by Max Delbrück (1940). The procedure is called a *one-step growth* experiment. Delbrück mixed a known number of T-even phages with a known number of potential host cells (bacteria). The mixture was incubated for a few minutes known as the *adsorption period,* during which the phages attach themselves to the host cells. The mixture was then greatly diluted, more than a thousandfold, so that the probability of further contact between phages and cells was made negligible. (Contact is made as a result of diffusion, aud dilution of the mixture reduces the concentration of both types of particles in the

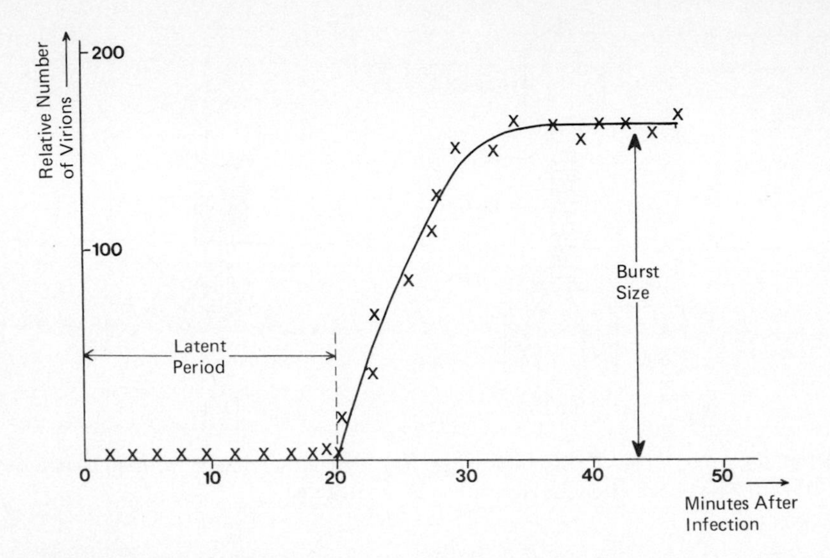

FIGURE 10-7. The one-step growth curve for a synchronized phage system. The curve represents the relative number of virions in a typical experiment as a function of time after infection of a host.

medium.) The result of this procedure was *synchronized infection*, i.e., all the individual infections occurred at essentially the same time. After the dilutions, samples of the medium were taken at regular intervals of time and *assayed* for phage content, i.e., the number of phages per unit volume of the medium were measured. The one-step growth process is, of course, applicable to studies of many kinds of viruses.

In his studies with T-even phages, Delbrück found that the phage concentration in the medium remained constant for about 20 minutes. This is called the *latent period,* which was followed by a rapid rise to a higher level where the concentration remained constant as long as no further bacteria were made available or if no readsorption could take place. Figure 10-7 is a graph of a typical one-step growth experiment with the concentration of phages normalized to an initial concentration of 1 phage per milliliter of medium. The ratio of the final concentration to initial concentration is called *burst size* for reasons that will become apparent. The characteristics of the one-step growth curve vary with different cell-virus systems and are dependent on temperature (lowing the temperature increases the latent period, for example).

By using the one-step technique, virologists have been able to detail the events in the replication cycle of viruses. Samples from the medium can be analyzed biochemically at any stage of the replication process, and the events of the process can be deduced. Since 1940 many clever experiments have provided us with a surprisingly complete picture of the viral replication process. We shall review some of the results here in an abbreviated form, using the T2 phage as a typical example of the replication process in general and of the replication of bacterial viruses in particular.

Adsorption and Penetration

The infective cycle of a virus begins with the coming together of a virion and a susceptible cell. The virions are sufficiently small that they are in constant *Brownian motion*, i.e., they are moved around by collisions with molecules of the medium that are in thermal motion. In other words, the virions diffuse until they make contact with a host cell.

The rocket-shaped phages attach themselves to the host cells by their tail fibers. This attachment is facilitated by the random diffusion process. Since the tail of a virion is less massive then the head, the tail moves about more rapidly than the head and therefore has a greater probability of making contact with a cell. The tail fibers of the phage attach themselves at very specific *receptor sites* on the cellular membrane. (A cell's surface may have different receptor sites for different phages, and a species of bacterium may undergo mutation that eliminates receptor sites for a particular phage while maintaining receptor activity for other types.)

The number of phages that can be adsorbed by a single cell is limited only by the space available on the surface of the cell membrane. Roughly 300 T-even phages can be adsorbed by a single bacterial cell of *E. coli*. The ratio of the number of adsorbed phages to the number of cells in a given medium is called the *multiplicity of infection*. In experimental practice, multiplicity of infection is determined by measuring the fraction of cells that are not infected and using the fact that, to a very good approximation, the phages are distributed among the cells according to the Poisson distribution,

$$P(n) = \frac{m^n e^{-m}}{n!} \tag{10-2}$$

where $P(n)$ is the fraction of cells infected by n virions. The absorption multiplicity m is determined from the $n = 0$ term of Equation (10-2):

$$P(0) = e^{-m} \tag{10-3}$$

where $P(0)$ is the fraction of cells that are not infected. Experimentally $P(0)$ is measured by plating equal volumes of cell suspensions, before and after infection, onto a nutrient culture medium. Counts of the number of colonies that grow reflect the number of cells that were not infected, i.e., $P(0)$. Then m can be calculated from Equation (10-3).

There is evidence that the attachment of the tail fibers to the cellular surface is initially accomplished by ionic bonding. Certain negatively charged areas on the cell membrane attract specific regions of the tail fiber that have a positive charge. The specificity of the adsorption process presumably depends on the complementary patterns of charge that exist on the cell and virus tail fibers.

The phage tail carries an enzyme called *lysozyme*, which, when brought into contact with the bacterial cell membrane, erodes a hole through it. If a large number of phages are attached to a single bacterium, there may be too many holes for the cell to repair; in such a case the cell dies before the viruses can use the cellular material to replicate themselves. Delbrück called this phenomenon "lysis from without," i.e., breakdown of the cell from an external source, a process in which viruses destroy cells without reproducing themselves.

Penetration is effected when the sheath about the phage tail contracts and the DNA in the head of the phage is extruded through the tail core into the body of the host cell. The contraction of the sheath is probably triggered by chemicals released from the cell through the hole in its membrane opened by the lysozyme. The DNA is accompanied into the host by a very small quantity of protein that is mixed with the DNA. The remainder of the virus protein, i.e., the phage ghost, remains ouside the cell membrane and does not participate further in the replication process.

The Latent Period

The one-step growth experiment does not tell us what is taking place during the latent period, the time after infection has taken place but before an increase in the number of viruses begins. In 1952 a method was devised for breaking open infected bacteria at different times during the latent period. It was found that the latent period can be divided into two phases lasting approximately equal times. The first half of the latent period is called the *eclipse*, because during that time the cells contain no virions. During the second half of the latent period, called the *maturation period*, complete virions accumulate within the bacterial cell. The virions present in the bacterial cells during the maturation period are not, of course, detectable in the one-step growth experiment.

ECLIPSE

In the eclipse phase, the phage DNA that has been injected into the bacterium assumes control of the chemistry of the cell. The DNA first acts as a template for RNA that produces enzymes that wrest control of chemical activity from the bacterial DNA. The phage DNA then begins replicating—and not only does the original phage DNA replicate, but the replicas replicate too. During the production of the numerous strands of phage DNA, the bacterial DNA is degraded by enzymes produced at the direction of the phage DNA. The pool of phage DNA that accumulates during eclipse contains DNA from three sources: the DNA made from nucleotides from the cytoplasm of the cell, the bacterial DNA, and the DNA of the original infecting phage.

A few minutes after infection the phage DNA that has accumulated begins to direct the synthesis of proteins (using, of course, the mRNA and ribosomes of the host cell). Toward the end of the eclipse phase, phage parts are synthesized and accumulated.

MATURATION

The maturation phase of viral replication starts with the *condensation* of the accumulated DNA, i.e., the phage DNA, which has been produced in long strands, begins to form into compact balls. It is thought that the very small quantities of protein found intermingled with the DNA of phage heads are probably responsible for the condensation. The condensed DNA then serves as a framework around which the protein shell of the phage is formed from the accumulated protein subunits.

There is formidable experimental evidence that the protein parts of a phage are assembled randomly during maturation. It is found, for example, that when bacteria are infected by both T2 and T4 phages, four kinds of phages are produced. Figure 10-8 illustrates the *phenotypic mixing* that occurs. The four types of progeny phages include not only T2 and T4 virions, but also include phages with genetic material of T2 and adsorption characteristics of T4 (that is, a T4 tail) and phages with T4 DNA and T2 tail parts. If these phages with mixed phenotypes are used to infect further bacterial hosts, but permitting only one phage to infect one cell, the progeny of the second infection produce the phenotype corresponding to genotype, i.e., each phage produced in the second replicative cycle has a T2 tail if the infecting DNA was T2 and similarly for the T4. Obviously, since only one kind of DNA infects the cell, it only codes for its own kinds of proteins.

Other experiments have shown that assembly of phage parts can occur *in vitro* (literally, in glass, i.e., in a test tube as opposed to *in vivo*, in the living system). These experiments show that assembly takes place when the system has been completely freed of all enzymes. This implies that assembly takes place spontaneously during maturation. All the facts of maturation seem to suggest that the predictions of Crick and Watson, cited earlier in this chapter, were entirely accurate.

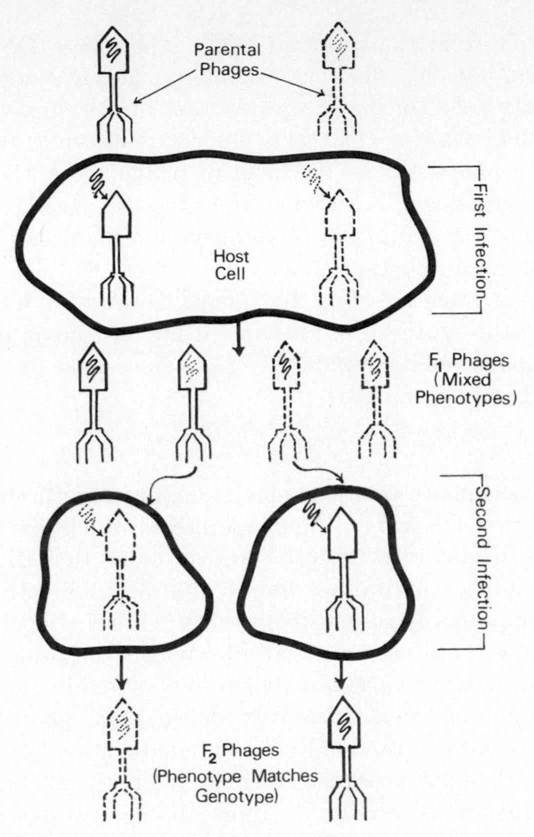

FIGURE 10-8. Phenotypic mixing. Infection of a host cell with two types of phages results in four types (including mixed parts) in the F_1 generation. A second infection by the virions of mixed phenotype results in virions whose phenotype match the genotype. This experiment demonstrates that the protein parts of a phage are randomly assembled during maturation.

Lysis

The replicative cycle of bacterial phages is completed when lysis occurs. Lysis is the breaking down of the cellular membrane of the host. During the latent period the host cells begin to swell, and the swelling continues as the replicative process continues until the cell literally explodes. The explosion releases the virions formed during maturation. Each progeny virion is then free and capable of infecting other host cells. Typically, between 100 and 200 virions are produced by the infection of a single cell.

The swelling and bursting of the host cell is due to osmotic pressure that results from alteration of the permeability of the cellular membrane. This alteration of permeability is accomplished by an enzyme—the same enzyme that is carried on the phage's tail. The synthesis of this lysozyme inside the host cell is directed by the phage DNA. Thus, the same enzyme that causes "lysis-from-without" also causes "lysis-from-within," which provides the means of escape for the phage that have been replicated. Any excess viral DNA that is not surrounded by a protein sheath is freed into the medium upon lysis of the host cell.

Replication of RNA Viruses

Some viruses do not contain DNA; such viruses are dependent on RNA to direct the replicative activities of the virus. We have learned in the central dogma (Chapter III) that RNA cannot be transcripted except through the use of DNA as template. In normal cells this is the case, and the production of RNA from the DNA template is mediated by RNA polymerase, an enzyme that is DNA-dependent, i.e., the enzyme action of RNA polymerase normally takes place because of its association with and attachment to DNA. In viruses containing RNA (and, of course, no DNA), the RNA is replicated with the aid of an enzyme called *replicase* that is an RNA-dependent RNA polymerase. In effect, RNA can serve as its own template with the assistance of replicase.

When RNA enters a host cell from an infecting virus, it begins to produce replicase using the host ribosomes to form the enzyme from amino acids in the host's cytoplasm. This enzyme in turn produces more viral RNA using the host's nucleotides. Each RNA molecule acts as messenger RNA in the production of proteins. Neither the host DNA nor the host RNA is involved in forming the progeny viruses.

RNA viruses are, in general, even more efficient than the DNA types. The proteins that they produce not only serve as the subunits for forming capsids but are responsible for causing lysis of the cell. Thus, the viral RNA needs only one gene to serve both purposes. RNA viruses therefore qualify as the simplest, most efficient organisms in nature.

References and Suggested Reading

Brooks, S. M. (1970). "The World of Viruses." A. S. Barnes, New York.

Crick, H. F. C., and Watson, J. D. (1956). Structure of small viruses. *Nature* (*London*) **177,** 473.

Delbrück, M. (1940). *J. Gen. Physiol.* **23,** (5), 643.

Fraser, D. (1967). "Viruses and Molecular Biology." Macmillan, New York.

Goodheart, C. R. (1969). "An Introduction to Virology." Saunders, Philadelphia, Pennsylvania.

Pollard, E. C. (1953). "The Physics of Viruses." Academic Press, New York.

Sigel, M. M., and Beasley, A. R. (1965). "Viruses, Cells, and Hosts." Holt, New York.

Smith, K. M., and Williams, R. C. (1958). *Endeavour* **17,** 12.

Valentine, R. C., and Pereira, H. G. (1965). *J. Molec. Biol.* **13,** 13.

UNIT III Biophysical Techniques

CHAPTER XI Techniques Using Wave Phenomena

In this chapter we shall be concerned with some of the techniques of biophysical study that involve wave phenomena—light waves, X-rays, ultraviolet and infrared radiation, and the wave characteristics of electrons. A great portion of the major advances in the understanding of living systems can be attributed to the development of those techniques that permit us to visualize the microscopic detail of organisms, cells, and cellular organelles that cannot be observed with the unaided eye. *Microscopy* is the field of physics concerned with the visualization of small objects. We shall consider some of the microscopic techniques that use both light and electrons. *X-ray diffraction* techniques have played a necessary and successful role in revealing the detailed structures of biological molecules. A brief discussion of X-ray diffraction will be included. Finally we shall consider one of the techniques used to identify biological chemicals and investigate chemical changes that occur in biological systems. This technique, *spectrophotometry*, involves measurements of the absorption characteristics of molecules primarily in the visible and ultraviolet regions of the electromagnetic spectrum.

Microscopy

A microscope is an optical device that effectively magnifies the size of an object to be observed. Without a microscope, an observer views an object in greatest detail by bringing the object as close to his eye as the eye can focus. Under these conditions, the focused image on the retina of the observer's eye is as large as possible. We have learned in Chapter VII that the minimum resolvable visual angle is about one minute of arc. At the near point of the eye (about 25 cm from the cornea), this means that the eye can resolve, or see as separate, objects that are about 0.03 mm apart. A microscope is a lens or system of lenses that effectively permits bringing objects closer to the eye while permitting the eye to focus the image on its retina. Some of the more powerful microscopes provide magnifications so great that the limiting factor in resolution is not visual acuity of the eye but the physical limitations imposed by the wave nature of the illuminating radiation.

Light microscopes generally use visible light to illuminate the object for viewing, although ultraviolet light has been used in some instances. In recent years high-energy electrons have been used to increase significantly the effectiveness of microscopes. We shall consider both light and electron microscopes, but first we shall review the optical properties of lens systems that are used in light microscopes. Most of the principles that apply to magnification and to the ultimate limitation of the microscopes are applicable to both types.

Light Microscopy

SIMPLE MICROSCOPE

The simplest optical microscope is a single convergent lens. For this simple microscope, the object is placed at or within the focal length of the lens. (The general terminology associated with lenses is in Chapter VII.) The constructions of Figure 11-1A and 11-1B show the comparison of images formed on the retina of a viewer's eye with and without a simple microscope. An object of height h placed 25 cm from the eye subtends an angle θ_1 at the unaided eye of Figure 11-1A. The distance from the lens to the eye is assumed to be negligible. A similar object viewed through a lens of focal length f placed as shown in Figure 11-1B produces a *virtual image* with height h' at a distance I from the lens. The image subtends an angle θ_2 at the eye. A *virtual image* is one that is formed by light rays that do not actually pass through the image (whereas a *real image* is formed by rays that pass through it).

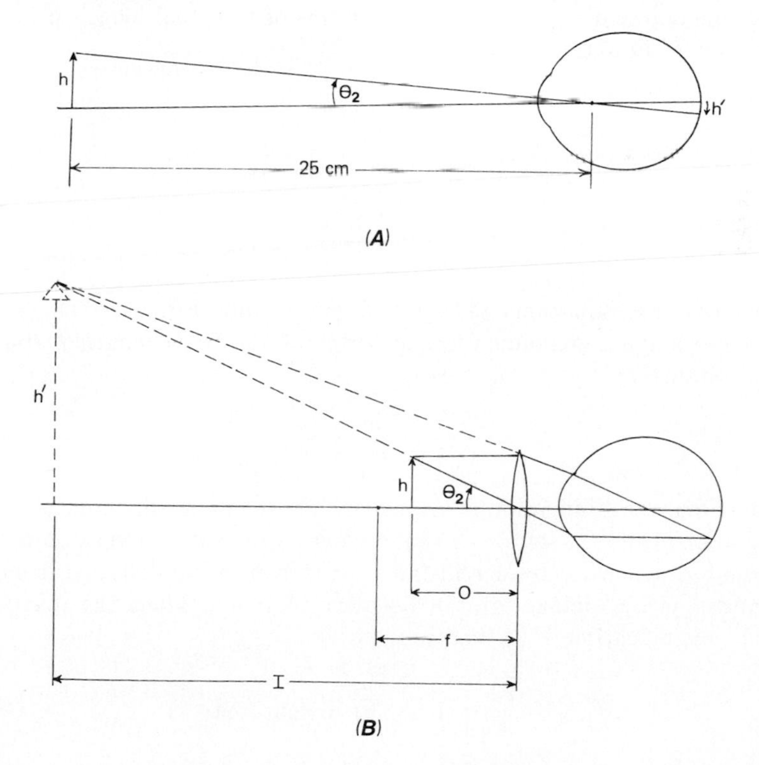

FIGURE 11-1. Geometric constructions showing the formation of images on the retina (A) with the unaided eye and (B) using a simple microscope.

The angular magnification M of a single lens used as a simple microscope is defined as the ratio of the angles θ_2 and θ_1, or

$$M = \frac{\theta_2}{\theta_1} \tag{11-1}$$

and since only small angles are involved, we may approximate θ_1 by $\tan \theta_1$ and θ_2 by $\tan \theta_2$. Then Equation (11-1) becomes

$$M = \frac{\tan \theta_2}{\tan \theta_1} = \frac{h'/I}{h/25} = \frac{25}{I}\frac{h'}{h} \tag{11-2}$$

The ratio of image size to object size is seen from the similar triangles of Figure (11-1b) to be

$$\frac{h'}{h} = \frac{I}{O} \tag{11-3}$$

and O, the object distance, is found in terms of the focal length using the lens formula [Equation (7-7)]:

$$\frac{1}{O} + \frac{1}{I} = \frac{1}{f}$$
$$\frac{1}{O} = \frac{1}{f} - \frac{1}{I} \tag{11-4}$$

By substituting Equations (11-3) and (11-4) into Equation (11-2), we obtain the angular magnification in terms of the focal length f and the image distance I:

$$M = \frac{25}{I}\frac{h'}{h} = \frac{25}{O} = \frac{25}{f} - \frac{25}{I} \tag{11-5}$$

The maximum magnification is achieved when the image distance is 25 cm. In Equation (11-5) I is set to -25 cm because of the sign convention used in the lens formula whereby a real image is at a positive distance from the lens and a virtual image is at a negative distance. Then the maximum angular magnification $M_{\max}$ becomes

$$M_{\max} = \frac{25}{f} + 1 \qquad (f \text{ in centimeters}) \tag{11-6}$$

The most comfortable viewing of an image is accomplished by placing the image an infinite distance away so that the eye is relaxed while observing the image. As $I \to \infty$ in Equation (11-5), the angular magnification for most comfortable viewing M_{com} becomes

$$M_{\text{com}} = \frac{25}{f} \qquad (f \text{ in centimeters}) \tag{11-7}$$

Notice that for strong lenses in which $f \ll 25$ cm, Equations (11-6) and (11-7) are nearly equal, or $M_{\text{max}} \cong M_{\text{com}}$.

The magnifying power of a microscope is usually expressed as the value of M in "number of times." For example, a lens of 2.5 cm focal length is said to have a magnification of 10 ×, meaning that the lens provides an increase in the angle subtended at the eye by a factor of 10.

COMPOUND MICROSCOPE

Two or more converging lenses can be arranged to form a *compound microscope*. The basic arrangement of two lenses as a compound microscope is shown in Figure 11-2. It consists of a very short focal length lens called

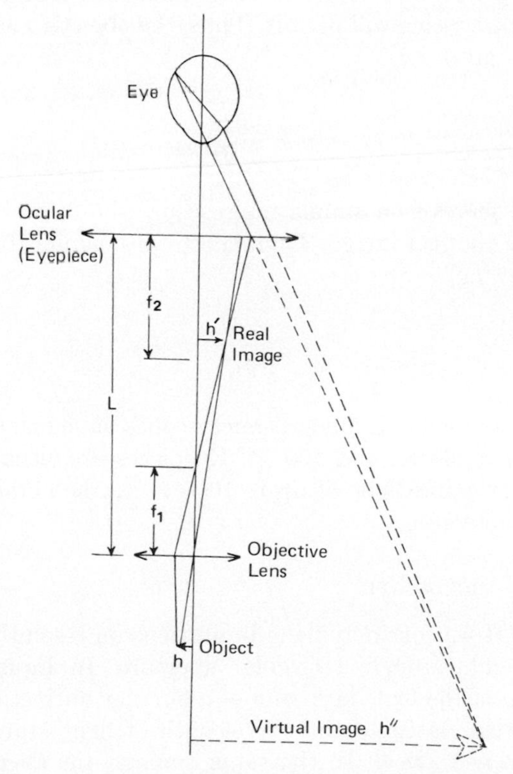

FIGURE 11-2. A schematic diagram of a compound microscope.

the *objective lens* placed close to the object and a short focal length lens called the *ocular lens* (or sometimes the eyepiece) placed near the eye.

The object is positioned just outside the focal length of the objective lens so that a real primary image is formed between the two lenses. The relative positioning of the two lenses is then adjusted so that this real image is slightly inside or coincident with the focal length of the eyepiece. The eyepiece then functions as a simple microscope, forming an enlarged virtual image using the primary real image formed by the objective lens as its object.

The lateral magnification m (approximately equal to the angular magnification) of a real image was given in Equation (7-9): $m = -I/O = h'/h$, where h is the object size, and h' is the size of the real image. I is the distance of the real image from the objective lens; O is the distance of the object from the same lens. If we assume that the object is very close to the focal point of the objective lens, we may approximate the object distance O as the focal length f_1 of the objective lens. Further, for all practical purposes, the image distance for the objective lens is made as long as the length L of the microscope housing will permit. Thus, the objective lens has a lateral magnification m given by

$$m \cong \frac{L}{f_1} \tag{11-8}$$

The eyepiece serves as a simple microscope whose focal length is f_2 to further magnify the real image. Then the total magnification is, for most comfortable viewing,

$$m \cong \frac{25L}{f_1 f_2} \tag{11-9}$$

Objective lenses found on modern microscopes have magnifications that are typically 10 ×, 30 ×, and 100 ×. Eyepieces are usually either 5 × or 10 ×. Thus magnifications of up to 1000 × can be obtained with compound light microscopes.

OIL IMMERSION MICROSCOPE

In Chapter VII we considered the limitations on resolution imposed by diffraction of light through a circular aperture. In Equation (7-7) the angular radius α of the first dark ring of a circular diffraction pattern (see Chapter VII) was related to the wavelength of light λ and the aperture diameter d by $\alpha = 1.22\lambda/d$. In the same manner, the aperture (diameter of the lens) of a microscope's objective lens determines the resolving power of the microscope, which is the measure of the microscope's utility. The *resolving power* R is definied as the reciprocal of α, or

$$R = \frac{d}{1.22\lambda} \tag{11-10}$$

This equation makes it clear that the resolving power can be increased (which is, after all, a primary goal of the science of microscopy) by either increasing the effective aperture d or decreasing the wavelength λ.

In an *oil immersion microscope,* the space between the object and the objective lens of a microscope is filled with a transparent oil, whose index of refraction n is greater than that of air. Synthetic oils ($n \cong 1.5$) are commonly used as the medium in oil immersion microscopes. Then the medium between object and objective lens have about the same index of refraction, and reflections from the lens are minimized. Because the focal length of high-powered objective lenses is very small, the immersion may be accomplished by placing a drop of oil onto the object, lowering the objective lens onto the oil, and raising the lens slightly.

The limiting aperture or diameter of a lens, which appears as d in Equation (11-10), is expressed as an angle rather than a length in the case of microscopes. A measure of the light-gathering power of a lens is the *numerical aperture* (NA), which is related to the half-angle of acceptance of light by the lens. If we consider a point P, illuminated from below and placed in front of a lens AB as shown in Figure 11-3a, we can see that the amount of light (the flux F) picked up by the lens from the cone QPR will be proportional to the solid angle subtended by that cone at P, i.e.,

$$F \propto \frac{(d/2)^2}{\overline{BP}^2} \propto \sin^2 a \tag{11-11}$$

where d is the diameter of the lens and a is the half-angle measured from

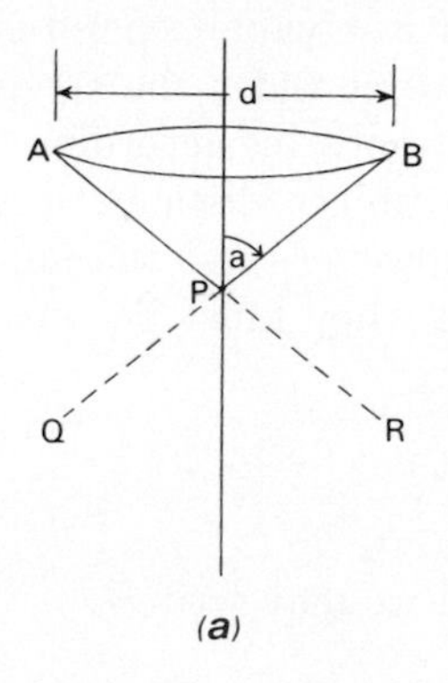

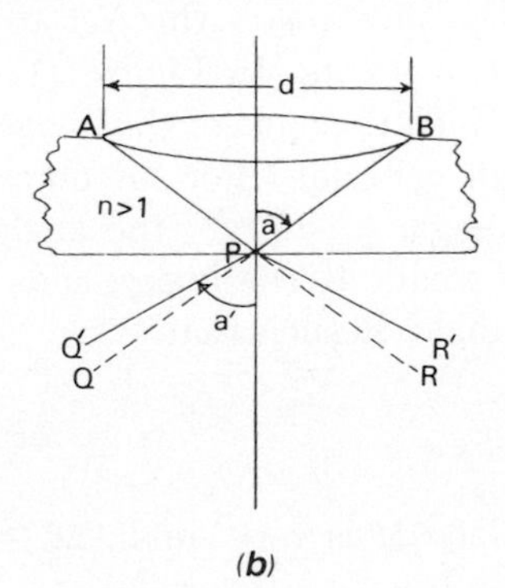

FIGURE 11-3. The light-gathering characteristics of lenses illuminated from below. (a) A dry lens. (b) An oil immersion lens with a medium of index of refraction $n > 1$ between the object at P and the lower surface of the lens.

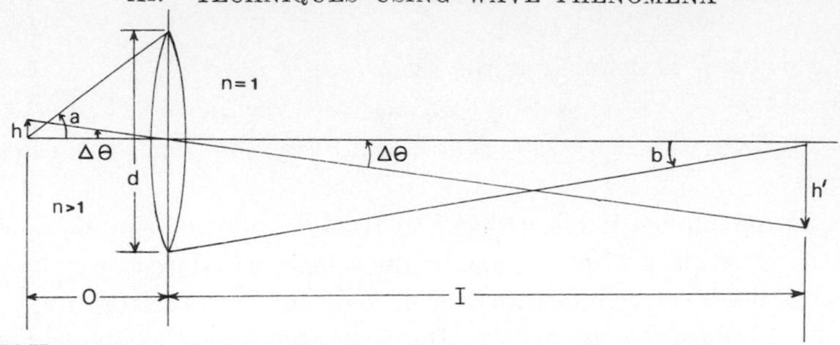

FIGURE 11-4. Construction for calculations of the resolution of a lens.

the lens axis to the line $\overline{PB}$. If now an immersion medium of refractive index n is placed between the point P and the lens, as in Figure (11-3b), the cone of rays entering the lens comes from the region within $Q'PR'$. The boundaries of the cone $Q'PR'$ are defined by the fact that the angle a' is such that Snell's law (see Chapter VII) is satisfied at the interface between the oil medium and air, i.e., such that

$$\sin a' = n \sin a \tag{11-12}$$

In this case, the amount of light F' entering the lens is proportional to $\sin^2 a'$ or

$$F' \propto n^2 \sin^2 a \tag{11-13}$$

The quantity $n \sin a$ is defined to be the numerical aperture, i.e.,

$$NA = n \sin a \tag{11-14}$$

The amount of light entering the lens is proportional to $(NA)^2$. In a microscope, the angle a is fixed by the geometry of the objective lens and its focal length, but the NA can be increased by filling the space between object and objective lens with oil of a relatively high refractive index.

We can now relate the NA to the minimum resolvable separation between two points. In Figure (11-4), suppose we consider the conditions under which the ends of the image h' are barely resolvable according to the Rayleigh criterion. For an objective lens, the object distance is short, so the angle a is large; the angle b is small, however, and the angle $\Delta\theta$ is very small. If the image ends are resolvable, they must be separated by an angle $\Delta\theta$ such that

$$\Delta\theta = \frac{1.22\lambda}{d} \tag{11-15}$$

Now, since $\Delta\theta$ is very small, $\Delta\theta \cong \tan(\Delta\theta)$, and we may write

$$\Delta\theta = \frac{h}{O} = \frac{h'}{I} = \frac{1.22\lambda}{d} \tag{11-16}$$

Helmholz has shown a general relation, now known as the *sine theorem*, between an object in a medium of index of refraction n and an image in air ($n = 1$):

$$nh \sin a = h' \sin b \tag{11-17}$$

Since the angle b is small, we may approximate $\sin b \cong \tan b$, so that the aperture d is related to b by

$$\frac{d}{2} = I \sin b \tag{11-18}$$

Then the minimum resolvable separation h' is obtained when the object height h is related to $\Delta\theta$ by

$$h = O\Delta\theta \tag{11-19}$$

and from Equation (11-16), $O = ih/h'$; $\Delta\theta$ is given in Equation (11-16). Incorporating these values into Equation (11-19) gives

$$h = I \frac{h}{h'} \frac{1.22\lambda}{d} \tag{11-20}$$

From Equation (11-17), we note $h/h' = \sin b/n \sin a$, and from Equation (11-18), we find $d = 2I \sin b$. Substituting these quantities into Equation (11-20) results in

$$h = I \frac{\sin b}{n \sin a} \frac{1.22\lambda}{2I \sin b} = \frac{1.22\lambda}{2n \sin a} \tag{11-21}$$

Since Equation (11-14) established that the numerical aperture is $n \sin a$, we may conclude that

$$\text{minimum resolvable separation} = \frac{1.22\lambda}{2(NA)} = \frac{0.61\lambda}{NA} \tag{11-22}$$

Because the eye is, in fact, capable of resolving points slightly better than the criterion established by Rayleigh, the equation used in practice is

$$\text{minimum resolvable separation} = \frac{0.5\lambda}{NA} = \frac{\lambda}{2(NA)} \tag{11-23}$$

Again, the smaller an object that can be resolved, the greater the resolving power (effectiveness) of a microscope. The resolving power R of Equation (11-10) has no units, yet it is common practice to speak of the reciprocal of the minimum resolvable separation as "resolving power," R.P., which has units of 1/length (or lines per cm, referring to the numbers of lines per cm on a grating that could be resolved). This "resolving power" is

$$\text{R.P.} = \frac{2(NA)}{\lambda} \tag{11-24}$$

Since the resolving power of a microscope increases with increasing NA, the use of oil immersion improves the resolving power.

CONSTRUCTION AND USE OF COMPOUND MICROSCOPES

Figure 11-5 shows some of the important optical features that are common to most compound microscopes. The mechanical arrangements for positioning of an object and for focusing of the images are not shown. The microscope proper is above the *stage,* a movable platform on which the sample to be observed is placed. The lighting arrangements are below the stage and consisting of the *substage condenser, substage iris,* mirror, *field iris, field condenser,* and lamp.

The field condenser is a lens that concentrates the light from the lamp onto the mirror, which reflects the light toward the object. The adjustable field iris is an adjustable diaphragm that limits the beam of the illuminating source so that it just fills the field of view of any objective.

The substage assembly is made up of a substage condenser and a substage iris. This condenser, another converging lens, focuses the illuminating beam onto the object so that the entire area of view of a particular objective lens is illuminated. The substage iris limits the cone of light reaching the

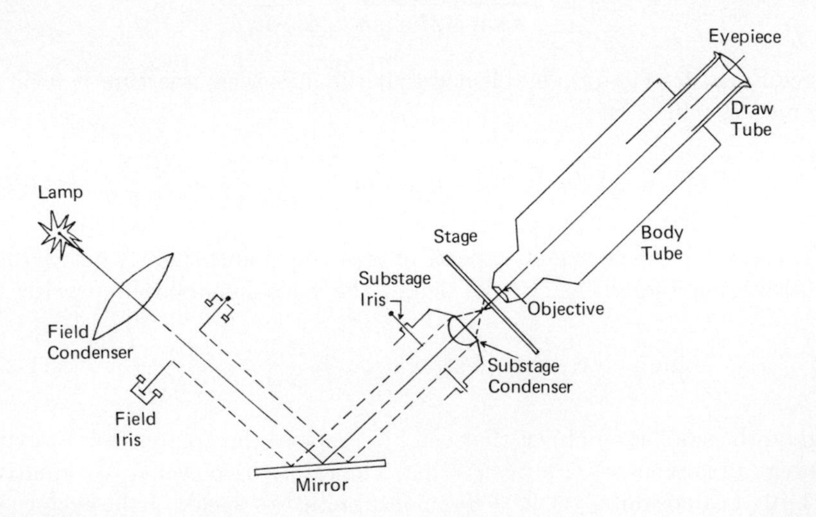

FIGURE 11-5. Schematic diagram of a complete compound microscope, including the lighting arrangement.

stage to the useful area in order to reduce reflections and unnecessary heating of the specimen.

The objective lens is usually one or three or more lenses of different magnification that are mounted on a revolving nosepiece at the lower end of the *body tube* of the microscope. Although we have discussed objective lenses as though each were a single lens, most quality microscopes have objectives lenses, each of which is actually a combination of lenses. The combination of lenses is designed to reduce spherical, chromatic, and other aberrations (see Chapter VII).

The eyepiece is housed in the *draw tube* of the microscope. Again, quality eyepieces are usually made up of more than one lens in order to minimize aberrations.

Proper illumination of an object is achieved by focusing the light from the lamp at the plane of the substage iris by adjustment of the field condenser. The substage iris is adjusted by removing the eyepiece so that the iris can be seen in focus on the back focal plane of the objective. The diameter of the substage iris is adjusted so that it is slightly smaller than the cone of light visible to the eye. Then an image of the field iris is formed in the plane of the object (on the stage) by positioning the substage condenser. Finally, the field iris is opened until it nearly encompasses the entire field of view.

The illumination just described is called *bright field* illumination. In many cases, however, greater contrast or more desirable emphasis on a particular part of a specimen can be achieved by *dark field* illumination. This effect most easily is achieved by placing an opaque disk in front of the condenser in the center of the light beam so that only the peripheral rays reach the stage. Special dark field condensers are available to achieve quality dark field illumination. The object is then illuminated only by light that falls onto it obliquely, and visibility is by reflection from surfaces that would otherwise be transparent.

Course and fine adjustments on the positioning of the body tube and draw tube allow precise focusing of the image seen by the observer.

It should be obvious that there are many variations in the construction of microscopes. For example, the entire illumination system is built into some microscopes. With any model there are many procedures that are necessary in order to obtain optimal viewing; these may be learned from the literature provided by the microscope's manufacturer.

PHASE CONTRAST MICROSCOPY

Many biological specimens, like living cells, are very difficult to study using the techniques of light microscopy discussed up to this point because they are composed of many parts that are all very nearly transparent. Some methods [e.g., fixing and differential staining (see Chapter I) and dark field illumination] are useful for contrasting some of the parts of cells under special circumstances. The *phase contrast* technique provides the means for visually contrasting materials within living cells, materials that are not differentially absorptive but do have other different optical properties.

As its name suggests, phase contrast microscopy depends on the differences in *phase* of portions of a light beam that have passed through different parts of a specimen. Two waves whose crests and troughs coincide are said to be in phase; when the crest of one occurs at the same time the trough of another occurs, they are 180° out of phase. Measurements of phase are based on 360° constituting a complete cycle of oscillation. Then phase differences, measured in degrees, specify the portion of a cycle that one wave leads or lags another. For example, waves exactly out of phase have a phase difference of 180°.

A phase difference is produced in two light waves of the same frequency if one wave has a different *optical path* length from the other. Optical path is the product of actual path length and index of refraction. Figure 11-6 illustrates how light passing through a thickness of a medium of higher refractive index than another beam may emerge with a phase that differs from that of the other beam.

Normal illumination of a specimen by a light microscope produces an absorption map of the specimen, i.e., contrast is provided between different portions of the specimen as a result of the differences in absorption properties. Phase contrast, on the other hand, produces an optical path map, in which the contrast is due to the different refractive indices among the various portions of the specimen.

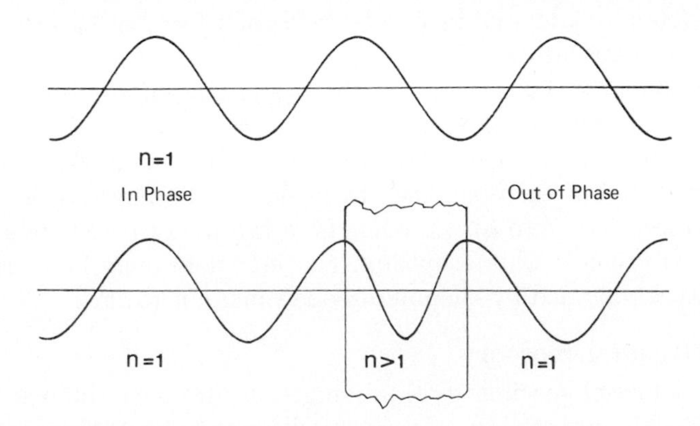

FIGURE 11-6. The introduction of phase shift between two waves by passing one of the waves through a medium of different index of refraction.

A schematic phase contrast microscope is shown in Figure 11-7. The phase variation process is begun by passing the direct light through an annular diaphragm below the substage condenser. This restricts the light illuminating the specimen to a hollow cone. After passing through the specimen, the light passes through the objective and then through a *phase plate* placed in the upper focal plane of the objective lens. The phase plate has a circular groove whose radius is designed so that when it is in position, the groove coincides with the direct image of the annular condenser diaphragm. The groove of the phase plate has a depth d such that the product of d and the index of refraction of the phase plate, i.e., the optical path length through the groove, is one-quarter of a wavelength less than through the other regions of the plate. The undeviated beam passes through this groove and is advanced in phase by 90° (a quarter of a cycle) over those rays that are diffracted or scattered by the specimen and pass through the thicker portion of the phase plate. On arrival at the image plane of the objective lens, the phase-shifted light interacts with the light diffracted by the specimen to produce the phase-contrasted image.

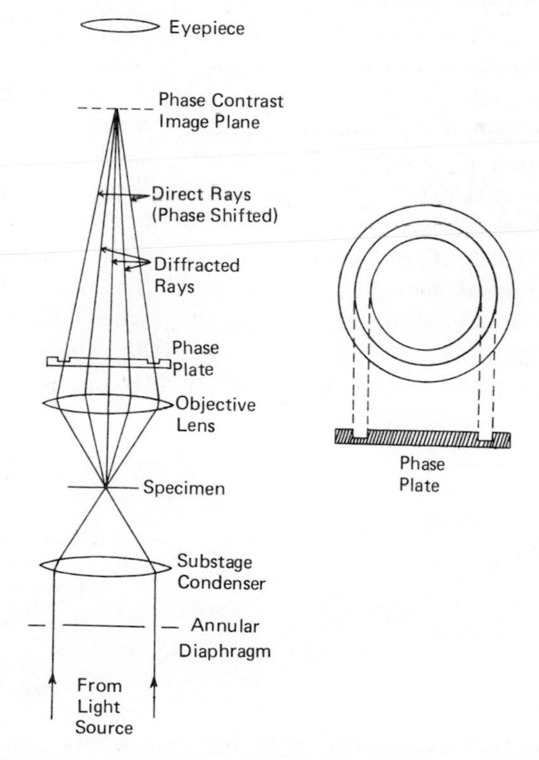

FIGURE 11-7. Schematic diagram of a phase contrast microscope.

How are phase differences in light waves converted into amplitude (and therefore intensity) differences so that the eye can visualize the contrasts produced? The answer is perhaps best understood by representing a light wave by a vector whose magnitude is proportional to the amplitude of the wave and whose angular orientation with respect to a reference wave is proportional to phase (which is why phase is measured in degrees). Figure 11-8A shows how a light wave, represented by a vector, is diminished in amplitude but not phase shifted in passing through a purely absorbing object that has the same refractive index as its surroundings. Figure 11-8B shows how a similar wave is phase shifted (but not diminished in amplitude) by passing through a phase object that has the same absorption as its surroundings. The observed intensity of light is proportional to the square of the wave's amplitude.

Suppose in Figure 11-9A the vector **OA** represents the average direct (undiffracted) light that appears in the image plane of the objective lens of a microscope *before* the phase plate is inserted. If the object is assumed to be perfectly transparent, but produces different phases at different elements of its surface, the amplitudes of light corresponding to each element will be the same and equal to the radius of the circle, OA. The vector **OB** is the total light at the image plane from an element of the object that has phase shifted the light through an angle ϕ but has not changed its amplitude. **OB** is the sum of the undiffracted light **OA** and the diffracted light **AB**. The light from all elements of the object would have the same amplitude (the radius of the circle) regardless of the phase shift ϕ.

If now the phase plate is positioned as in Figure 11-7 so that the direct rays are advanced in phase by 90° compared to the diffracted rays, the vectors are related as shown in Figure 11-9B. The vector **OA** has been phase shifted 90° and now appears as **O′A**, which when added vectorially to the diffracted light **AB**, results in the vector **O′B**. The amplitude of the light in the image plane, i.e., the magnitude of **O′B**, has become dependent

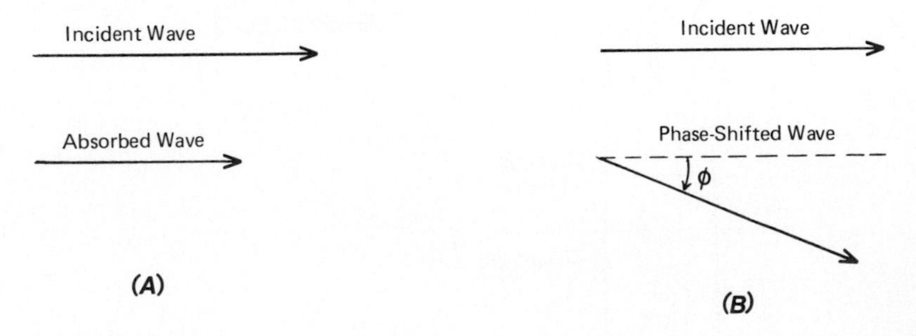

FIGURE 11-8. Vector representations of light waves before and after passing through (A) an absorbing object and (B) a phase object.

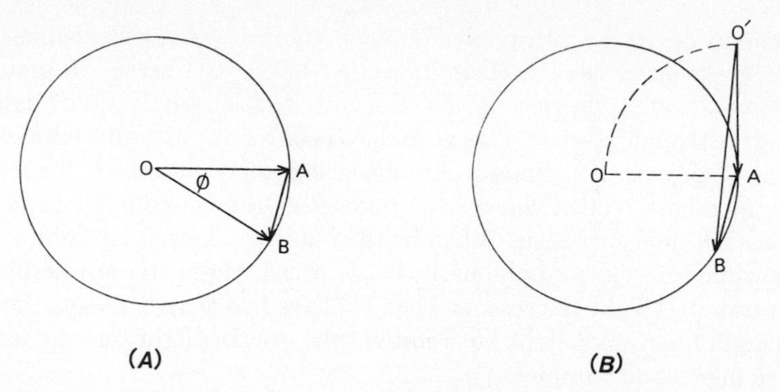

FIGURE 11-9. Diagram illustrating the production of amplitude contrast from variation in phase. (A) Before the insertion of a phase plate into a phase contrast microscope, **OB** is the sum of the average direct (undiffracted) light **OA** and the diffracted light **AB**. **OB** has the same amplitude and therefore the same intensity as **OA**. Thus, **OB** is independent of phase shift φ. (B) With the phase plate in place, the direct rays are advanced by 90°, and **O′A** added vectorially to **AB** results in **O′B**, which then becomes dependent on phase shift φ. In this case, phase differences produced by elements in the object are transformed into amplitude (intensity) differences at corresponding elements of the image.

on ϕ. Thus, the phase difference produced by elements of the object are transformed into amplitude, and therefore intensity, differences at corresponding elements of the image.

SOME SPECIALIZED MICROSCOPIC TECHNIQUES

Many variations on and adaptations to light microscopy have been developed for studies of biological materials. *Ultraviolet microscopes* were developed for two reasons: to take advantage of the shorter wavelength light, which, as we have shown earlier, improves the resolving power over microscopes using visible light, and to utilize the highly absorptive properties of many cellular components, like proteins and nucleic acids, in the ultraviolet region of the spectrum. There are, however, many problems that accompany the advantages of ultraviolet microscopy. Images must be visualized by photographic or photoelectric processes. Refracting microscopes cannot focus effectively over a large range of ultraviolet wavelengths. These problems can be dealt with, but only at considerable expense and expenditure of time. Thus, ultraviolet microscopy is in the realm of specialized research and does not provide the simple utility of the more common forms of light microscopes.

Fluorescence microscopy has in recent years become a valuable tool for studying biological systems. Fluorescence refers to the visible radiation emitted by some atoms after the electrons of the atom have been excited to high energy levels by ultraviolet or even short wavelength visible light. As the electrons return to their unexcited level by a series of jumps to successively lower energy levels, longer wavelength visible light is emitted. Although most biological molecules are not naturally fluorescent, or only weakly so, a technique has been developed in which fluorescent groups of atoms (called *fluorescent antibodies*) can be coupled to certain locations within a specimen. When irradiated with an exciting light source, the specimen becomes luminous in those areas where the antibodies are concentrated. A light microscope that is filtered so that it passes the long wavelength fluorescent light but removes the exciting light may be used to observe fluorescent samples.

Electron Microscopy

Although it is generally assumed that the purpose of a microscope is to magnify an object, it should be clear at this point that the limiting consideration in the usefulness of a microscope is resolution. It serves no useful purpose to increase arbitrarily the size of an image after details of the image can no longer be distinguished. We have discussed the limitations imposed on resolution by diffraction properties of apertures in Chapter VII and in this chapter. In particular, we have seen that the measure of resolving power of any microscope depends on the ratio of dimensions of the aperture to the illuminating wavelength. Efforts to reduce the wavelengths used in light, or electromagnetic wave, microscopy have met with only limited success. A significant breakthrough in microscopy occurred with the advent of the *electron microscope* in 1931. The concept of electron microscopy, using the wave nature of particles as the illuminant in a microscope, had advanced microscopy to a level beyond light microscopy that compares with the advance of the first microscopes over the naked eye. Today the electron microscope probably represents the most useful single tool in the biological sciences.

ELECTRON WAVES

The source of illumination in an electron microscope is an *electron gun*, which consists of a metal filament that is heated and from which electrons are ejected by *thermionic emission* and an accelerating anode, a positively charged baffle that attracts the electrons that have been "boiled off," giving them very high velocities. Some of the high-speed electrons continue through an opening in the anode; this electron beam is used to "illuminate" a sample under examination. The difference in potential across which the electrons fall in exciting the electron gun usually ranges between 40 and

100 kV in modern electron microscope systems. This means that the electrons possess energies up to 10^5 *electron volts* (eV). One eV is equivalent to 1.6×10^{-19} joule.

In 1925 de Broglie speculated that particles in motion have a wavelength associated with them such that

$$\lambda = \frac{h}{p} \tag{11-25}$$

where λ is the wavelength of the "matter wave"; h is Planck's constant, introduced in Chapter VII; p is the momentum of the particle. The validity of Equation (11-25) was experimentally demonstrated using electrons as the illuminating particles in interference experiments by Davisson and Germer (1927). Notice that as the momentum p of the particle increases, its associated wavelength becomes shorter. At the high speeds that can be attained by electrons from electron guns, speeds that are significant fractions of the velocity of light, the relativistic momentum is given by

$$p = \frac{m_0 v}{(1 - v^2/c^2)^{1/2}} \tag{11-26}$$

where m_0 is the electron's rest mass (9.1×10^{-31} kg), v is the electron's speed, and c is the speed of light. The relativistic kinetic energy (KE) is expressed in terms of the same parameters by

$$KE = \frac{m_0 c^2}{(1 - v^2/c^2)^{1/2}} - m_0 c^2 \tag{11-27}$$

One can use Equations (11-25), (11-26), and (11-27) to compute the wavelengths of electrons used in electron microscopes. An electron with kinetic energy of 60 keV (60,000 eV) has a wavelength of approximately $5 \times 10^{12}\ m$ = 0.005 nm. Comparison of this value with the wavlength of visible light of 500 nm shows that the wavelengths associated with electrons are about 100,000 times shorter than those of light waves. This decrease in wavelength represents a theoretical advantage in resolution of five orders of magnitude! Unfortunately, as we shall see subsequently, not all of this theoretical advantage can be realized in practice.

ELECTRON LENSES

Beams of electrons can be focused optically by *electron lenses* in much the same way that light is focused by glass lenses. Two types of electron lenses are used in electron microscopes—*electrostatic lenses* and *magnetic lenses.*

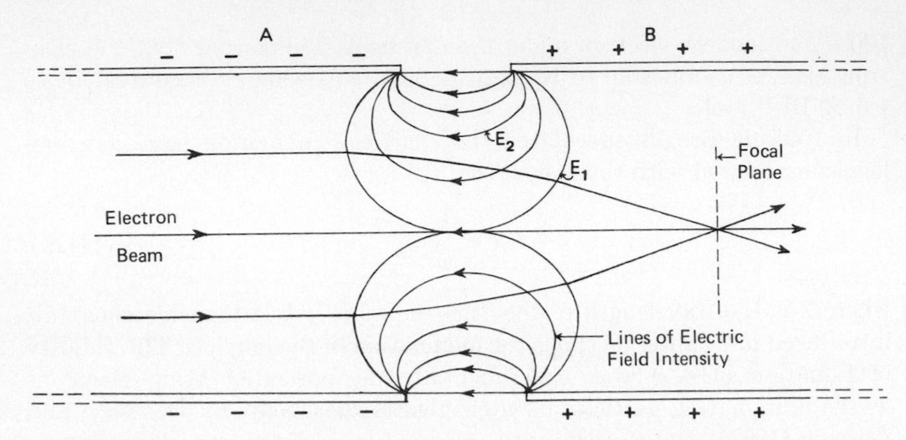

FIGURE 11-10. An electrostatic lens. The drawing is a cross section of two cylinders, A and B, each at a different potential, so that an electric field configuration, shown in a plane through the cylinder axis, exists in the gap region between the cylinders.

The Electrostatic Lens. A pair of hollow cylindrical electrodes arranged as shown in Figure 11-10 and held at different electrical potentials can serve to converge a beam of electrons. The lines of electric field intensity, i.e., the points having the same magnitude of electric field intensity E, across the gap between the two cylinders in the plane of the page are shown in Figure 11-10. Without a detailed analysis, we can understand how such an arrangement can function as an electron lens. The electric field associated with a line passing near the axis, like E_1, is weaker than the field associated with a line farther from the axis, like E_2. If cylinder B is at a higher electrostatic potential than cylinder A, electrons on the cylinder axis traveling from A to B will experience an acceleration along the axis from A to B. These electrons speed up without changing direction. Electrons off the axis, traveling in the same direction, experience an acceleration with a component toward the center of the tube as they approach the region of the gap. There is, therefore, a focusing effect while the electrons are in cylinder A because the electrons farther from the axis encounter stronger fields and receive greater acceleration toward the axis than those nearer the axis. Once an electron passes the center of the gap, it experiences accelerations away from the axis; this is a defocusing effect. The defocusing effect is considerably less than the focusing effect for two reasons: First, the fields nearer the axis where the electrons are defocused are weaker than the fields nearer the cylinder wall where the electrons were focused; second, the electrons are moving faster during the defocusing period because they were sped up during the focusing phase. Hence, the defocusing force acts for a shorter period of time.

The net result of the action of an electrostatic lens is that a parallel beam of electrons is focused at a point in the same way that a converging optical lens focuses a parallel beam of light at its focal distance. By reversing the potential on cylinders A and B, the electrostatic lens becomes a divergent lens with a parallel beam of electrons diverging from a point on the axis at the focal distance from the gap. In further analogy to geometric optics, an electrostatic lens may suffer from aberrations, or failure to focus electrons perfectly, because of distortions in the electric field configuration or variations in velocity among the electrons within the beam. Electrostatic lenses can form images of objects illuminated by the electron beam just as optical lenses form images of objects illuminated by light.

The Magnetic Lens. Magnetic lenses are more commonly used in electron microscopes than are electrostatic lenses. The function of a magnetic lens is perhaps understood most easily if we first consider the trajectory of an electron in the uniform field of a *solenoid*, a hollow cylinder formed of many turns of wire through which current may flow.

In Figure 11-11, a long (compared to its diameter) solenoid carrying current I produces a magnetic flux density $\mathbf{B}$, whose magnitude inside the solenoid (but not near the ends) is given by

$$B = \mu_0 n I \qquad (11\text{-}28)$$

where n is the number of turns of wire per unit length and μ_0 is the permeability of free space (a constant equal to 1.26×10^{-6} weber/ampere meter). The field is uniform throughout the interior of the solenoid and is directed along the axis of the solenoid. Suppose an electron with velocity $\mathbf{v}$ directed at an arbitrary angle θ from the direction of the magnetic flux density $\mathbf{B}$ is at some point within the solenoid, as shown in Figure 11-12. The component of $\mathbf{v}$ parallel to $\mathbf{B}$ has magnitude $v \cos \theta$. The magnitude of the force F exerted on an electron of charge $-e$ by a magnetic field $\mathbf{B}$ is

$$F = -evB \sin \theta \qquad (11\text{-}29)$$

where θ is the angle between $\mathbf{v}$ and $\mathbf{B}$, and $\mathbf{F}$ is perpendicular to both $\mathbf{v}$ and $\mathbf{B}$. We may consider the forces and their resulting effects on each velocity

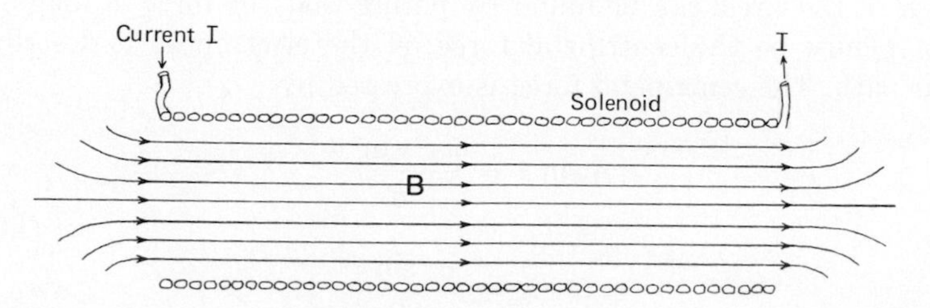

FIGURE 11-11. The magnetic flux density inside a long solenoid. Far from the ends of the solenoid, the magnetic flux density is uniform.

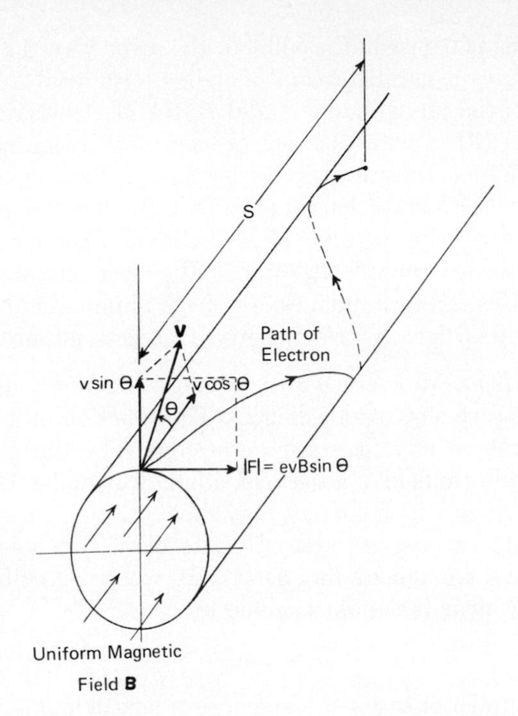

FIGURE 11-12. The path of an electron in a uniform magnetic field. An electron, arbitrarily directed, follows a helical path.

component of the electron separately. The component of **v** parallel to **B** has the magnitude $v \cos \theta$; here θ is zero, and there is no force on the electron associated with this component of its velocity. The velocity component $v \sin \theta$ is perpendicular to **B** and produces a force on the electron that is of magnitude $Bev \sin \theta$ (here θ is 90° and perpendicular to both **v** and **B**, as shown in Figure 11-12). In fact, the force on the electron continues to be mutually perpendicular to **B** and to the direction of $v \sin \theta$, so the electron would move in a circle if its velocity had only the component $v \sin \theta$. The radius R of the circle can be found by noting that the force of magnitude $evB \sin \theta$ must be the centripetal force on the electron as it describes a circular path. The centripetal force is expressed by

$$evB \sin \theta = \frac{mv^2 \sin^2 \theta}{R}$$

$$R = \frac{mv \sin \theta}{eB} \tag{11-30}$$

where m is the electron mass. The time t required for the electron to complete one revolution of the circle is found from

$$t = \frac{2\pi R}{v \sin \theta} = \frac{2\pi\, mv \sin \theta}{eB\, v \sin \theta} = \frac{2\pi m}{eB} \tag{11-31}$$

The electron is progressing along the axis of the solenoid at a constant speed $v \cos \theta$ at the same time it is executing circular motion in the plane perpendicular to the axis. The resultant trajectory of the electron is a helix. The pitch of the helix, i.e., the distance s the electron travels along the axis while making one revolution, is given by

$$s = (v \cos \theta)\, t = \frac{2\pi m}{eB} v \cos \theta \tag{11-32}$$

This result means that electrons that initially travel parallel to the axis and are scattered by an object at an angle θ to the axis will form an image at a distance s from the object. The scattered electrons must be limited to small angles θ from the axis to produce a sharp image since s is proportional to $\cos \theta$; for small θ, however, one may write $\cos \theta = 1 - \frac{1}{2}\theta^2 + \cdots \cong 1$, so s is approximately independent of θ if θ is restricted to small angles. Notice that a solenoid used as just described forms an image but does not produce magnification.

Now we can consider how a thin magnetic lens can be constructed to image and magnify. If the windings of a solenoid are surrounded by an iron casing, except for a small circular opening on the axial side of the casing, a nonuniform magnetic field similar to that shown in Figure 11-13 will be produced inside the solenoid. An electron traveling at an angle to the axis will experience a force tending to rotate it about the axis in an azimuthal direction. The insets of Figure 11-13 show how the azimuthal component of velocity $\mathbf{v}_{\mathrm{Az}}$ (out of the plane of the page) causes a force $\mathbf{F} = -e\mathbf{v}_{\mathrm{Az}} \times \mathbf{B}$ on the electron toward the axis. Since the magnitude of the magnetic flux density $\mathbf{B}$ is greater farther from the axis, the electrons farthest from the axis experience the greatest force toward the axis; thus focusing is achieved. It is a characteristic of all magnetic lenses that the image is rotated because of the helical trajectory of off-axis electrons in the magnetic field.

Further analysis of the magnetic lens would demonstrate their ability to magnify an object that is judiciously positioned. Equation (11-32) warns us that the distance s of the image plane from an object is proportional to $\cos \theta$; this fact means that magnetic lenses have an aberration analogous to spherical aberration in a glass lens. This defect is minimized just as in geometric optics—by limiting the beam to paraxial electrons, i.e., those electrons for which $\cos \theta$ is very small. Any variation in the speed of

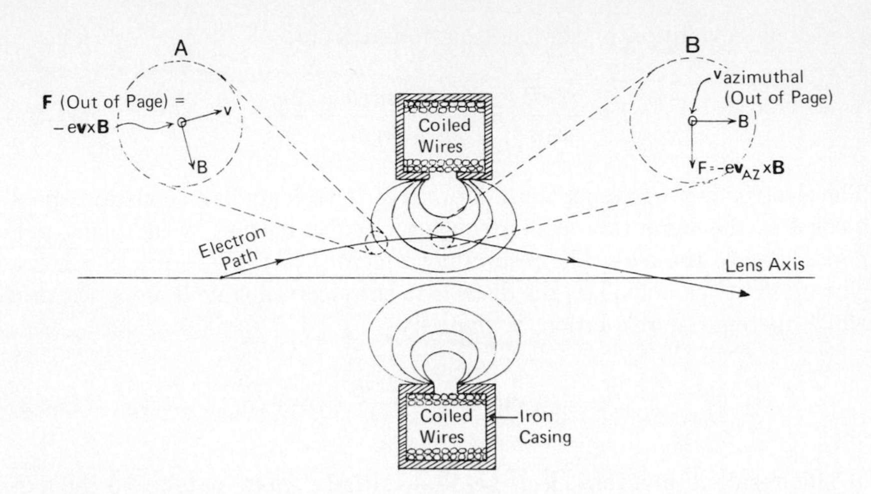

FIGURE 11-13. A magnetic lens. The nonuniform magnetic fields near the openings of the casings containing many turns of wire bend electrons that pass the lens off the lens axis. The inserts show (A) the production of an azimuthal component of velocity in the electron and (B) the production of a focusing force due to the azimuthal component of velocity.

electrons within the illuminating beam causes imperfect focusing that is analogous to chromatic aberration in lenses using light.

MICROSCOPE CONFIGURATION

An electron gun and electron lenses may be positioned so that an object illuminated by the electrons produce a highly magnified image, as shown in Figure 11-14. Normally three magnetic lenses are used in modern electron microscope systems. The first lens, the condenser lens, forms a parallel beam of electrons from the beam that diverges from the electron gun. The sample to be examined is placed very near the first focal plane of a second lens, the objective lens. A magnified image is formed at a plane near the focal plane of a third lens, the *projection lens*, which forms a real image on a luminescent screen or photographic film plate. The magnification is controlled by varying the current through the coils of the magnetic lenses.

The design of an electron microscope represents a compromise in choosing the numerical apertures of the lenses. Too small a NA, i.e., restricting θ to small angles to reduce spherical aberration, results in electron flux insufficient to provide reasonable contrast in the image; too large a NA results in loss of detail through spherical aberration. The NA is usually chosen to lie between 0.01 and 0.001. When used with the wavelength

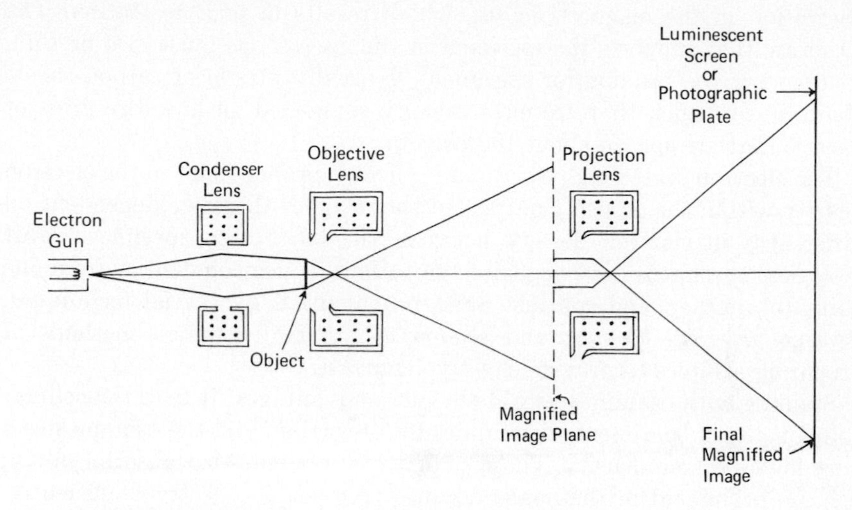

FIGURE 11-14. Production of a magnified image by a series of magnetic lenses comprising an electron microscope.

associated with 60 keV electrons, i.e., $\lambda = 0.005$ nm, Equation (11-23) gives the minimum resolvable separation that can be obtained when the $NA = 0.005$ (this means $n \sin \theta \cong \theta = 0.005$ radians, since $n = 1$ in the evacuated regions through which the electrons travel):

$$\text{Minimum resolvable separation} = \frac{\lambda}{2(NA)} = \frac{0.005 \text{ nm}}{0.01} = 0.5 \text{ nm} \quad (11\text{-}33)$$

This result is very nearly the resolvable separation that can be achieved in practice. Separations of 0.6 to 1 nm are often obtained in practice when excellent samples are used.

SPECIMEN PREPARATION

The observation of specimens using electrons as the illuminant places severe restrictions on the characteristics of both the specimen and the environment of the electron beam. The electron beam is not absorbed by the specimen but is *scattered*. Since electrons are not very massive, they are easily scattered even by the molecules in air; for this reason the chamber housing the electron optical system, including the specimen, must be evacuated.

Specimens used in electron microscopy must be very thin. Since high resolution is desirable, the samples must be thin enough that the electron beam loses negligible energy in negotiating the specimen; otherwise variations in velocity occur in the electron beam, producing "chromatic" aberration in the magnetic lenses, which results in poor resolution. The structure that supports the specimen in the microscope must also be thin. In practice, the backing for specimens is usually plastic or carbon sheets about 10 nm thick that are mechanically supported on fine wire grids or meshes that are spaced about 100 per cm.

The electron scattering power of an object is proportional to the electron density within the object. Contrast in the image is therefore dependent on differences in electron density between the parts of a specimen. Most biological specimens have negligible variations in electron density from one point to another, and contrast must be enhanced by special techniques. *Osmium tetroxide staining* and *shadowing* are two common methods of preparing samples to provide effective contrast.

Staining with osmium tetroxide has two advantages. It fixes the cellular components of specimens with minimum distortion, and the osmium has a very high electron density. Those cellular components that absorb osmium become prominent in the image because they scatter electrons efficiently.

Shadowing, a technique frequently used in electron microscopy, produces great contrast between objects on the specimen's surface and gives a three-dimensional effect to the image. Shadowing is accomplished by depositing thin layers of heavy metal atoms onto the surface of a specimen by evaporating the metal from hot filaments in a vacuum. Figure 11-15 illustrates how the evaporation is accomplished and how the shadowed sample appears. Since the electrons in an electron microscope cannot penetrate the dense electronic sheaths around the heavy atoms, the electrons that reach a photographic plate have passed through regions where no metal was deposited; the resulting picture, an electron *photomicrograph*, appears dark in regions where no metal was deposited. A photomicrograph of a shadowed sample looks as if an object on its surface were casting shadows due to a light source that was placed where the evaporation filament was. Figure 11-16 shows examples of electron photomicrographs of stained and shadowed preparations.

Since only thin samples can be directly observed by an electron microscope, a special technique is required to view the surfaces of thick specimens. A *replica* of a surface is made by depositing a thin layer (a few hundred angstroms) of plastic on the surface to be studied and then stripping off this thin "negative" from the surface. The replica may be shadowed before being used as the object in the electron microscope.

X-Ray Diffraction

X-rays are electromagnetic waves that originate in the electronic structure of atoms. X-rays typically have wavelengths between 0.1 and 1 Å;

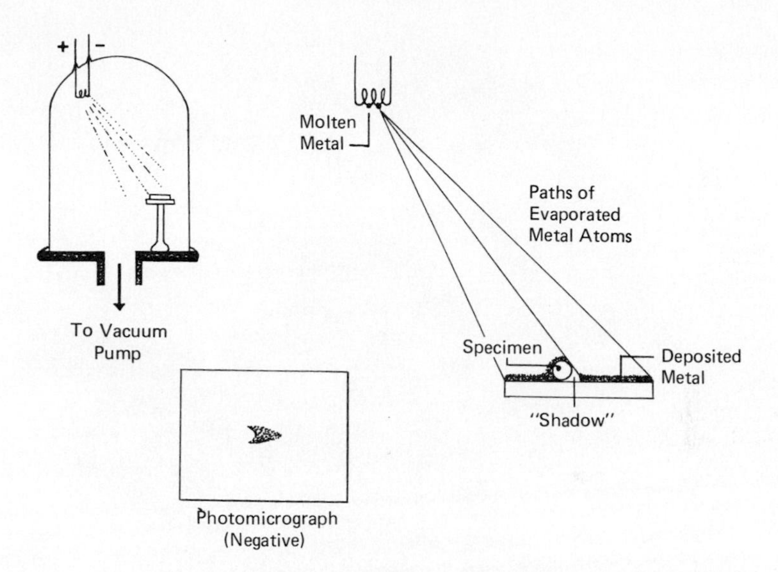

FIGURE 11-15. Evaporation technique for shadowing a specimen for an electron microscope. Thin metal wires are melted by passing current through the coil in a vacuum. As the coil temperature is increased, the molten metal evaporates, and the metal atoms follow essentially straight-line paths. As the metal deposits on the sample holder, shadows are formed in regions where no metal is deposited.

such wavelengths permit them to be diffracted by "gratings" composed of atoms in solid crystals. Analysis of X-ray diffraction patterns from crystals can yield details of the structure of polyatomic molecules. The mathematical techniques required to acquire such information are beyond the scope of this book, but the basic principles of X-ray diffraction will be discussed briefly.

The techniques of X-ray diffraction have been used to determine the detailed structure of DNA and of several proteins. Such contributions have been the basis for significant progress in better understanding biological systems at the molecular level.

Here we shall consider the production and physical characteristics of X-rays. Then the elementary principles involved in the determination of spatial relationships between atoms of a crystal will be discussed.

Production of X-Rays

X-rays are produced when high-speed electrons strike atoms. Figure 11-17 represents the basic elements of an *X-ray tube*. An electron gun, similar to that used in electron microscopes, provides a beam of high-

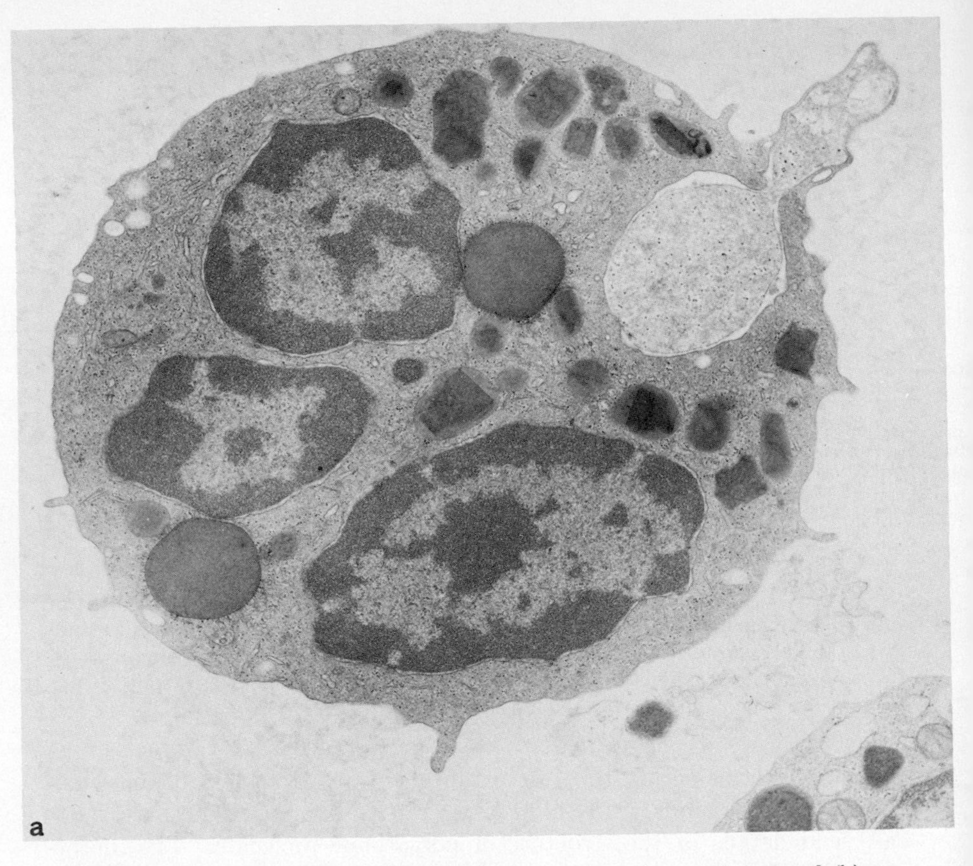

FIGURE 11-16. Photomicrographs illustrating (a) osmium staining, and (b) carbon-platinum shadowing. The stained specimen is a granular white blood cell (see Chapter VIII) which has three large nuclei, two circular lipid droplets, and crystalline granules visible. This phagocytic cell is shown ingesting a fragment of a cell. The very small, darkly stained spots are ribosomes. The shadowed specimen is a cluster of four bacteria, *Neisseria gonorrhoeae,* whose surface texture and three-dimensional arrangement are visualized by the shadowing. Courtesy of Carey Callaway, Pathology Division, Center for Disease Control, Atlanta, Georgia.

energy electrons (typically greater than 40 keV), which is directed at a metal target. The region inside the tube is evacuated. Electromagnetic radiation is emitted from the target, and the useful beam of X-rays is formed by surrounding, except for a small hole, the glass tube with material such as lead or steel that is highly absorptive to X-rays.

When the high-energy electrons strike the target material, two effects are responsible for the radiated electromagnetic waves. If a charged

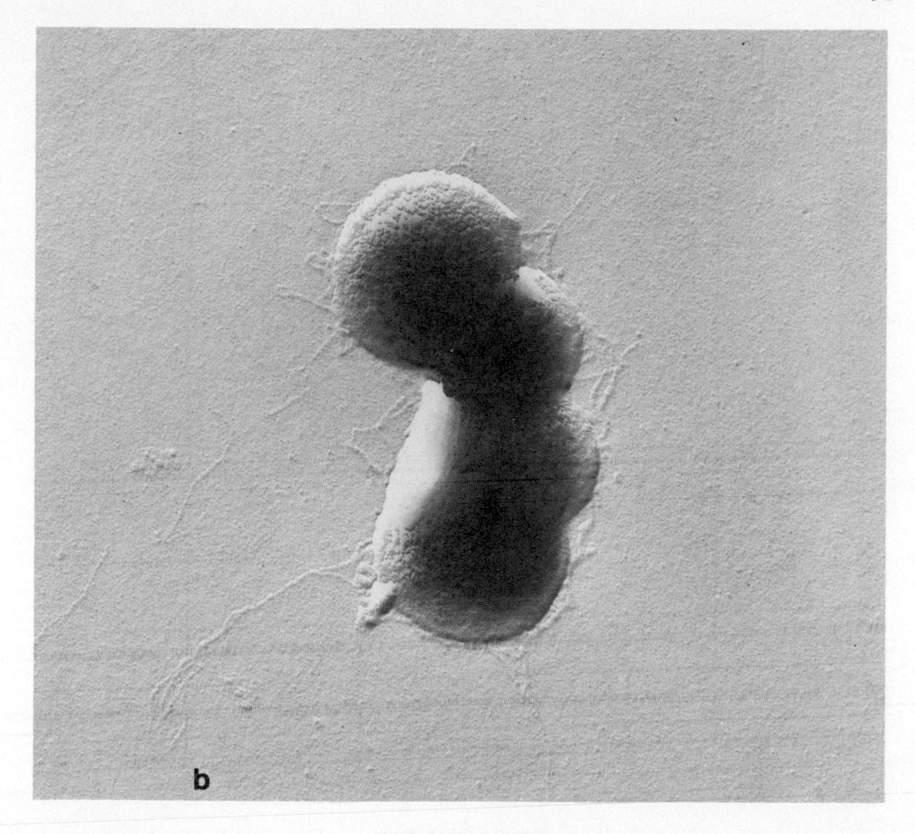

FIGURE 11-16b

particle changes velocity, i.e., accelerates or decelerates, it radiates energy away from it as an electromagnetic wave (this is the phenomenon that occurs in transmitting antennas). In striking the target, electrons are rapidly decelerated, and a *continuous spectrum* of wavelengths is radiated. These radiations, caused by "braking" the electrons, are called *bremsstrahlung*, or braking radiation. Since no electron has kinetic energy in excess of that acquired in accelerating across the difference in potential between the filament and anode of the electron gun, there is a maximum kinetic energy that the electrons may have. The maximum energy (in eV) is numerically equal to the accelerating potential (in volts) across the electron gun. Conservation of energy requires that no X-rays can be emitted with greater energy than the maximum kinetic energy of the electrons. Since the energy E of a photon is $hf = hc/\lambda$, the maximum kinetic energy of the electrons corresponds to a minimum wavelength of

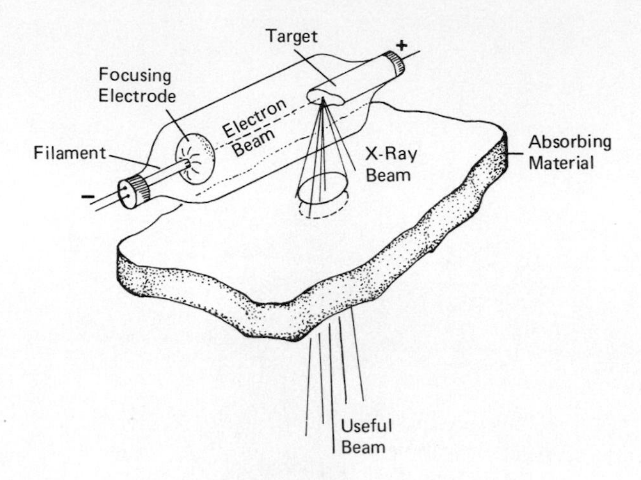

FIGURE 11-17. An X-ray tube.

the X-rays:

$$eV = \frac{hc}{\lambda_{\min}}$$

or (11-34)

$$\lambda_{\min} = \frac{hc}{eV}$$

where V is the accelerating voltage and h, c, and e are physical constants (Planck's constant, the speed of light, and the charge on an electron, respectively). When the numerical values of the constants are substituted into Equation (11-34), it is found that when V is in volts, λ is expressed in angstrom units according to

$$\lambda_{\min}\ (\text{Å}) = \frac{1.24 \times 10^4}{V\ (\text{volts})} \cong \frac{12{,}345}{V} \tag{11-35}$$

Figure 11-18 is a representation of the intensity of radiation from an X-ray tube as a function of wavelength. The *bremsstrahlung* is cut off at the short wavelength end of the spectrum according to Equation (11-35). There is no long wavelength limit to the continuous spectrum.

Superimposed on the continuous radiation spectrum of Figure 11-18 are spikes—high X-ray intensities that occur at particular wavelengths. The X-rays in these spikes are collectively called *characteristic X-rays* because the wavelengths at which they occur are characteristic of the target material. The characteristic X-rays originate in the electronic structure of

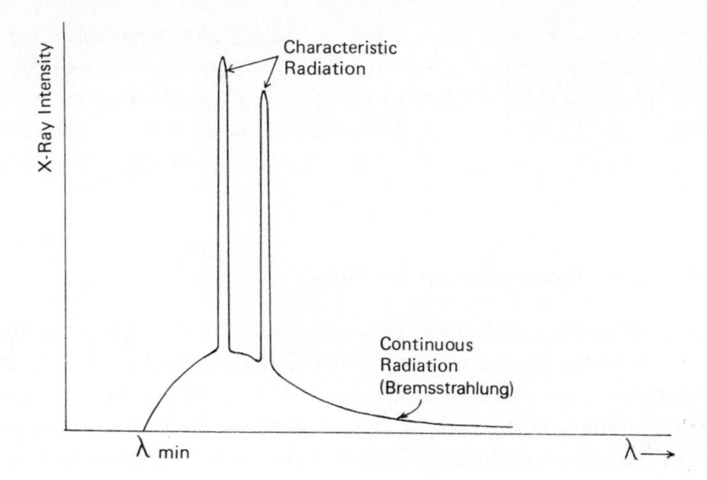

FIGURE 11-18. The intensity of X-rays emerging from the target of an X-ray tube as a function of wavelength. The characteristic radiation spikes are superimposed on the continuous radiation.

target atoms. Figure 11-19 pictures the Bohr model of a target atom and illustrates how characteristic X-rays are emitted when the electrons from higher energy levels (farther from the nucleus) fall to lower levels to fill vacancies caused when orbital electrons have been knocked out by the high-energy electrons of the electron gun. The orbits within an atom have been traditionally labeled K, L, M, N, . . ., starting from the innermost orbit and proceeding outward. X-rays produced as a result of vacancy in the K-shell are called K X-rays, and L, M, N, . . ., X-rays are similarly named.

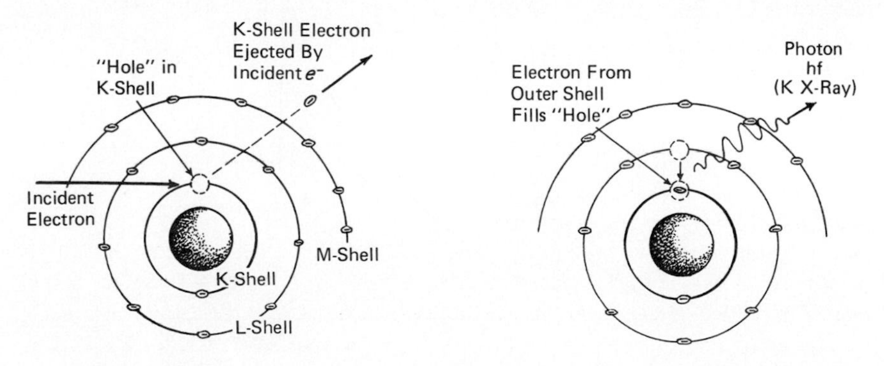

FIGURE 11-19. The production of a characteristic X-ray. An electron strikes and ejects an electron from an atom, leaving a hole in one of the electron shells (the K-shell in this case). An electron from a higher shell (the L-shell in this case) fills the hole in the lower shell and emits an X-ray in the process.

The characteristic X-rays from a copper target, for example, are at different wavelengths from another target, say tungsten, because the energy difference between orbits is different for different species of atoms. A photon is emitted from the atom as an electron jumps to a lower orbit, and the energy ($= hf$) of the photon is the difference in the energies associated with each orbit.

Detection and Absorption of X-Rays

X-rays can interact with matter by being scattered or by being absorbed. For the moment we shall consider absorption. X-rays may be absorbed by giving up their energy to free electrons from atoms, thereby creating ions. Some X-ray detectors depend on the current produced by ions created as the X-rays pass through a gas; others use the electrons freed by the X-rays to produce photons of light that are detected in photomultiplier tubes, which in turn produce electrical pulses that can be counted electronically. The latter type detector, called a *scintillation counter,* is very efficient, capable of detecting very weak fluxes of X-rays.

The intensity I of an X-ray beam passing through matter decreases exponentially with distance x according to

$$I = I_0 e^{-\mu x} \tag{11-36}$$

where I_0 is the intensity of the beam incident on the matter and μ is called the *absorption coefficient* of the particular absorbing material. The coef-

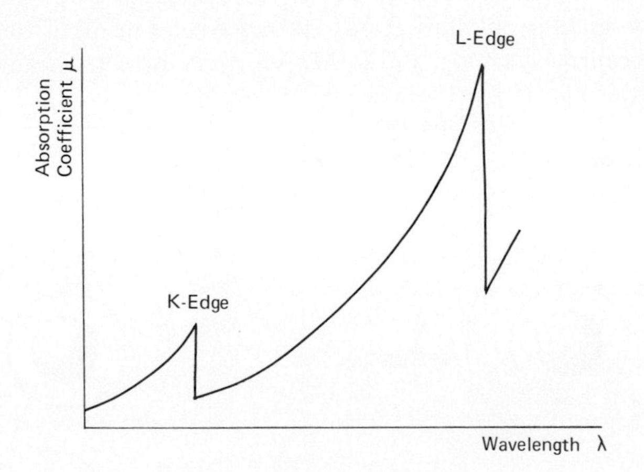

FIGURE 11-20. Absorption spectrum of a typical X-ray absorber material. The absorption coefficient μ as a function of wavelength shows sudden changes, called edges, at wavelengths corresponding to the photon energies capable of freeing electrons from the K-shell (K-edge), L-shell (L-edge), etc., of atoms in the absorber material.

ficient μ is itself dependent on wavelength (or photon energy) of the X-rays being absorbed (see Figure 11-20). In fact, *absorption edges,* or discontinuities in μ, occur in a material at wavelengths corresponding to X-ray photon energies sufficient to eject electrons from a particular atomic shell. Thus, there is a K-edge, an L-edge, etc., in the absorption coefficient of an absorber composed of atoms of one element.

Since the absorption edges occur at different wavelengths for different elements, a judicious selection of absorbing materials can be made to provide an effective filtering system for the X-ray beam emerging from an X-ray tube. Figure 11-21 shows how a carefully selected filter material, placed in the beam from an X-ray tube, passes a relatively *monochromatic*

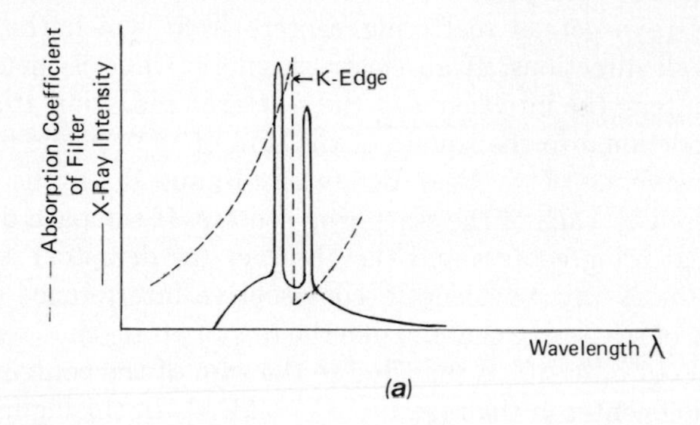

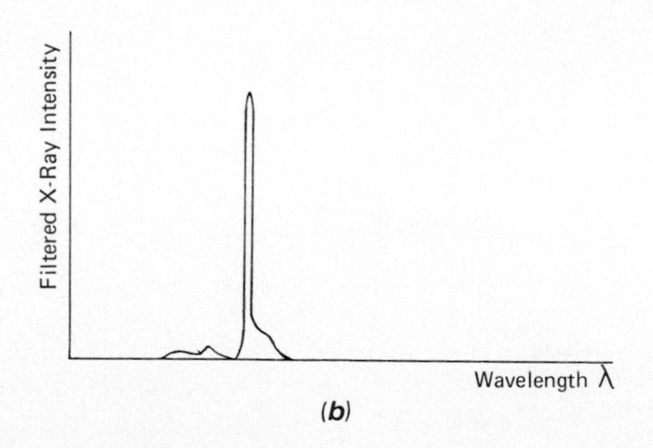

FIGURE 11-21. Production of essentially monochromatic X-rays by interposing an appropriate absorber in the beam of an X-ray tube. (a) The absorber is chosen so that one of its absorption edges lies between the two principal characteristic peaks of the emitted beam. The absorption is large except in the wavelength region of one characteristic wavelength. (b) The resultant filtered intensity as a function of wavelength is nearly monochromatic.

(single wavelength) beam of X-rays. One characteristic spike in the radiation from the target occurs at a wavelength at which the filter material has a low absorption coefficient; the other occurs where the filter absorbs efficiently. At other wavelengths, the filter absorbs a considerably greater fraction of the beam intensity. The availability of essentially monochromatic X-rays is necessary in order to implement the diffraction techniques that we shall now consider.

Determination of Crystal Parameters

X-rays interact with matter by scattering. The atoms of a crystal irradiated by X-rays act as scattering centers from which the X-rays are scattered in all directions. If an electromagnetic wave is incident on two scattering centers, the intensity I of the scattered radiation at any point in space is proportional to the square of the sum of the electric field strength due to each wave, i.e., $I \propto (\mathbf{E}_1 - \mathbf{E}_2)^2$, where $\mathbf{E}_1$ and $\mathbf{E}_2$ are the fields of the waves scattered by each of the scattering centers. If the path difference between the scattering centers and the observer (or detector) is an integral multiple of the X-ray wavelength, constructive interference occurs. In a regular array of scattering centers, like the row of atoms shown in Figure 11-22, the scattered intensity is that due to the sum of the contributions from each scattering center in the row, i.e., $I \propto (\Sigma \mathbf{E}_i)^2$. In this figure, where the

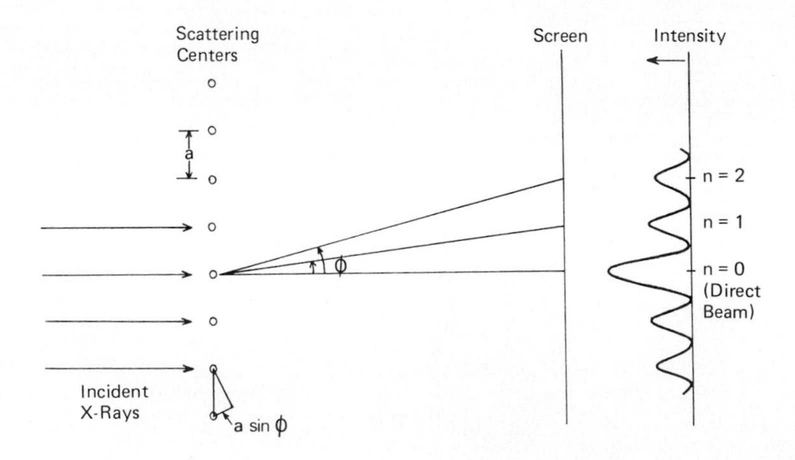

FIGURE 11-22. Scattering of an X-ray beam from a single row of atoms. Constructive interference causes intensity maxima to occur on the screen at angles ϕ such that $n\lambda = a \sin \phi$, where a is the distance between scattering centers and λ is the wavelength of the incident X-rays.

separation between scattering centers is a, constructive interference occurs only at particular angles ϕ such that

$$n\lambda = a \sin \phi \qquad n = 0, 1, 2, \ldots \tag{11-37}$$

As indicated in Figure 11-21, $a \sin \phi$ is the difference in path length for X-rays scattered by adjacent atoms.

A three-dimensional array of scattering centers may be analyzed similarly by summation of fields from individual scattering centers, though this procedure becomes quite complex. A simple procedure, introduced by Bragg (1912), analyzes the three-dimensional problem in terms of reflections from the various planes formed by the array of scattering centers. This simplified analysis gives the same results as the more physically satisfying, but more complex, scattering analysis. The Bragg analysis can be seen from Figure 11-23 to be obtained by considering a crystal composed of planes of atoms separated by a distance d. Plane waves incident on the crystal at an angle θ measured from one of the sets of planes are considered to be reflected from those planes at an angle equal to θ. The path difference between adjacent rays in the figure is $\overline{CB} + \overline{BD}$, and since $\overline{CB} = \overline{BD} = d \sin \theta$, the path difference is $2d \sin \theta$; the same path difference exists between every pair of rays. The reflected waves from many planes constructively interfere when the path difference $2d \sin \theta$ is an integral member of wavelengths, i.e., when

$$n\lambda = 2d \sin \theta \qquad n = 1, 2, \ldots \tag{11-38}$$

Equation (11-38) is *Bragg's law*, which is satisfied for all planes spaced d

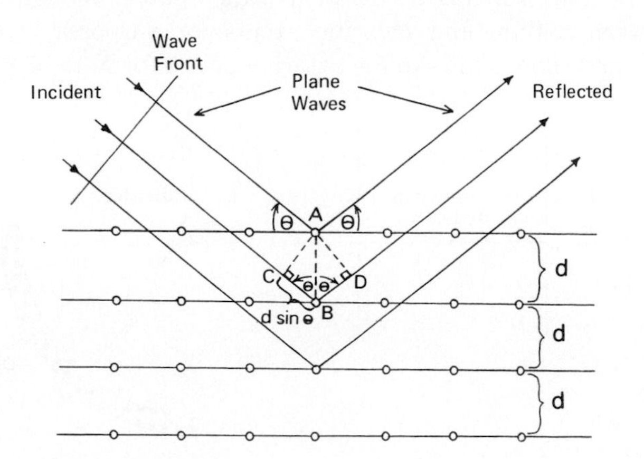

FIGURE 11-23. Bragg analysis of the reflection of plane waves of X-rays from planes of atoms separated by a distance d.

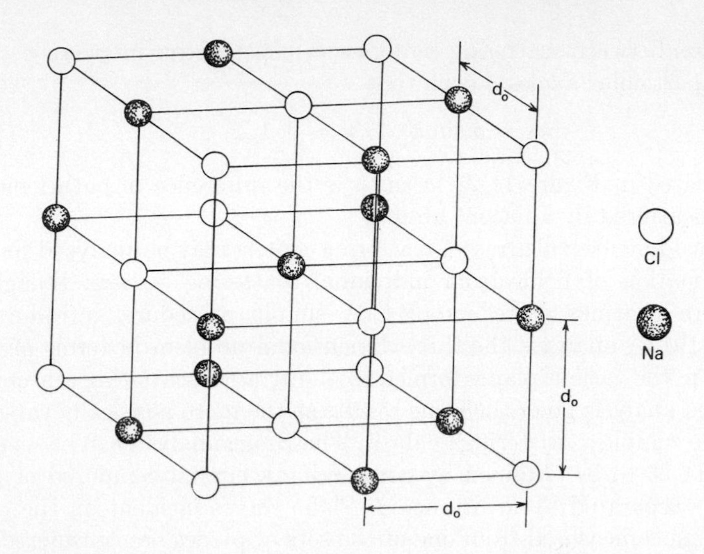

FIGURE 11-24. The structure of sodium chloride. This crystal is cubic, and the lattice spacing d_0 is the separation between nearest neighbor atoms.

apart. An intense reflection of X-rays from the crystal when Equation (11-38) is satisfied is called a *Bragg reflection*, and the whole number n that specifies each intense reflection is called the *order* of that reflection.

To illustrate the use of Bragg's law in obtaining information about crystal structure, let us consider a crystal of NaCl in which the atoms are spaced a distance d_0, called the *lattice constant*, from their nearest neighbor atoms. This structure, shown in Figure 11-24, is called cubic. We ignore the difference between sodium and chlorine atoms for purposes of diffraction because both have about the same scattering power for X-rays. Suppose we

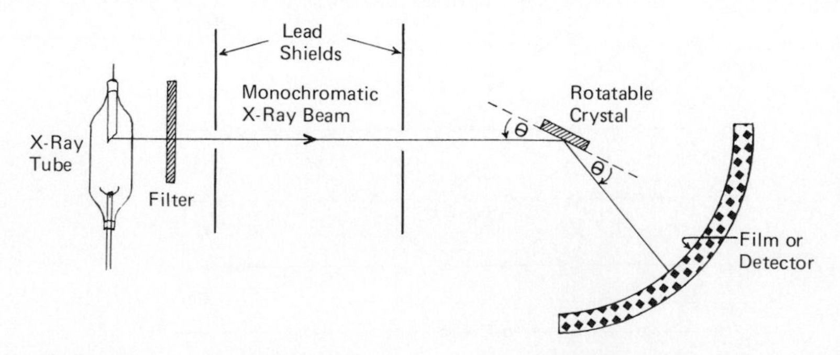

FIGURE 11-25. Irradiation of a crystal with X-rays. Bragg reflection occurs when θ is such that the Bragg condition $n\lambda = 2d \sin \theta$ is satisfied. The reflected maxima of X-ray intensity are recorded on film or other suitable X-ray detector.

irradiate a crystal of NaCl with essentially monochromatic X-rays of known wavelength λ, as illustrated in Figure 11-25. If the crystal is rotated slowly about an axis out of the page, θ will change, and Bragg reflections will be recorded on the photographic film (or other detectors) at appropriate angles. One series of angles that produces Bragg reflections will satisfy Bragg's law for $n = 1, 2, \ldots$, and provide the value of $d = d_0$. There will be other series of reflections that arise from other planes of atoms within the crystal. Figure 11-26 shows other planes in a cubic crystal; each set of these planes is separated by distance d_1, d_2, etc., all of which are less than d_0. We may note from Equation (11-38) that θ becomes larger as d becomes smaller, so the smallest angles at which Bragg reflection occurs corresponds to reflections from the planes separated by d_0. The intensity of the reflections from the planes separated by d_0 is greater than those from other planes because the density of scattering centers is greater.

Samples that can be obtained only in very small crystals, i.e., in powder form, can be studied in much the same way as larger single crystals. If a collimated beam of monochromatic X-rays is allowed to fall on a powder sample, as shown in Figure 11-27, it will encounter all possible orientations of atomic planes within the powder. Bragg reflections occur at every angle ϕ [$\phi = 2\theta$, where θ is the angle used in the Bragg law, Equation (11-38)]. Since the powder particles may have any azimuthal orientation about the incident pencil of X-rays, a circle appears on the detecting film corresponding to each Bragg angle. In this way, Bragg reflections from all sets of atomic planes in all orders are obtained on a single photograph, called a *Laue powder pattern.*

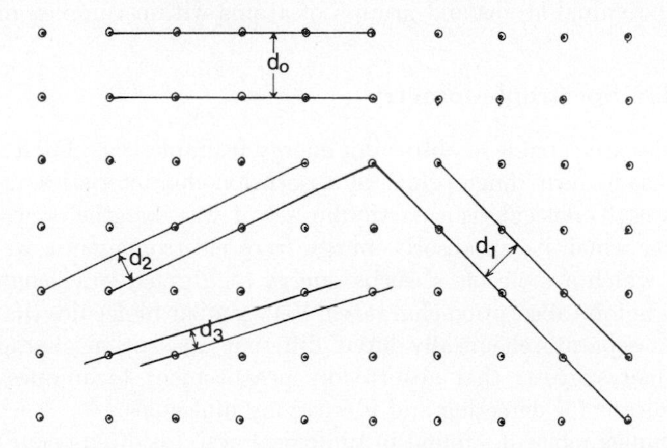

FIGURE 11-26. Bragg planes in a cubic crystal. As the separation between planes becomes smaller, the density of scattering centers in the planes becomes smaller.

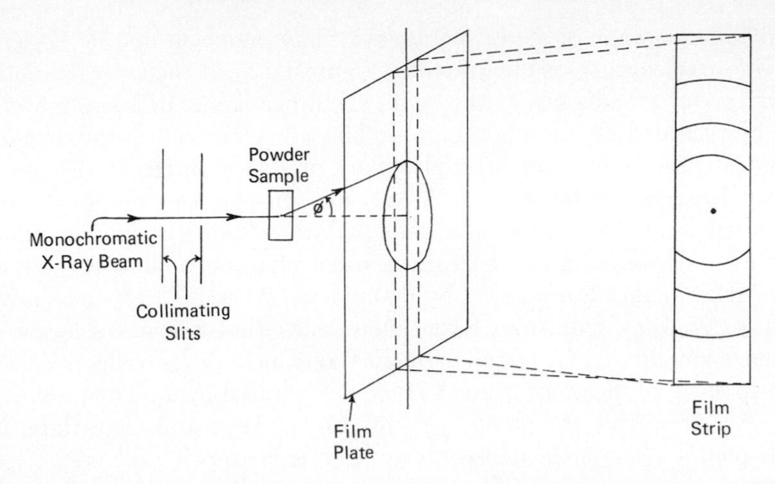

FIGURE 11-27. A schematic diagram of an X-ray powder diffraction apparatus. In a powder sample, all possible orientations of the atomic planes are presented to the X-ray beam, and Bragg reflections from each set of planes are recorded as arcs on a film strip. The display on the film strip is called a Laue powder pattern.

The foregoing examples illustrate how monochromatic X-rays can be used to determine one parameter of the simplest crystal forms. Similar parameters associated with more complex crystal systems may be determined by these and similar methods. But X-ray diffraction techniques involving far more involved analysis are available to specialists in this area of research, and these techniques, together with great patience and skill on the part of investigators, have yielded detailed information about the placement of individual atoms and groups of atoms within complex molecules.

Absorption Spectrophotometry

Molecules are capable of absorbing energy from photons. Each species of molecule has its own "fingerprint" of absorption characteristics, i.e., associated with each molecule is a particular set of wavelengths or wavelength ranges over which it can absorb energy from electromagnetic waves. The degree to which a molecule absorbs energy at a given wavelength is also part of its unique absorption character. Very similar molecules that may be difficult to separate chemically have different absorption characteristics, and this fact suggests that absorption measurement techniques offer an effective means for detecting and identifying molecules.

The complex molecules found in biological systems often occur in chemically similar groups, like the various heme proteins, the considerable number of chlorophylls, etc. The techniques of absorption measurement provide a means of distinguishing between molecules of such groups. Absorption techniques have been developed and refined in recent years; now the presence of even short-lived intermediate molecules in a series of reactions can be detected in this way. Changes in molecular structure that occur with changing temperature can also be detected by modern absorption measurement systems.

The absorption technique that has proved most valuable in studies of biological molecules is absorption spectrophotometry. We shall now consider some aspects of instrumentation, function, and operation of spectrophotometric devices.

Instrumentation

Most biological molecules absorb electromagnetic energy most readily in the visible (λ = 350 to 750 nm) and ultraviolet (λ = 1 to 350 nm) regions of the spectrum. Typical instruments used in the study of biological materials are designed to study absorption in all of the visible region and the adjoining ultraviolet region extending to wavelengths of 200 nm. Infrared instruments are available that extend the wavelength range upward from visible light to 1000 nm.

It is perhaps instructive to distinguish between a number of optical devices that are used in these spectral regions in order to clarify any further references to them:

A *spectroscope* is an instrument that presents the various wavelengths in the visible range so that they may be observed with the eye.

A *spectrometer* measures the wavelengths of emitted or absorbed electromagnetic radiations.

A *colorimeter* is an instrument that, by means of optical filters, selects particular wavelengths (colors) of visible light from a beam of white light, splits the colored light into two beams, and passes the two beams through different samples, whose absorption characteristics in the selected color band are compared.

A *photometer* is any device capable of comparing either the electromagnetic absorption or emission characteristics of two samples.

A *spectrophotometer* is a combination of a monochrometer and a photometer.

The instruments used in absorption spectrophotometry are extremely varied in the details of construction and operation. For our purposes we may consider a "typical" spectrophotometer, composed of a source of electromagnetic radiation, a monochrometer, a cell chamber containing the samples under observation, a detector, and a recording or display device.

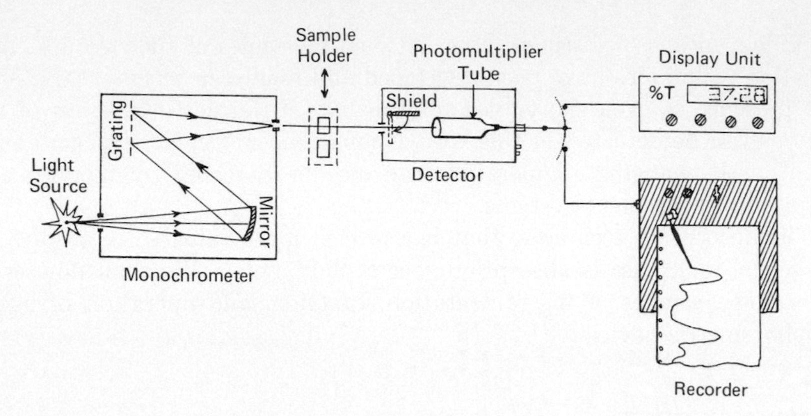

FIGURE 11-28. A schematic representation of a spectrophotometer. The components of the instrument are discussed in the text.

Figure 11-28 is a schematic illustration of the components that comprise a spectrophotometer.

LIGHT SOURCE AND POWER SUPPLY

Sources of electromagnetic radiation in the spectral regions of interest here are usually confined to two types, *discharge tubes* and *heated filaments*. Both types are usually incorporated into a modern spectrophotometer. Discharge tubes emit radiations that are primarily *line spectra* (discrete wavelengths), at the longer wavelengths but provide a continuous distribution of wavelengths in the ultraviolet region. In most studies, a source of continuous wavelengths is desirable, but line spectra are useful in some specialized studies such as fluorescence studies, in which the illuminating radiation must be filtered out. Sources with line spectra are also useful in the calibration of the monochrometer. Heated tungsten filaments (a less technical term for this type source is "light bulb") are used to provide visible continuous-wavelength radiation. Heated carborundum rods are often used as infrared sources whose wavelength distributions are continuous.

Very stable power supplies are required in spectrophotometers. Fluctuations in the voltage on power lines supplying the voltage across light sources can cause variations in light intensity that are greater than the differences in intensity due to the absorption that is being measured. The temperature T of a filament is proportional to the power consumed in heating the filament, so the temperature is proportional to the square of the voltage V across the filament. The *Stefan–Boltzmann law* tells us that the intensity I of electromagnetic radiation from a radiating object is proportional to the fourth power of the temperature. Then the light intensity is proportional to V^8. This means that relatively small voltage variations can cause significant light intensity variations. In certain spectrophotometric studies, such as the measurements of enzyme concentrations, voltage fluctuations across the light source must be less than 3 parts in a million.

MONOCHROMETER

The monochrometer in a spectrophotometer accepts light from the source through an entrance slit, selects a narrow wavelength band of radiation, and directs this essentially monochromatic light through an exit slit.

Figure 11-29 shows the basic elements of a *prism monochrometer*, one of two commonly used wavelength selectors. Light from the entrance slit is collimated (made into a parallel beam) by a lens and is introduced into a prism. The prism is made of glass or quartz depending on the wavelength range of the light being used (quartz is relatively transparent farther into the ultraviolet range than glass). A prism material exhibits different indices of refraction for different wavelengths of light, a phenomenon known as *dispersion*. In passing through the prism, longer wavelengths are deviated less than shorter ones. A second lens concentrates the dispersed beam that emerges from the prism onto the exit slit. The particular wavelength band of light leaving the exit slit is controlled by rotating the prism about an axis out of the page.

A *grating monochrometer*, the essential features of which are illustrated in Figure 11-30, uses a *diffraction grating* to separate spatially the light with different wavelengths. A diffraction grating is a surface on which alternate transparent and opaque slits are spaced closely. When the spacing between

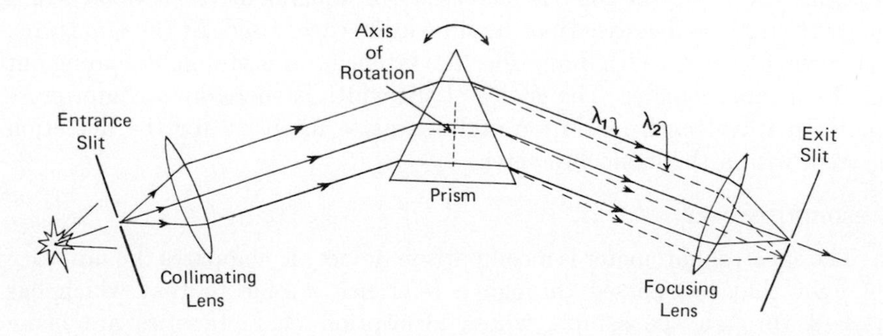

FIGURE 11-29. A rotating prism monochrometer. Rotation of the prism about the axis shown causes different wavelengths of light to emerge from the exit slit.

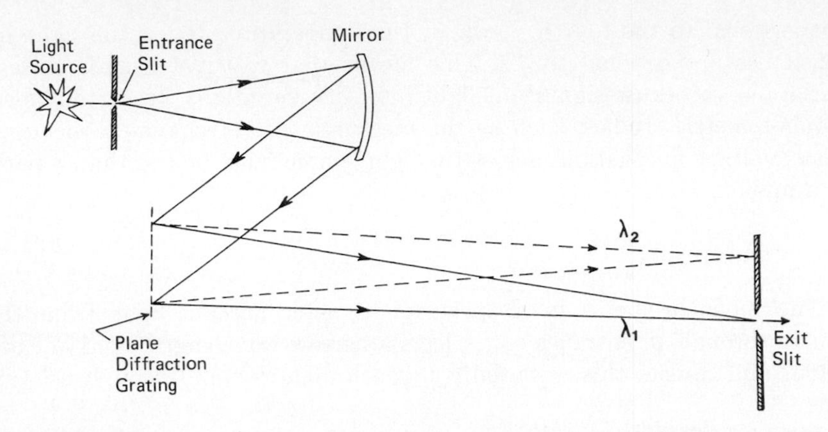

FIGURE 11-30. A grating monochrometer. Different wavelengths of light constructively interfere at different points along a plane where the exit slit is positioned to select a particular wavelength.

transparent slits of the grating is b, light incident on the grating will be diffracted at each slit, and light rays of wavelength λ that leave the grating at an angle θ from the normal to the grating will be in phase when the condition

$$m\lambda = b \sin \theta \qquad m = 0, 1, 2, \ldots \tag{11-39}$$

obtains. A converging lens or mirror brings the in-phase parallel rays together at the focal plane of the lens, where the rays constructively interfere. Thus, different wavelengths, emerging from the grating at different angles, are separated spatially. Rotation of the grating about an axis out of the page brings different wavelengths to the exit slit.

Since the beam of light inside the monochrometer at the exit slit of either type monochrometer has a continuous distribution of wavelengths across the slit, the purity of color is increased (or equivalently, the band width of wavelengths is decreased) as the slit width is decreased. At the same time, narrowing the slit width diminishes the intensity of useful light coming out of the monochrometer. The choice of slit width is therefore a compromise between wavelength resolution and adequate intensity for the detection system within the monochrometer.

ABSORPTION CELL

The spectrophotometer is a comparison device. It compares the intensity of light that has passed through a reference sample to that which has passed through the sample whose absorption characteristics are being examined. Typically two sample holders, one containing solvent alone and the other containing solvent plus the molecules under study, are successively placed in the beam from the exit slit of the monochronometer. The sample holders, called *cuvettes*, are containers of glass or quartz with two parallel faces (of known separation), which are oriented perpendicularly to the light beam.

DETECTOR

The light transmitted through the sample and reference material is passed into a detecting system, usually a photomultiplier tube but sometimes a thermopile or photoconductive cell—all devices that convert the energy of the light into an electrical signal. The circuitry associated with the detector in a spectrophotometer is usually designed so that when the light has passed through the reference sample, the detector output indicates 100% transmission. Similarly, when the light is completely blocked off from the detector, the output of the detector is adjusted to read "zero." Then, with appropriately linear circuitry, the detection system is capable of measuring transmission of light through the sample directly in percentage of light transmitted.

In modern systems the detector output is usually displayed in digital readout devices directly in percent transmission or some other convenient measure of absorption (some of the common measurement parameters of spectrophotometry are discussed in a subsequent section of this chapter). The output of the detector is sometimes connected to writing recorders that are driven in synchrony with the rotation of the grating (or prism) in the monochrometer; in this case a permanent record of the sample's absorption characteristics is obtained as a function of wavelength.

Physical Parameters of Absorption Spectrophotometry

The object of a spectrophotometric measurement is to obtain the ratio of the light intensity I_0 passing through a reference sample (or "standard") containing only a solvent to the intensity I passing through a similar sample that contains the same solvent *and* the molecules under study. The absorption of light is described mathematically by the same equation used for the absorption of X-rays:

$$I = I_0 \, e^{-\mu x} \tag{11-40}$$

where x is the thickness of the sample, and μ is the absorption coefficient —exactly the same as Equation (11-36). Spectroscopists refer to this relationship as *Lambert's law*. Again we should note that, just as in the case for X-rays, the absorption coefficient μ is a function of wavelength λ, i.e., in general, μ is different at different wavelengths. The more general form of Lambert's law is

$$I = \int I_{0_\lambda} e^{-\mu(\lambda)x} \, d\lambda \tag{11-41}$$

when I_{0_λ} is the intensity of the incident light lying within the wavelength range between λ and $\lambda + d\lambda$. Equation (11-41) is difficult to use because μ is a function of λ. In practice, narrow bands of wavelength are used so that μ is essentially constant throughout that range or is at worst a simple function of λ.

It is very important that the solutions used as samples in spectrophotometers contains low concentrations of the absorbing materials under study. At low concentrations of the absorber, the absorption coefficient μ is proportional to the concentration, or

$$\mu = \beta c \tag{11-42}$$

in which c is the concentration in moles/liter of absorbing molecules and β is a constant, the *molar absorption coefficient*. Equation (11-42) is called *Beer's law*. At high concentrations of absorbing molecules, both Lambert's law and Beer's law fail; consequently, it is important to check experimentally the validity of these relationships before extended measurements are begun on a sample solution.

Experimentally, μ can be determined from measurements of light intensity, since from Equation (11-40) we may obtain

$$\mu = \frac{1}{x} \ln \left(\frac{I_0}{I} \right) \tag{11-43}$$

It is more customary, however, to measure a quantity D defined by

$$D = \log_{10} \left(\frac{I_0}{I} \right) \tag{11-44}$$

D is called the optical density, and by comparing Equation (11-43) and (11-44), we see that μ and D are related by

$$\log_{10} \left(\frac{I_0}{I} \right) = \frac{1}{2.3} \ln \left(\frac{I_0}{I} \right)$$

$$\tag{11-45}$$

$$\mu = \frac{2.3D}{x}$$

The values of μ and β can be specified in many ways, in terms of various units, depending on the units chosen for length x and concentration c.

Table 11-1 Common Terms and Symbols Used in Spectrophotometry

Term	Symbol	Defining equation	Units
Transmittance	τ, T	$\tau = 100\left(\frac{I}{I_0}\right)$	percent
Optical density	D, o.d., E	$D = \log_{10}\left(\frac{I_0}{I}\right)$	pure number
Extinction coefficient	ϵ, κ	$\epsilon = \frac{D}{x}$	x in cm, ϵ in cm^{-1}
Molar extinction coefficient	ϵ_{mol}	$\epsilon_{mol} = \frac{D}{cx}$	c in mole/liter, ϵ in liter/mole cm
Millimolar extinction coefficient	ϵ_{mM}	$\epsilon_{mM} = \frac{D}{cx}$	c in millimoles liter, ϵ_{mM} in liter/millimole cm
Absorption coefficient	μ	$\mu = \frac{1}{x}\ln\left(\frac{I_0}{I}\right)$	x in cm, μ in cm^{-1}
Molar absorption coefficient	β	$\beta = \frac{\mu}{c}$	c in mole/liter, β in liter/mole cm

Many forms and symbols are used in current literature to report results of absorption measurements. Table 11-1 is a compilation of a number of common terms and symbols in present usage. The symbols used in the defining equations of the table are consistent with those we have been using in this section.

Specialized Spectrophotometric Systems

Almost every individual spectrophotometer arrangement used in research is in some way specialized to serve the particular demands of particular experiments. Some of the refinements and specializations are relatively common. Two types of special systems will be briefly described here.

FLOW SYSTEMS

In order to make absorption observations on molecules involved in very rapid chemical reactions (i.e., those having reaction times of less than a millisecond), it is necessary that the reagents be mixed and observed before the reaction has progressed very far. In general, rapid observations cannot be made using cuvettes, in which mixing times usually exceed one second.

Flow systems, an example of which is shown in Figure 11-31, are arrangements that bring reactants together immediately before the mixed solution enters the path of the light beam in the spectrophotometer. The reagents A and B are in separate tubes and are mixed at rates controlled by the weights W placed on the plunger. If the liquids flow at a constant rate, the absorption of light is constant. A number of absorption measurements are usually made, each at a different flow speed. Such a set of measurements is equivalent to a set of observations of absorption at different times after mixing and provides information about how the reaction proceeds as a function of time. Very rapid reactions can be studied in this manner, even those whose half-times (the times in which half the reagents have reacted chemically) are of the order of one millisecond.

If a reaction is not too fast (for example, those having half-times greater than 30 milliseconds), the flow of the solutions can be stopped after mixing is completed. Then changes in absorption by the mixture are recorded

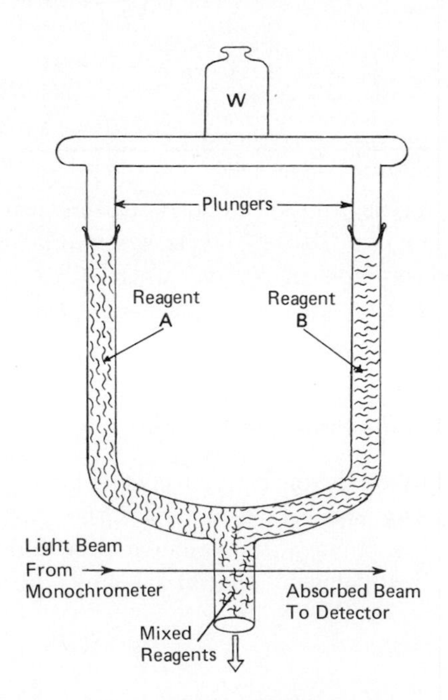

FIGURE 11-31. A flow system for spectrophotometric measurements of reagents as a function of time. The time between the mixing of the reagents and the observation of the absorption characteristics of the mixture is controlled by varying the weight W.

directly as a function of time. This use of the flow system is called the *stopped-flow method.*

SPLIT-BEAM AND DUAL-BEAM SPECTROPHOTOMETERS

If, for any reason, light intensity changes are introduced during the course of a spectrophotometer run, errors in absorption measurements may result. A *split-beam* spectrometer eliminates this source of error by causing the light beam from the monochrometer to be split so that the beam rapidly alternates between two paths, one of which passes through the reference sample and the other of which passes through the sample under study. The detector output then gives rapid point-by-point pairs of signals, the difference between which can be calibrated to give the difference in absorption between reference and sample under study during very short time intervals. Thus, any variation in light intensity from the source is shared by both samples being compared, and no error due to light intensity variations is introduced into the absorption measurement.

Split-beam spectrophotometers offer a further advantage. As an example, two suspensions of bacteria, one in the presence of oxygen and one deprived of oxygen, could be compared to each other instead of comparing each to a standard. The absorption spectrum thus obtained is a difference spectrum and indicates the absorption characteristics of molecules present in one, but not the other, of two suspensions. Such systems are used to study metabolites of anaerobic bacteria and provide one of the means by which bacteria are classified or identified.

A *dual-beam* spectrophotometer is used to detect specific changes in absorption that might occur during a reaction, changes that are not of a general, nonreactive nature, like the swelling or settling of suspended particles. In the dual-beam instrument, the light from the source alternately passes through two monochrometers, each of which provides a beam of different wavelength. The wavelengths are chosen so that the absorption characteristics of the specific products of the reaction being studied are affected at only one of the wavelengths. The output beams of both monochrometers are passed through the sample. The changes in absorption due to nonspecific changes in the sample can then be subtracted out of the resulting absorption data, leaving the desired absorption characteristics associated with the specific change.

References and Suggested Reading

Engström, A. (1950). Use of soft x-rays in the assay of biological materials. *Progr. Biophys.* **1**, 164.

Fraser, W. D. B. (1953). The infrared spectra of biologically important molecules. *Progr. Biophys.* **3**, 47

Harrison, G. R., Lord, R. C., and Loofbourow, J. R. (1948). "Practical Spectroscopy." Prentice-Hall, Englewood Cliffs, New Jersey.

Heywood, V. H., ed. (1971). "Scanning Electron Microscopy." Academic Press, New York.

Hiskey, C. F. (1955). Absorption spectroscopy. *In* "Physical Techniques in Biological Research" (G. Oster and A. W. Pollister, eds.), Vol. 1, p. 74. Academic Press, New York.

Koehler, J. K., ed. (1973). "Advanced Techniques in Biological Electron Microscopy." Springer-Verlag, Berlin and New York.

Lonsdale, K. (1949). "Crystals and X-Rays." Van Nostrand-Reinhold, Princeton, New Jersey.

Rhodes, J. E. (1949). Demonstrating the phase contrast principle. *Amer. J. Phys.* **17**, 70.

Richards, O. W. (1955). Fluorescence-antibody method. *In* "Analytical Cytology," 2nd ed. (R. C. Mellors, ed.), p. 1. McGraw-Hill, New York.

Osterberg, H. (1955). Phase and interference microscopy. *In* "Physical Techniques in Biological Research" (G. Oster and A. W. Pollister, eds.), Vol. 1, p. 378. Academic Press, New York.

Swift, J. A. (1970). "Electron Microscopes." Barnes & Noble, New York.

White, G. W. (1966). "Introduction to Microscopy." Butterworth, London.

Wischnitzer, S. (1970). "Introduction to Electron Microscopy," 2nd ed. Pergamon, Oxford.

CHAPTER XII **Techniques Using Nuclear Phenomena**

Nuclear physics has made significant and unique contributions to the biological sciences. The radiations from certain atomic nuclei and the distinguishing characteristics of certain forms of atoms provide the tools with which living systems can be examined in aspects that are not available through normal chemical methods. In this chapter we shall examine some of the techniques by which nuclear phenomena have been applied to the understanding of biological systems. We shall first summarize some basic concepts of nuclear physics, particularly those that emphasize the nature and production of nuclear radiations. We shall briefly touch on the instrumentation employed in assaying, or measuring, nuclear radiations and then proceed to *tracer techniques*, the methods by which a particular species of atomic nucleus, i.e., a specific *nuclide*, can be incorporated in a living system and later located, thereby supplying information about the biochemical pathway of the nuclide. Finally, we shall touch on some of the high points of a continuously advancing area of biophysical research that is usually referred to as "radiation biology." This area is concerned with the effects—destructive, mutational, and therapeutic—that nuclear and atomic radiations produce in living systems.

Fundamental Properties of Atomic Nuclei

All matter is composed of atoms. The modern concept of the atom identifies the massive central core of the atom as the *nucleus*, surrounded by orbital electrons. Since we shall be concerned here primarily with nuclear phenomena, the terminology and systematics of appropriate aspects of nuclear physics will be reviewed briefly.

Nuclear Terminology

Atomic nuclei are composed of positively charged *protons*, each of which is a hydrogen nucleus, and uncharged *neutrons*. The mass of a proton and that of a neutron is nearly the same. The number of protons in a given atom determines the *atomic number* Z of that atom. Atoms of different atomic number are different elements, and the chemical properties of an atom are determined by its atomic number. The *atomic mass number* A is the total number of protons and neutrons in the nucleus of an atom. Thus, every nuclide is an atomic form with a nucleus having a specific Z and A. A nuclide is uniquely specified by its Z an A, but a nuclide is symbolically represented by using the chemical symbol of the element (which is determined by Z) with Z specifically expressed by a left subscript and A expressed by a left superscript, e.g., ${}^{1}_{1}H$, ${}^{18}_{8}O$, ${}^{31}_{15}P$. Note that subtraction of the subscript from the superscript gives the number of neutrons in the nuclide. There are, for example, 10 neutrons in ${}^{18}_{8}O$.

Nuclides that have the same Z but different A are said to be *isotopes* of the element characterized by Z. For example, ${}^{16}_{8}O$, ${}^{17}_{8}O$, and ${}^{18}_{8}O$ are isotopes of oxygen, each of which has a nucleus containing 8 protons and 8, 9, and 10 neutrons, respectively. *Isobars* are nuclides with the same A but different Z, like ${}^{40}_{18}A$ and ${}^{40}_{20}Ca$.

Some nuclei are of particular importance in nuclear phenomena and have special names. The deuteron (d) is the nucleus of a heavy isotope of hydrogen (${}^{2}_{1}H$) and consists of one proton and one neutron. The *alpha particle* (α) is the nucleus of a helium atom (${}^{4}_{2}He$) and is composed of two protons and two neutrons. Another heavy isotope of hydrogen, tritium (${}^{3}_{1}H$), has a nucleus called a *triton* (t) with one proton and two neutrons.

Nuclear particles, protons and neutrons, are positively charged and neutral, respectively. There is therefore a repulsive force between the positive charges that tends to disrupt a nucleus. A *nuclear force* is the cohesive force between *nucleons* (a generic term for nuclear particles, i.e., a nucleon is either a proton or a neutron). Nuclear forces bind the nucleons together despite the mutually repellent electrical forces between protons. The energy required to keep the nucleus together is called the *nuclear binding energy*. In some isotopic forms of nuclides, the binding energy is sufficient to hold the nucleus together for an indefinite period of time. These isotopes are said to be *stable*. Examination of a table of nuclides shows that the stable isotopes among the lighter elements are those for which the number of neutrons is roughly equal to the number of protons. Actually, the "line of stability" passes through nuclides for which the ratio of neutrons to protons increases slowly as Z increases so that the neutron to proton ratio for very heavy stable elements, like Pb, is about 1.5. It is also found that isotopes with even Z are in general more stable than those with odd Z; elements with even Z usually have more isotopes than the elements with odd Z, which seldom have more than two isotopes.

Among the many empirical facts that have been garnered in studies of nuclear systematics, one will be of particular interest for our purposes. When the relative isotopic composition of elements is studied, it is found that the relative abundance of the isotopes of nearly every element is essentially constant. This fact is true regardless of the source from which the element is obtained. For example, all naturally occurring sulfur is found to be composed of 95.02% ${}^{32}S$, 0.75% ${}^{33}S$, 4.21% ${}^{34}S$, and 0.02% ${}^{36}S$. Even elements comprising extraterrestrial materials, such as meterorites and rocks from the moon, have the same isotopic content as terrestrial samples. This means that alteration of the natural isotopic content of any element constitutes "tagging" or *labeling* of that element. Isotopes introduced into a normal system to label an element of that systems are called *tracers*. The inclusion of labeled atoms into a species of molecule results in a labeled molecule.

Any nuclide that does not have sufficient binding energy to keep its nucleus together indefinitely is said to be *unstable*. Unstable nuclides adjust themselves to stable configurations by releasing energy in the form of nuclear radiations, either particles with kinetic energy or electromagnetic waves (photons) with their associated energy, hf, where h is Planck's constant and f is the frequency of the wave. The process by which unstable nuclides transform themselves to lower energy states by radiation is called *radioactive decay*, a process we shall consider further.

Radioactivity

An excess of protons or neutrons from the proportion that characterizes the stable isotope of an element causes the nucleus to be unstable. The redistribution of numbers of nuclear particles is the basis of radioactive decay. Nuclei achieve stability by undergoing one or more of the following processes:

1. The transformation of neutron into a proton. This process is the release from the nucleus of a *negatron*, an electron of nuclear origin, often called a negative *beta ray* (β^-).
2. The transformation of a proton into a neutron. This process consists of either releasing from the nucleus a *positron*, a positive beta ray (β^+), whose mass and magnitude of charge are identical to that of an electron, or capturing an extranuclear electron (usually from the K-shell, whence the name *K-capture* for this process).
3. The emission of α-particle from the nucleus.

All of the above processes may be accompanied by the emission of electromagnetic waves—*gamma rays* (γ)—when the waves originate in the nucleus, and X-rays when they result from extranuclear electron transitions between orbits (see Chapter XI) resulting from the holes produced in inner shells by electron capture. In some cases, in lieu of the emission of nuclear γ-rays, the excess nuclear energy is released by exciting electrons out of their extranuclear orbits. This process, *internal conversion*, produces an electron and the accompanying X-rays that result from filling the orbital vacancy created by the ejected electron.

DECAY LAWS

Since nuclear radiations arise primarily because of nuclear constitution, the rate of nuclear reconstitution is not usually altered by either chemical or physical means. The decay rates for processes involving atomic electrons, electron capture, and internal conversion can be affected slightly by the chemical environment to which the radioactive material is exposed. Radioactive decay is a purely statistical process. This means that we may assume that the rate of decay for radioactive atoms at any time is proportional to the number of such atoms and is independent of the age of any particular atom. Then the number of atoms N of a radioactive isotope is related to the change in N, i.e., the number decaying ΔN, in a time Δt by

$$\frac{\Delta N}{\Delta t} = -\lambda N \tag{12-1}$$

Equation (12-1) means that the time rate of change of the number of atoms is proportional to the number present, and λ is the constant of proportionality called the *disintegration constant.* For small time intervals dt, we may write Equation (12-1) as

$$\frac{dN}{dt} = -\lambda N$$

or

$$\frac{1}{N}\frac{dN}{dt}\,dt = -\lambda\,dt \tag{12-2}$$

which can be integrated directly:

$$\int \frac{1}{N}\frac{dN}{dt}\,dt = -\int \lambda\,dt$$

$$\ln N = -\lambda t + C \tag{12-3}$$

If we assume the number of undecayed atoms is N_0 at $t = 0$ and evaluate Equation (12-3) at $t = 0$, we have

$$\ln N_0 = C$$

$$\ln\left(\frac{N}{N_0}\right) = -\lambda t \tag{12-4}$$

or

$$N = N_0 e^{-\lambda t} \tag{12-5}$$

which is the exponential decay law for radioactive transformation of a single isotope.

A useful characterization of a radioactive isotope is its *half-life.* The intensity of radioactivity, or the "activity" is dN/dt (the number of decays per unit time), which is $dN/dt = -\lambda N$. The activity is half of its initial value when $N/N_0 = \frac{1}{2}$. The half-life, $\tau_{1/2}$, is then the time required for the intensity (and therefore the number of undecayed atoms) to be reduced to half its initial value; $\tau_{1/2}$ is expressed in terms of λ by substituting $N/N_0 = \frac{1}{2}$

into Equation (12-4):

$$\ln \tfrac{1}{2} = -\lambda\tau_{1/2}$$

$$\ln 2 = \lambda\tau_{1/2}$$

$$0.693 = \lambda\tau_{1/2} \tag{12-6}$$

or

$$\tau_{1/2} = \frac{0.693}{\lambda}$$

The half-life of a radioactive isotope is determined experimentally by measuring the activity as a function of time.

For a simple decay, a linear plot results if the log (base 10 is easiest) of the activity is plotted against time, as illustrated in Figure 12-1. When

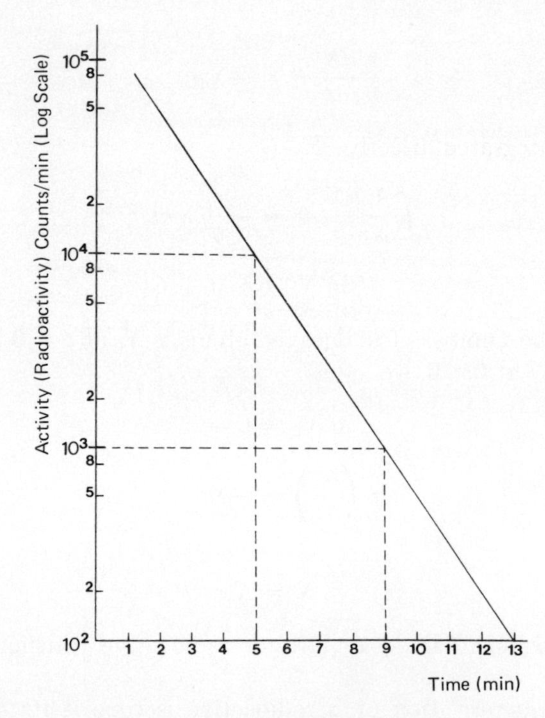

FIGURE 12-1. A radioactive decay curve. The decay of a single nuclide appears as a straight line when the logarithm of the activity is plotted as a function of time. The coordinates indicated by dotted lines in the illustration are used in the text to show that this particular curve corresponds to a nuclide whose half-life is approximately 1.2 seconds.

base 10 is used, Equation (12-4) may be rewritten as

$$2.31 \log_{10} \frac{N}{N_0} = -\lambda t$$

or

$$2.31 \log_{10} N = -\lambda t + 2.31 \log_{10} N_0$$

The slope of the logarithmic plot is then

$$\frac{\Delta \log_{10} N}{\Delta t} = -\frac{\lambda}{2.303}$$

As an example, the values indicated by the dotted lines in Figure 12-1 may be used giving

$$\frac{\log_{10} 10^4 - \log_{10} 10^3}{5 - 9} = \frac{4 - 3}{-4} = -\frac{1}{4} = -\frac{\lambda}{2.303}$$

so that $\lambda = 0.576 \text{ min}^{-1}$ for the plot in this figure. Equation (12-6) shows that this value of λ corresponds to a half-life of about 1.2 seconds.

Rather complex decay chains sometimes occur in which parent atoms decay into daughter atoms, which are also radioactive and decay with a disintegration constant different from that of the parent. The daughter may in turn decay, and so on. The radiation characteristics of such chains may be analyzed by methods not too much more complex than we have used for a single isotope. Although we shall not pursue the analysis here, isotopes in such chains are useful as tracers in biological systems and are commonly employed as tracers.

The units in which radioactivity is measured are based on the number of disintegrations occurring per second. The *curie,* chosen because it represents the activity of 1 gm of radium, is 3.7×10^{10} disintegrations per second (dps). This is a very large unit for practical purposes, and the *millicurie* and *microcurie* are usually encountered. Often it is more significant to specify the *specific activity* of the sample, which is expressed in disintegrations per second per gram of substance.

PRODUCTION OF RADIOSOTOPES

There are about forty naturally occurring radioisotopes, most of which occur among elements with large atomic number. Hundreds of useful isotopes are prepared by bombarding target atoms with protons, deuterons, neutrons, or other nuclear projectiles. When the resulting isotopes are radioactive, the process is said to produce artificial radioactivity.

Cyclotrons are usually used to produce radioisotopes that result from bombardment of target atoms by protons or deuterons. Nuclear reactors are sources of intense neutron fluxes, and the majority of radioisotopes used as biological tracers are prepared at reactor facilities.

Nuclear reactions are expressed in equations similar to chemical reactions. For example,

$$^{14}_{7}\mathrm{N} + n \rightarrow p + ^{14}_{6}\mathrm{C} \tag{12-7}$$

describes the production of $^{14}_{6}\mathrm{C}$, a radioactive isotope of carbon, by the bombardment of nitrogen with neutrons. Equation (12-7) indicates that a proton is produced along with $^{14}_{6}\mathrm{C}$. A shorthand notation is usually used to indicate nuclear reactions; the target material is indicated first, followed by parentheses with the bombarding particle and emitted particle (or wave) enclosed in that order, and the product nuclide is indicated after the parentheses. The reaction of Equation (12-7) is indicated by the following:

$$^{14}_{7}\mathrm{N}(n, p)^{14}_{6}\mathrm{C} \tag{12-8}$$

One concept of nuclear reactions considers the reaction as a two-step process in which the target atom and bombarding particle merge to form an intermediate nucleus (or compound nucleus), which later breaks up to form the product nucleus and emitted particle. In this model, it is assumed that the impinging particle distributes its energy more or less uniformly among all the particles of the intermediate nucleus, whose total energy content is raised. Since the excitation energy is divided among the nucleons of the intermediate nucleus, no one particle (on the average) has enough energy to escape the nucleus. Disintegration of the intermediate nucleus takes place when sufficient energy for escape is concentrated in one particle. The intermediate nucleus does not "remember" which of its constituent particles was the impinging particle. The amalgamation process is assumed to take place in about 10^{-20} second, whereas the emission process takes about 10^8 times longer. The emitted particle does not take away all of the excitation energy, and the product nucleus remains in an excited state, from which it ultimately drops to lower energy states by emission of light particles such as β^-, β^+, or γ-rays. This relaxation of the product nucleus to a lower energy state takes place with comparatively long characteristic times (a second to millions of years) and constitutes the radioactivity associated with the product nucleus. $^{14}_{6}\mathrm{C}$, for example, reverts to the original target nucleus, $^{14}_{7}\mathrm{N}$, by β^- decay, and the half-life of the decay is 5600 years.

Isotopic Tracer Methods

The use of tracer techniques in the study of biological systems depends on labeled elements, molecules, or atomic groupings. The labeling of biosystems is accomplished by incorporating into living systems either radioactive isotopes or stable, but rare, isotopes. The labeled material must later be identified and assayed in some manner to provide useful information. Tracer methods depend on the fact that living organisms, in the chemical conduct of their lives, cannot distinguish between different isotopes of an element. (There are rare exceptions like hydrogen, where the mass ratios of its isotopes are extraordinarily high and can be distinguished, for example, by the rates at which they diffuse across cellular membranes.) In general, labeled isotopes are as easy to incorporate into living systems as natural materials. Another fact, already mentioned, permits the accurate assay of tracers. The relative abundance of isotopes of a given element in its natural occurrence, regardless of the source of the element, is constant.

In our considerations of tracer methods, let us first indicate briefly what is involved in the assay of isotopic tracers.

Assay

When biological systems have been labeled, either with radioactive isotopes or with relatively rare stable isotopes, the determination of the quantity of labeling atoms at particular sites within the system is called the isotopic assay. In the case of radioactive isotopes, assay is accomplished by radiation detectors—either electronic instruments or photographic films. Stable isotopes are generally assayed by means that can distinguish between the small mass difference of the isotopes of one element. One such technique is mechanical centrifugation, the discussion of which will be postponed until Chapter XIII. Here we shall consider the assay of radioactive isotopes.

RADIATION DETECTORS

A large number of devices are available that detect nuclear radiations. Here we shall consider only a few representative types.

One of the oldest and simplest nuclear detectors is a gas-filled container, shown in Figure 12-2a, with a central electrode (usually a thin wire) maintained at a positive potential relative to the conducting wall of the container. When ionizing radiation passes through the tube, positive ions and electrons will be produced in the filling gas. In general, the electrons will be attracted to the positive center wire, and the positive ion will move to the

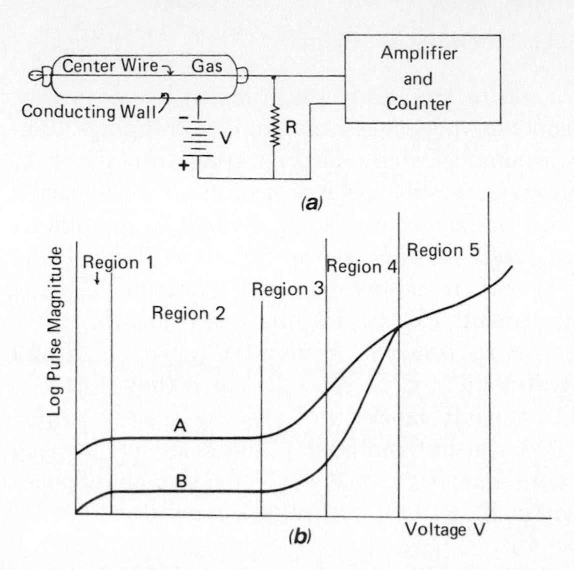

FIGURE 12-2. (a) A gas-filled radiation counter. (b) Curves illustrating the magnitude of pulses produced across the resistor R as a function of voltage V for ionizing events producing a large number of initial ion pairs (curve A) and a large number of initial ion pairs (curve B).

outer wall. The total charge passing through the resistor R of Figure 12-2a due to a single incident ionizing particle depends on the dc voltage V maintained between the central wire and counter wall and on the energy and type of particle causing ionization. Figure 12-2b shows schematically the magnitude of voltage pulses across R as V increases for two typical ionizing particles. Curve A represents the response to a highly ionizing particle, such as an α-particle, and B represents that of a relatively slightly ionizing particle, such as a β-particle. Or, alternatively, both curves might represent the same kind of particle, but curve A would represent a more energetic particle than curve B. In region 1, recombination of the ions and electrons takes place to some extent, and not all of the original ion pairs are detected. In region 2, essentially all ion pairs produced by the detected particle are collected. This is the *ion-chamber region*. Region 3 is called the *proportional region*, and here the electrons are accelerated by the high electric fields near the center wire so that they collide with gaseous atoms and form more ion pairs, which tend to cascade into the production of more ion pairs, etc. In region 4, the difference between effects produced by different initial ionizing particles diminishes until, in region 5, called the *Geiger–Mueller* region, even a minimally ionizing particle produces a very large pulse. Although occasionally the gas-filled detectors are used as proportional counters or ionization chambers, they are most frequently used as Geiger–Mueller counters, which are, in spite of a number of more sophisticated radiation detectors that have been developed, still useful because they are simple, inexpensive, and reliable radiation detectors.

Photographic film plates are exposed by nuclear radiations, and under many circumstances, where the location of the radiation source is of primary interest, this method of detection is simple and effective.

Modern assay systems usually use one of two types of detectors that have been highly developed for sensitivity and resolution (the ability to distinguish between radiations of different energies). We have already mentioned scintillation counters as X-ray detectors (Chapter XI), which typically consist of NaI crystals that absorb radiation and produce photons, which are transformed into an electrical pulse by a photomultiplier tube. Solid state detectors made of germanium and silicon, drifted (i.e., diffused) with lithium, have excellent resolution. These solid state detectors function very much like proportional counters in which a solid medium has replaced the gas.

Figure 12-3 shows a comparison of the response of a scintillation counter and a Ge (Li-drifted) detector to a single γ-ray produced by a ^{137}Cs source. This isotope of cesium emits a single γ-ray of energy 662 keV. In each of the curves, the *photopeak* appears centered about 662 keV. As the name suggests, this peak is the result of the γ-ray being absorbed in the detecting crystal by giving up its entire energy to one of the electrons freed from an atom of the absorbing crystal. In the scintillation counter, this electron, called a photoelectron, produces *luminescence* (the emission of photons of visible or ultraviolet light) to which the crystal itself is transparent. This light is collected by the photomultiplier tube and converted to an electrical pulse whose amplitude is proportional to the energy of the incident γ-ray. In the Ge detector, the photoelectrons cause charge "carriers" (conduction electrons and "holes," as they are called by solid state physicists) to be formed in numbers proportional to the energy of the incident γ-ray. An electric field, maintained by a battery across the semiconductor (Ge), sweeps the electrons away through an electrical circuit that produces a pulse. This output pulse is also proportional to the energy of the totally absorbed incident γ-ray. The pulses produced in both detectors at lower amplitudes than the photopeak are due to Compton scattering, or degradation of the photon incident energy by collisions with electrons in the detecting crystal, in which only a part of the photon energy is transferred to an electron.

Both the scintillation detector and the solid state detector have very high speeds of response compared to gas-filled tube detectors discussed earlier. The excellent resolution associated with the newer devices (particu-

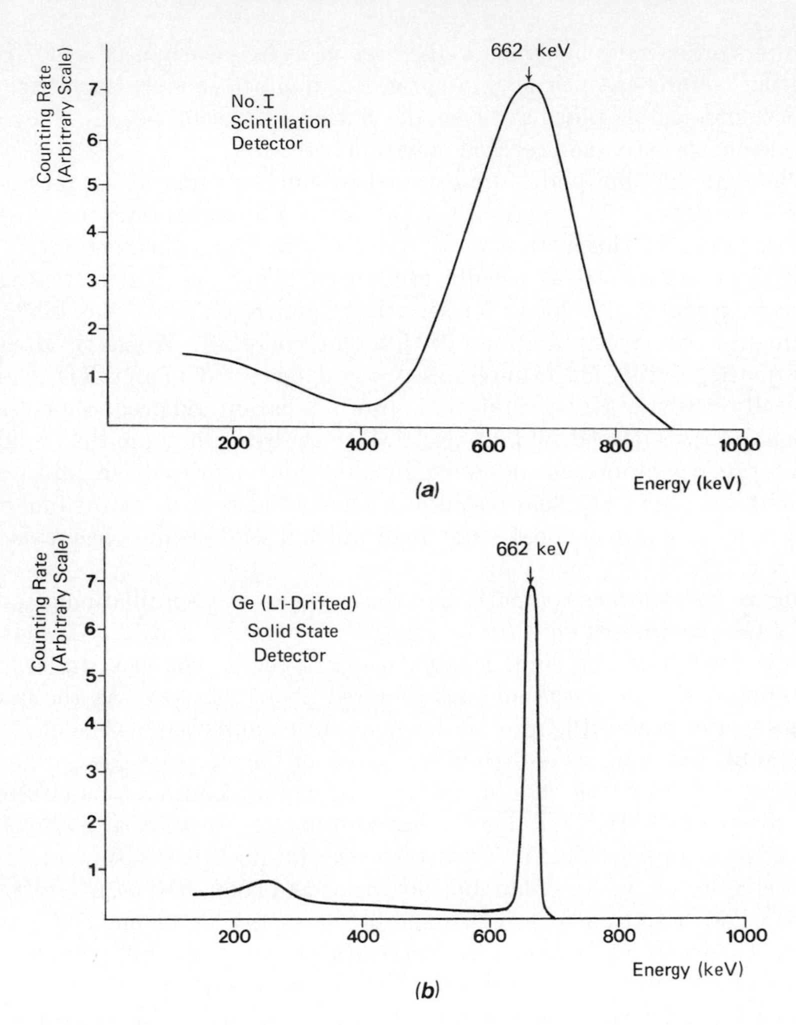

FIGURE 12-3. Spectra obtained using a ^{137}Cs source, which has a single γ-ray with an energy of 662 keV: (a) us'ng a NaI scintillation detector and (b) using a Ge (Li-drifted) solid state detector.

larly the solid state detectors) permits the assay of a specific radioisotope even when a large number of radiating isotopes are present. The scintillation counter can be constructed geometrically so that very weak sources can be placed entirely within the scintillation crystal; the radiation is then collected throughout the entire 4π steradians (i.e., the complete spherical solid angle) about the source, and the efficiency of detection is considerably enhanced. Solid state detectors can be constructed so that they are physically small—small enough that they may be introduced inside living animals to conduct *in vivo* assays.

CORRECTIONS ASSOCIATED WITH RADIOACTIVE ASSAY

Although many assays of nuclear radiation consist only of relative measurements of radiation intensity, it is important to recognize a variety of conditions that can affect the outcome of an assay. Some of the factors we shall consider are due to the characteristics of the detectors, some are due to the nature of the source of radiation, and still others are due to geometric and physical arrangements associated with the source and detector together.

Counter Efficiency. The absolute efficiency of a detector is very difficult to determine. Since efficiency is a measure of the ratio of the number of counts recorded to the number of disintegrations that occur, it involves the detector-source geometry. One geometric factor depends on the fact that the radiations from a radioactive source are emitted throughout 4π steradians, and the solid angle intercepted by the detector is inversely proportional to the square of its distance r from the source. In other words, radiation intensity obeys the *inverse square law.*

Other factors affecting counting efficiency depend on the detailed properties of a specific detector, like its absorption characteristics for each kind and energy of radiation and the geometry of the detector itself. Fortunately it is not often necessary to know the absolute efficiency of a detector.

Coincidence. The time required for a radiation counting system to detect a radiation particle, record the event, recover, and be ready for the next event is short but significant. At relatively high counting rates, more than one event may occur during this time. Such events are said to be coincident, and only one of the coincident events will be recorded. As the number of counting events per unit time increases, the error (missing counts) due to coincidence increases.

Background. Natural radioactivity and cosmic rays (radiations from high-energy particles incident on the earth's atmosphere from extraterrestrial sources) cause additional counts to be introduced into any assay. Such counts that are not from the source being measured are called *background* and must be subtracted from the total count recorded during an assay. Although counting rates associated with the background are usually small, they obviously become important in assays of weak radiation sources.

Self-Absorption. As the amount of radioactive material comprising a source is increased, the net activity of the source increases. At the same time, however, radiations exiting the source must pass through more of the material, which absorbs some of the radiations. Thus, increasing the thickness of a source causes *self-absorption* of radiations within the source material. After a limiting source thickness is reached, it is necessary to increase the specific activity of the sample in order to increase the measured counting rate. Unless very thin samples are used in an assay, it is necessary to make a correction for source self-absorption.

Scattering. Anytime an object is brought near the detector-source system, there is a possibility that radiation from the source may be scattered into the detector and increase the measured counting rate. Similarly, a sample containing a source that is being assayed may produce scattering.

Metabolic Tracer Studies

The tracer method depends on introducing labeled molecules into living systems, later separating the chemical components of the system, and determining by isotopic assay the amount of the labeled material in the various biochemical fractions of the system. Metabolic tracer studies are concerned with what biochemists call *intermediary metabolism,* i.e., the determination of the pathways and ultimate fate of particular molecules involved in one or a series of metabolic reactions. Before tracer techniques were developed, it was not possible to distinguish between carbon, for example, introduced into an organism in the form of glucose from the cellular carbon already present in tissues and organelles. There was no effective means for following the metabolic history of the glucose once it was introduced into the living system. Tracer technology has permitted us the detailed examination of the most complex metabolic pathways.

Adequate labeling for metabolic studies is accomplished only by compliance with a number of criteria:

1. The label isotopes must adhere to the molecule or functional group into which it is originally incorporated. For example, labeled –OH groups or –H groups tend to exchange with similar groups in aqueous media independently of any metabolic processes. Such labeled groups must be avoided in metabolic studies.
2. The initial concentration of tracer must be sufficient to withstand repeated dilution during successive metabolic processes. Labeled isotopes may be diluted by factors of several hundreds when incorporated into a living system.
3. The half-life of a radioactive tracer isotope used in metabolic studies must be long enough that the decay does not remove too much of the tracer before it can be extracted and assayed. An isotope such as $^{14}_{6}C$, with a half-life of 5600 years, has essentially constant activity throughout the

course of any experiment. On the other hand, $^{11}_{6}C$ decreases in intensity by a factor of about 1000 in 10 half-lives, about 3.5 hours.

4. The intensity of radiation from the tracer must not affect the metabolic processes being studied. Intensely ionizing radiation sources used as labels may damage the organism under study (radiation damage to living systems is considered later in this chapter). Indeed, care must be taken that very intense radioactive isotopes do not cause damage to the experimenter.

Metabolic tracer studies are made easier by the fact that certain elements specific to particular biological molecules are available as convenient isotopic tracers. Phosphorus, which has a radioactive isotope (^{32}P) with a half-life of 15 days, is found in nucleic acids but not in the amino acids. On the other hand, sulfur, which has the radioactive isotope ^{35}S with a half-life of 80 days, is in the amino acids methionine, cystine, and cysteine, one or more of which is found in practically every protein. But no sulfur is found in nucleic acids. These tracers have been especially useful in studies of protein synthesis, replication of DNA, and transcription of RNA.

The labeling of the media in which organisms grow permits the use of many clever schemes to determine metabolic pathways. The understanding of the mechanics of viral replication within the cytoplasm of its host cell was aided by studies in which the host bacteria were grown on "hot" media rich in ^{32}P. The nucleic acids of the viruses, formed within the first generation host cells, were assayed for ^{32}P. In this manner it was determined that the viral nucleic acids were formed almost wholly from phosphorus incorporated into the host after viral infection occurred. Thus, it was learned that the viruses did not use the preexisting phosphorus compounds (nucleic acids) of the host to replicate their own genetic material.

One of the most important metabolic tracer studies, performed by Calvin and others, led to the understanding of the carbon pathways in the photosynthesis process. Radioactive ^{14}C was introduced into an alga by way of CO_2, and the biochemicals in the plant were assayed at various time intervals to determine how the carbon had been transferred from one compound to another. Such detective work with metabolites led to a rather complete understanding of photosynthesis (see Chapter II).

The enormous opportunities for tracer studies of metabolism (not all of which have been pursued) have been exploited for nearly half a century. Several volumes would be required to summarize the work adequately.

Physiological Tracer Studies

Physiologists, though their interests often coincide with those of biochemists, are not so much interested in specific enzyme systems of metabolism as in the mechanisms that regulate them and the ways in which they relate to overall cellular economy. Tracer techniques afford the physiologists unique opportunities to study permeability, absorption, and distribution of cellular materials, nutrients, and fluids into and through cellular interfaces.

The problems of cellular membrane permeability are ideally suited to tracer methods. In many cases where there is no *net* transfer of a metabolite across a membrane (but diffusion is taking place, nevertheless), labeled samples of the metabolite provide a singular means of studying the diffusion process. The permeability of cellular membranes to numerous ions has been established by tracer techniques. These techniques are similarly effective in the investigation of active transport, where the incorporation of certain materials against considerable concentration gradients may depend on "carrier" molecules for which no concentration gradient exists (see Chapter IV).

The flow of ions across nerve and muscle membranes has been clarified considerably by tracer studies of Na^+ and K^+ ions. An interesting technique has been used to permit observation of outward flow of these ions across axon membranes of neurons. In this method, an isolated axon is irradiated with known fluxes of neutrons, after which an assay of radioactive Na and K ions is conducted. Knowledge of the activation efficiency of neutrons in creating radioisotopes of Na and K then permits estimates of the quantities of stable isotopes originally present in the sample. Assays of the Na and K radioisotopes before and after the neuron is stimulated provide information that is related to the actual transport of the ions in the pulse transmission process (action potential propagation). What is presently known of the "sodium pump" is primarily due to results of studies using this type of tracer technique, often called "activation analysis."

The absorption or excretion of materials under steady state conditions is particularly amenable to tracer studies. Ingestion of a substance from a medium into which labeled elements have been incorporated is proved if radioactivity is demonstrated in the metabolizing organism after the organism has been withdrawn from the medium and washed. Similarly, excretion of materials can be demonstrated by suspending organisms containing labeled isotopes in a nonradioactive medium if, when the medium is later examined, it shows appropriate radioactivity. Such simple absorption and excretion experiments were among the first performed using radioactive tracers. These techniques were used to demonstrate the uptake of numerous trace elements in plants. Subsequently, similar techniques were found to be effective in studying both the uptake and distribution of a wide variety of trace elements in extremely small quantities in higher animals.

The distribution, localization, and site of synthesis of metabolites can be determined by tracer techniques. For example, labeled iodine compounds have been used to investigate the details of the concentration of iodine in the thyroid gland. Rather complex experiments with ^{32}P tracers have elucidated the role of the liver in producing the phospholipids that appear in the bloodstream. And through the use of Na and K tracers the distribution routes of plant nutrients have been deduced.

One special technique, *radioautography,* has been particularly useful to physiologists. The distribution of labeled materials introduced into a living organism can later be visualized if sections of the organism (or, in some cases, the whole intact organism) are placed directly upon a photographic film, which the radiations expose. The image on the film formed in this manner is called a *radioautograph.* Since a radiation source emits in all directions, thin samples, pressed closely against the film plate, are necessary if good resolution is to be achieved. Figure 12-4 is an example of an radioautograph. It is a whole-body section through a pregnant mouse that had been injected with ^{14}C-labeled glucose 5 minutes before being sacrificed. The lighter areas are those in which the film was more exposed, and the lighter areas therefore indicate the regions in which the glucose has been concentrated. Although radioautography is an expedient means for localizing tracers, it should be remembered that the technique provides neither information about how the material arrived at its observed site nor in what chemical form it exists there.

Radiocarbon Dating

The determination of the age of a once-living artifact is not exactly a tracer technique but is related to such techniques. The most effective method used for this purpose is *radiocarbon dating,* a technique that depends on assay of a naturally occurring radioisotope—not on artificially labeled samples. The technique provides important biological information concerning the age of fossil animals and plants. Apart from its contributions to biology, radiocarbon dating has become a particularly valuable tool of archeologists, geologists, and historians.

Every living thing exchanges carbon with the atmosphere. Plants take in carbon in the form of CO_2 during photosynthesis, and animals ingest either plants or other animals that have consumed plants. Therefore, during the life of an organism, its carbon composition (with respect to relative isotopic abundance) is the same as that of the atmosphere. The CO_2 in the atmosphere contains a fixed proportion of stable ^{12}C and radioactive ^{14}C, a β-emitter with a half-life of 5600 years. The ^{14}C is formed by cosmic radiation in the upper atmosphere and all the evidence available suggests that the fixed ratio of ^{14}C to ^{12}C in the atmosphere is essentially

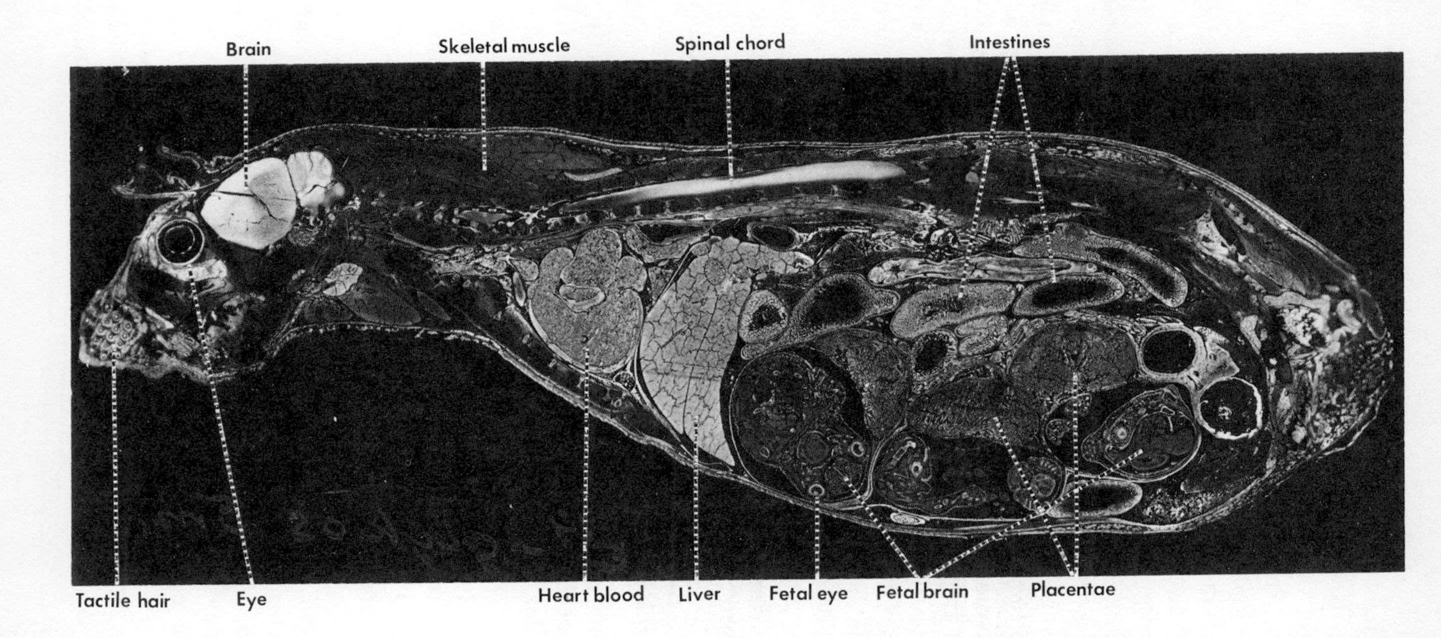

FIGURE 12-4. An autoradiograph of a section through a pregnant mouse that has been injected with glucose labeled with radioactive ^{14}C. The mouse was injected 5 minutes before being sacrificed. The lighter areas are those most exposed by the radiation from ^{14}C, and it can be observed that the glucose has been concentrated mostly in the mother's liver, heart, stomach, and brain, but not in those of the fetus. There appears to be some uptake of glucose in the fetal bones. Courtesy of Professor Sven Ullberg, Uppsala University.

constant throughout the earth. In life, an organism has the same proportion of ^{14}C as the atmosphere, but upon the death of the organism, the exchange of carbon between the atmosphere and organism ceases. From that time onward the ^{14}C concentration in the organism decays, halving itself every 5600 years.

Bone, wood, paper, linen, fossils—all are materials that were at one time living. Samples such as these can be dated by comparing their ^{14}C activity to that of the atmosphere. Specimens living over 25,000 years ago have been dated using this technique, but only by very carefully conducted measurements. Since the specific activity of ^{14}C in the atmosphere is about fifteen disintegrations per minute per gram, counting rates of less than one count per minute must be accurately measured to date samples of one gram mass that are 25,000 years old.

The techniques of counting the very low activities required in radiocarbon dating have been highly developed. Specimens may be prepared by burning, after which the carbon residue in the form of soot is used to line the inside of the wall of a proportional counter. Sometimes CO_2 gas, containing the specimen's carbon, is prepared and used as the filler gas of a detector tube. Because of the low counting rates, it is often necessary to count for months to accumulate a significant number of counts compared to the background. Indeed, it is necessary to reduce the natural background count by careful screening or shielding. Even shielding does not screen out the very penetrating cosmic rays. Clever *anticoincidence* arrangements of detectors are used in conjunction with appropriate electronics to eliminate most cosmic ray background. Figure 12-5 illustrates one way in which

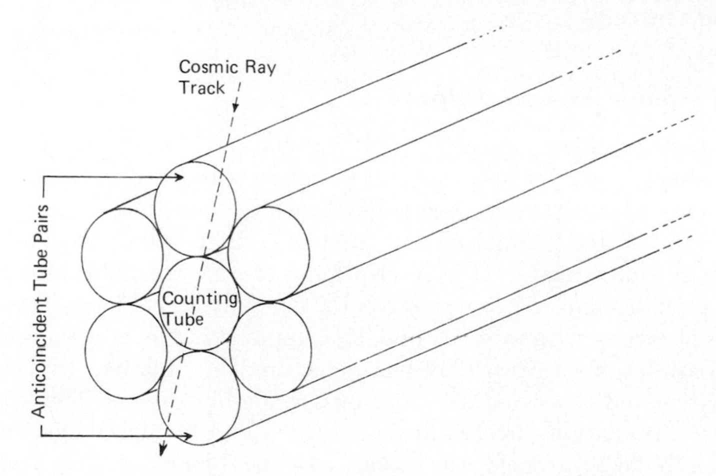

FIGURE 12-5. A counting tube surrounded by anticoincidence counters. When any opposing pair of coincidence tubes is fired by a cosmic ray, the resulting count is electronically eliminated fiom those detected by the counting tube alone.

pairs of anticoincidence detectors can be arranged around the ^{14}C counter so the cosmic rays passing through the ^{14}C counter have a high probability of firing a pair of anticoincidence detectors. The associated electronics eliminates the counts due to radiations that fire a pair of anticoincidence detectors "simultaneously," i.e., within a predetermined very short span of time.

The efficacy of radiocarbon dating has been confirmed repeatedly by carefully comparing ^{14}C data with known ages of specimens such as redwood trees (whose rings may be counted) or mummies of particular pharoahs whose death dates are accurately known.

Effects of Radiation on Living Systems

Nuclear radiations produce ions in passing through matter. When such ionizing radiations pass through living material, they release energy in very localized regions, causing damage out of proportion to the energy involved. Radiation biology is the study of the sequence of events that follows the absorption of energy from ionizing radiation by living cells. This field of study has, of course, received impetus from the development of nuclear weapons and nuclear power sources and the accompanying concern for the effects of their radiations on biological systems. Although radiation biology includes the therapeutic uses of ionizing radiations as well as the techniques by which the cellular systems themselves can be investigated using ionizing radiations as probes, we shall be concerned here with the physical damage associated with the absorption of ionizing radiations by cells.

Dose–Response Relationships

The basic problem associated with radiation biology is to describe quantitatively the effects of ionizing radiations absorbed in living matter in terms of the physical processes—like heating, ionization, excitation, and dissociation of the biomolecular structures—that take place following absorption. Unfortunately, the descriptions presently available are speculative, imprecise, and very suggestive that biophysical science is not yet capable of coping with systems of such complexity. The general approach to the basic problem consists of characterizing the radiation by type and by quantity and observing its effect on specific biosystems. The type and quantity of radiation directed at or absorbed in a biological system is, in radiation biology, termed the *dose*, and the biological, physical, and chemical effects on the living system constitute the *response*.

IONIZING RADIATION DOSIMETRY

The measurement of ionizing radiation is called *dosimetry*. The most common dosimetric technique involves the determination of the amount of ionization produced by radiation within a specified volume of air. This technique measures *exposed dose*, as opposed to *absorbed dose*. Obviously, the absorbed dose is the biologically affective radiation, but often only the amount of radiation directed at the biological system, i.e., the exposed dose, is measurable.

The basic unit of radiation quantity is the *roentgen* (R), defined as that quantity of X- or γ-radiation that produces one esu (electrostatic unit; 1 coulomb $= 3 \times 10^9$ esu) of charge as ions of either sign in air at standard conditions (0°C and 760 torr). Thus, the roentgen is a measure of exposed dose of photons; the unit is not applicable to particulate radiations. Most studies in radiation biology are concerned with the effects of X-rays and γ-rays, however, and the roentgen is an appropriate measure for such cases. The *milliroentgen* (mR) is often a convenient subunit.

The common unit of absorbed dose, the *rad*, is not restricted to particular kinds of radiation. The rad is defined as 100 ergs of energy absorbed in one gram of any medium from any type of ionizing radiation. Although this unit seems completely reasonable for radiobiologic studies, the absorbed dose is difficult to measure. The measurement is usually accomplished by relating absorption in a given medium to that in a gas—not a very accurate or satisfactory procedure.

Further units have been introduced to account for the fact that the biological effectiveness of radiation may not depend only on quantity of ionization and absorption. The *roentgen equivalent man*, or *rem* for short, is that amount of any type of radiation that produces the same biological effects as an absorbed dose of 1 rad of X-rays from an X-ray tube across which 200 kV is impressed. Notice that the rem cannot be measured directly by physical instrumentation; some biological effect must be observed and correlated with a measurement of dose obtained by physical methods. Now add to this array of units one more, the *relative biological effectiveness* (abbreviated RBE), through which the rem is related to the rad according to

$$\text{rem} = \text{RBE} \times \text{rad}$$

The RBE is simply a numerical factor that indicates the specific biological effectiveness of any radiation. For example, the RBE of fast neutrons with respect to mammalian mortality is approximately unity, while the same fast neutrons have an RBE of about 20 with respect to producing opacity in the lens of the mammalian eye.

It should be apparent that many of the experimental measurements of radiation biology are far from fundamental. Since many factors, difficult to recognize and all but impossible to completely control, are involved, especially careful examination of experimental conditions and results are recommended before one draws conclusions from such measurements.

POINT-HEAT THEORY

The energy incident on living systems from nuclear radiation sources—even intense ones—is exceedingly small; yet this small energy often produces profound effects on the absorbing organism. Absorbed energy is ultimately degraded into heating the tissues involved. However, a million roentgens causes a typical tissue culture to incur a temperature rise of less than 1°C, not a significant change in itself. The *point-heat theory* considers the possibility that the energy of ionizing radiation is absorbed as heat in very specific and localized regions, which sometimes are regions critical to the function of the cell or cellular subunit involved.

As an example, if it is supposed that a moderate energy of 100 eV is absorbed by a single molecule of molecular weight 10,000, we can (by assuming that microscopic physical properties may be applied microscopically) estimate the temperature rise of the molecule

$$100 \text{ eV} \times 1.6 \times 10^{-19} \frac{\text{joule}}{\text{eV}} \times \frac{1 \text{ cal}}{4.18 \text{ joule}} = 3.8 \times 10^{-18} \text{ cal} \qquad (12\text{-}9)$$

is the heat absorbed by the molecule. The mass to be heated, i.e., the mass of the molecule, is

$$10^4 \frac{\text{gm}}{\text{mole}} \times \frac{1 \text{ mole}}{6.02 \times 10^{23} \text{ molecules}} = 1.66 \times 10^{-20} \frac{\text{gm}}{\text{molecule}} \qquad (12\text{-}10)$$

If the specific heat of the molecule in question is assumed to be about the same as that of water, i.e., 1 cal/gm °C, the temperature rise ΔT of the molecule is estimated to be

$$\Delta T = \frac{3.8 \times 10^{-18} \text{ cal}}{1.66 \times 10^{-20} \text{ gm} \times 1 \text{ cal/gm °C}} = 229\text{°C} \qquad (12\text{-}11)$$

This local rise in temperature, according to point-heat theory, is expected to initiate destructive action at the site of absorption.

Of course, there is no reasonable way to estimate how the heat generated at a point site will be dispersed nor to know how quickly the dispersion will take place. Because of the many uncertainties associated with this theory, radiobiologists usually interpret the effects of absorbed radiation in terms of hits by the radiation on targets at sensitive sites within cells.

TARGET THEORY

The supposition that there exist specific "target" volumes, vital to cellular function and capable of producing biological change in cells when receiving a "hit" by ionizing radiation, has led to the widely used *target theory*. This theory supposes the existence of key molecules in living tissue, upon which direct action of ionizing radiation is the predominant mechanism producing molecular change. These particular molecules, or some special region within them, are the targets in studies interpreted according to this theory.

Target theory is usually interpreted in terms of survival of cells or small organisms (like bacteria or viruses) after exposure to ionizing radiation. Sometimes this concept is extended to survival of function, as in the case of enzymes, which can be rendered inactive by radiation. The target theory in its simplest form considers the probability of survival of a member of a set of irradiated cells or organisms, i.e., the probability that a vital target within the cell or organism survived with function intact after irradiation. The average number of hits is the product of target volume V and the number of effective hits per unit volume H. The probability $P(n)$ of n hits occurring within the volume V is expressed by the Poisson distribution (encountered earlier in Chapter IV),

$$P(n) = \frac{e^{-HV}(HV)^n}{n!} \qquad (12\text{-}12)$$

The target volume has sustained no hits when $n = 0$, and the survival of the volume V is a certainty. In this case, $P(0) = e^{-HV}$, so that the probability of survival is e^{-HV}. The number of organisms surviving N out of an original number N_0 is expressed as the *survival ratio*, N/N_0, given by

$$\frac{N}{N_0} = e^{-HV} \quad \text{or} \quad \ln\frac{N}{N_0} = -HV \qquad (12\text{-}13)$$

This logarithmic form of the survival ratio is found to coincide with experimental results of irradiation when the number of hits HV is expressed in terms of dose. In other words, it is found experimentally that the survival ratio is related to dose D by

$$\ln\frac{N}{N_0} = -kD \qquad (12\text{-}14)$$

where k is a constant. Then, since Equations (12-13) and (12-14) show that

$$kD = HV \qquad (12\text{-}15)$$

the effective target volume can be determined if the dose D can be related to H, the number of effective hits per unit volume. Such relationships are relatively easy to obtain in "single-hit" organisms, those in which function is eliminated by one hit on one target. Higher on the scale of life, however, where inactivation of the organism by a single hit becomes less likely, a more complex "multi-hit" theory is used.

Survival data, even when interpreted in terms of sensitive volumes like single molecules, do not tell us what the molecules are. Nor do they give us information on the ability of organisms to repair damage wrought by radiation. In dealing with the effects of radiation on higher life forms, no theoretical treatment has been satisfactory, and we are forced to rely on empirical data almost exclusively.

Expression of Radiation Injury

When living systems are exposed to significant doses of ionizing radiation, no immediate change or damage is observable. After a period of time, called the *latent period*, the appearance of change (i.e., damage) follows irradiation. Most expressions of radiation damage are attributed to alteration of metabolic processes. Metabolites—maybe enzymes, maybe nucleic acids, or maybe amino acids—can be changed by radiation. We have seen repeatedly how rigid the cell's requirements—both chemical and structural—are for these metabolites. Defective structure and consequent defective function ultimately follow sufficient radiation exposure, and only when such changes become apparent can we perceive the "radiation effect." Thus, the chain of events following damage to metabolites proceeds imperceptibly during the latent period.

RADIATION DAMAGE AT THE CELLULAR LEVEL

Experiments on animals and plants strongly suggest that the nucleus of a cell is far more *radiosensitive*, i.e., susceptible to damage by ionizing radiation, than is the cytoplasm. Further, the most radiosensitive region of the nucleus is the genetic material, the chromosomes.

In Chapter III, we considered a number of chromosomal rearrangement processes that can occur when breakage occurs along the length of the chromosome. In the same chapter we described the rare process of mutation, the sudden discrete change in genetic material that results in permanent change in the expression of genes. Genes mutate spontaneously at a fixed rate; the rate is different for different genes, but the rate of spontaneous mutation for each gene is fixed. A mutagen, which we noted earlier to be any agent that produces mutations, increases the frequency of spontaneous mutations.

Ionizing radiation is a mutagen and therefore produces an increased frequency of mutations; the changes incurred by radiation in chromosomes are identical with those that occur spontaneously. Many of the types of chromosomal change that can occur result from abrupt breaks in the DNA chain within the chromosome. Indeed, some of the chromosomal changes discussed in Chapter III (deletion and reciprocal translocation, for example) require that two breaks occur on the same chromosome. One effect of ionizing radiation on DNA is therefore the complete dissociation of nucleotides along the DNA chain. In addition, ionizing radiation can cause point mutations on genes by alteration of the nitrogenous base on a nucleotide. This is effected by ionization of a base so that, during synthesis of DNA, "forbidden" pairs (like guanine-thymine or adenine-cytosine) can form, producing heritable changes in the genetic code. Such a process is called *base change.* On occasion, sufficient numbers of ionizations occur within a base that it becomes completely dissociated from the DNA molecule; this process is called *base deletion*, and the genetic effect is again a coding error.

Mutation frequency is, of course, related to dose, and over wide ranges of doses the relationship is linear. Because of the low mutation rates that accompany small doses, it is difficult to establish the linear relationship at small doses, and the question of the existence of a minimum, or threshold, dose has been debated over the years. It seems reasonable to assume that mutation frequency is proportional to dose at any dose level because any ionizing radiation is capable of producing a base change in DNA.

The dose rate, i.e., the time rate at which radiation is administered, *apparently* does not affect mutation frequency. This means that whether a given dose is provided in a single time period of short duration or occurs over an extended time, the total number of mutations is the same. This result is expected because a gene mutation is permanent. Chromosome repair is possible and has been demonstrated to occur, particularly at low dose rates. In such a case, when a chromosomal break occurs, the two pieces may later rejoin in the original configuration; the effect is the equivalent of a chromosomal repair job. However, *permanent* chromosomal changes produced by any fraction of the total dose accumulate with those produced by any other fraction. Although some mutations are reversed (i.e., restored to the original configuration), this process is itself a bona fide change in the structure of the chromosome, though it provides the appearance of dependence of mutation rate on dose rate.

Although the radiosensitivity of cellular cytoplasm is far less than that of the nucleus, cellular death may occur after exposure of the cytoplasm to high radiation levels. It has been suggested that the sensitive regions of the cytoplasm are the mitochondria, which provide most of the cellular energy supply.

Different tissues of an organism exhibit different radiosensitivities. A general principle (to which quite a few exceptions are demonstrable), called the *law of Bergonie and Tribondeau*, expresses the dependence on radiosensitivity on the number of undifferentiated cells contained in the tissue and the degree of mitotic activity in the tissue. Thus, the more primitive (i.e., less specialized, or less differentiated) cells are and the faster they divide, the more likely they are to experience radiation damage.

RADIATION DAMAGE TO WHOLE ORGANISMS

Here we shall restrict our considerations to the irradiation of mammals and assume that the whole animal receives the radiation. The most significant effect of whole-body irradiation of mammals is the shortening of life span of the organism. Essentially no dose of radiation, however large, produces instantaneous death, but doses that produce death within about 30 days of exposure are termed *immediately lethal.* Some effects of irradiation, called *late effects,* may not become apparent until many years after exposure.

As the dose is increased from zero to beyond a level that begins to kill animals, a larger fraction of the irradiated population dies, and the survival time among those that die becomes shorter. A single-exposure irradiation of mammals results in a survival curve whose pattern of response is the general shape shown in Figure 12-6. In the dose range from about 200–1000 R, survival time depends on dose and decreases from several weeks

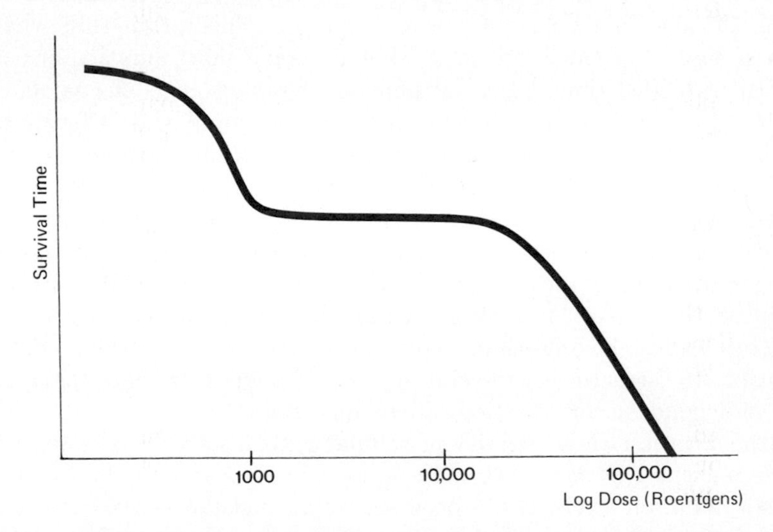

FIGURE 12-6. The mean survival time in mammals after they have received a single dose of whole-body radiation as a function of dose.

to a few days. In the plateau region extending roughly from 1000 to 10,000 R, response is independent of dose; any dose in this region produces death in mammals in 3 or 4 days. Above doses of 10,000 R, the survival time again decreases as dose increases until survival time becomes hours and finally a few minutes.

Of course, there is considerable variation about the mean survival time among individuals within an irradiated group. Because of this statistical variation, it is usual to describe the immediately lethal response to a single exposure of whole-body irradiation as the dose required to kill a certain percentage of irradiated animals within a certain number of days. Thus $LD_{70/30}$ specifies the lethal dose for 70% of animals within 30 days of irradiation. Often the time is not specified; LD_{50} simply refers to the lethal dose for half the exposed animals.

Figure 12-7 illustrates the fractional survival of a typical mammal population (rats in this case) after having been exposed to a single dose of radiation as a function of radiation dose. Notice that 900 R is sufficient to be fatal to 100% of the animals, and this dose is $LD_{100/30}$.

The clinical symptoms that occur after exposure to doses of 100 R or more are commonly called *radiation sickness*; these symptoms—nausea, vomiting, fatigue, and loss of appetite—typically appear within a few hours after exposure and persist for one or two days. A number of *syndromes* (a syndrome is a group of clinical symptoms occurring together that characterize a particular abnormality) are associated with various dose

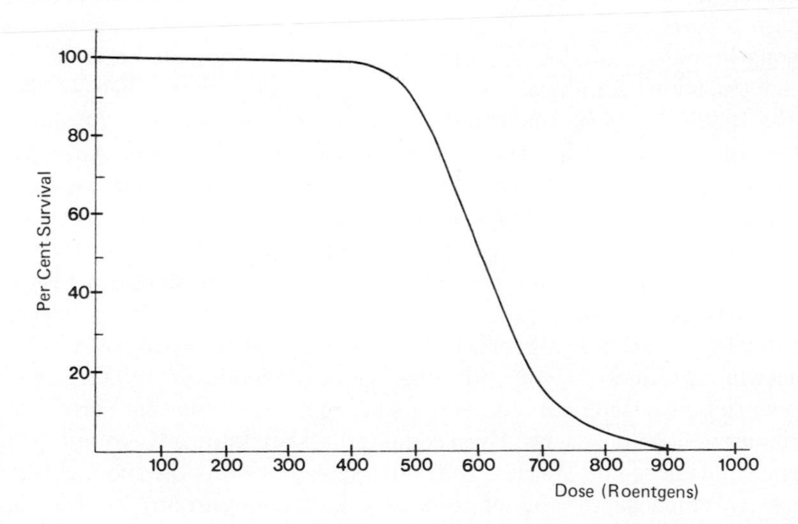

FIGURE 12-7. A typical representation of the percentage survival of a population of rats as a function of dosage of single-dose, whole-body irradiation.

ranges. In the dose range $LD_{0/30}$–$LD_{100/30}$, i.e., in what is called the lethal dose range, the radiation sickness syndrome is apparent, and death results from failure of the *hemopoietic system* (the system responsible for manufacture of cellular components of blood). Animals exposed to lethal doses but that die after 30 days usually die as the result of secondary infections, to which they have a considerably reduced resistance.

When the dose range exceeds 2 $LD_{100/30}$, i.e., twice $LD_{100/30}$, the *gastrointestinal syndrome* occurs. In this case, the lining membranes of the intestinal tract are seriously damaged, causing bleeding and diarrhea. Bone marrow damage accompanies these symptoms. Death occurs in 3 to 4 days after irradiation. At still higher doses, greater than 2 $LD_{100/30}$, or in general greater than 5000 R, the *central nervous system syndrome* appears. Convulsions, coma, and other evidences of central nervous system failure appear in addition to the symptoms seen after lesser doses. Death is presumed to occur from failure of the central nervous system in its control of vital functions. Death occurs relatively quickly in this dose range, but, as we have noted, death is not instantaneous even at doses of 100,000 R.

LATE EFFECTS

Even though all biological effects of radiation are due to the transfer of energy to cells during irradiation, many effects are not apparent for very long periods of time. Indeed, some genetic changes resulting from radiation damage may not be actually expressed for generations.

We have already mentioned the fact that life span is shortened by exposure to radiation. This fact is demonstrable in animal populations. Among humans, radiologists, who are exposed continually to small radiation doses, have been shown to have shorter life spans than other physicians. In the 1930's it was assumed that a dose of 1 R per week would produce no detectable effects during the lifetime of an exposed person. After World War II, the acceptable dose was reduced to 0.3 R per week. At present, 0.1 R per week is considered safe. Since 1960, no detectable shortening of life span among radiologists has been apparent, probably because they are now quite careful to expose themselves only to doses considerably below the accepted danger level.

Ionizing radiation is a general carcinogen, i.e., it is capable of producing cancer in any kind of tissue. Although certain chemicals and abrasives are known to be carcinogenic, they are usually rather specific for certain types of tissues or certain species. Even some viruses are known to produce cancer in specific tissues and species. But ionizing radiation is the most general of known carcinogens, capable of producing malignancy in any kind of tissue.

Fertility, the ability of spermatozoa to fertilize ova, is affected by ionizing radiation. Single doses above 500 R may produce permanent sterility. Lower doses produce temporary sterility, and doses as low as 30 R can reduce the sperm count and reduce fertility. Neither sex drive nor sexual capacity is affected by radiation-induced infertility.

Radiation exposure of the gonads (either testes or ovaries) can rarely reach levels in humans where fertility is affected without producing radiation syndromes. Nevertheless, very small doses to the gonads are capable of producing mutations in gametes, which can transmit the defects to later generations. A gonadal dose of somewhere between 50 and 100 R results in doubling the natural mutational frequency. Genetic injury is the most important form of radiation damage from a biological (as opposed to individual) point of view because it inflicts on the species a genetic burden, the import of which is difficult to ascertain but is almost certainly undesirable.

References and Suggested Reading

Andrews, H. L. (1961). "Radiation Biophysics." Prentice-Hall, Englewood Cliffs, New Jersey.

Calvin, M. (1962). Path of carbon in photosynthesis. *Science* **135**, 879.

Eisenbud, M. (1963). "Environmental Radioactivity." McGraw-Hill, New York.

Francis, G. E., Mulligan, W., and Wormall, A. (1959). "Isotopic Tracers," 2nd. ed. Oxford Univ. Press (Athlone), London and New York.

Glasser, O. E., Quimby, E. H., Taylor, L. S., and Weatherwax, J. L. (1944). "Physical Foundations of Radiology." Harper (Hoeber), New York.

Glasstone, S., ed. (1957). "The Effects of Nuclear Weapons." U.S. At. Energy Comm., (U.S. Gov. Printing Office), Washington, D.C.

Gude, W. D. (1968). "Autoradiographic Techniques." Prentice-Hall, Englewood Cliffs, New Jersey.

Johns, H. E. (1961). "The Physics of Radiology." Thomas, Springfield, Illinois.

Kamen, M. D. (1957). "Isotopic Tracers in Biology," 3rd ed. Academic Press, New York.

Libby, W. F. (1956). Radiocarbon dating. *Amer. Sci.* **44**, 98.

Pizzarello, D. J., and Witcokski, R. L. (1967). "Basic Radiation Biology." Lea & Febiger, Philadelphia, Pennsylvania.

Quimby, E., and Feitelberg, S. (1963). "Radioactive Isotopes in Medicine and Biology." Lea & Febiger, Philadelphia, Pennsylvania.

Rogers, R. A. (1967). "Techniques of Autoradiography." Elsevier, Amsterdam.

Siri, W. E. (1949). "Isotopic Tracers and Nuclear Radiations." McGraw-Hill, New York.

CHAPTER XIII

Techniques Using Mechanical and Electrical Phenomena

Some of the oldest and most useful techniques of biophysical research will be considered in this chapter. *Centrifugation* and *sedimentation* are mechanical processes by which materials suspended in a liquid medium are separated or compacted. The techniques that have been developed in these areas have become routine in biology, physiology, and biochemistry laboratories. There are innumerable electromechanical devices that record quantitative measurements of physiological movements and forces. In this chapter we shall consider some of the basic techniques that have historically linked physics and instrumentation to the physiology of living systems.

We shall inquire into the physical principles underlying each of these techniques and consider some typical applications of each technique to the study of biological molecules and systems.

Centrifugation and Sedimentation

A *centrifuge* is a mechanical device that spins one or more sample holders containing particles suspended in a fluid medium. Those devices capable of spinning at very high angular speeds are called *ultracentrifuges*. Some commercial ultracentrifuges are capable of providing rotational speeds up to about 75,000 rpm. Such speeds are achieved by carefully balancing rotors mounted on low-friction bearings; occasionally even air friction against the rotor is eliminated by mounting the rotor in an evacuated chamber. The purpose of such high-speed rotating mechanisms is perhaps best understood by considering some fundamental mechanical principles that apply to particles suspended in a liquid.

Some Physics of Centrifugation

One might expect that particles, such as large biological molecules, suspended in water would settle out onto the bottom of a container because of the attraction of gravity. A simple calculation demonstrates the necessity for an "artificial" gravitational force of considerable magnitude, in order that such sedimentation take place. Suppose we consider the difference in potential energy possessed by molecules of molecular weight 10,000 suspended in a liquid and separated by a vertical distance of one centimeter. This potential energy difference, due to position in the gravitational field, is

$$E_p = mgh \tag{13-1}$$

where g is the acceleration due to gravity (9.8 m/sec^2), and h is the vertical separation (1 cm in this case) of the particles. The mass m of a particle is found from the molecular weight by noting

$$m = \frac{10^4 \text{ gm/mole}}{6.02 \times 10^{23} \text{ molecules/mole}} = 1.66 \times 10^{-20} \frac{\text{gm}}{\text{molecule}} \times \frac{1 \text{ kg}}{10^3 \text{ gm}}$$

$$= 1.66 \times 10^{-23} \frac{\text{kg}}{\text{molecule}} \tag{13-2}$$

Then the gravitational potential energy difference between the molecules separated by one centimeter is

$$E_p = mgh = 1.66 \times 10^{-23}\,\text{kg} \times 9.8\,\frac{m}{\text{sec}^2} \times 10^{-2}\,m = 1.63 \times 10^{-24}\,\text{joule} \tag{13-3}$$

In Chapter IV we learned that the average thermal energy $\bar{E}_t$, i.e., the energy associated with random motion of particles at a temperature T (°K), is related through the Boltzmann constant $k = 1.38 \times 10^{-23}$ joule/°K to the temperature according to

$$\bar{E}_t = kT \tag{13-4}$$

Then at room temperature (about 300° K), the energy of thermal agitation is

$$\bar{E}_t = 1.38 \times 10^{-23}\,\frac{\text{joule}}{°\text{K}} \times 300°\,\text{K} = 4.14 \times 10^{-21}\,\text{joule} \tag{13-5}$$

Comparison of Equation (13-3) and (13-5) shows

$$\frac{\bar{E}_t}{E_p} = \frac{4.14 \times 10^{-21}\,\text{joule}}{1.63 \times 10^{-24}\,\text{joule}} = 2540 \tag{13-6}$$

which indicates that the agitation energy is more than 2500 times as great as the gravitational potential energy. The thermal randomization of the molecules (often called *Brownian motion*) is far more effective than the settling influence of gravity, and the gravitation effect is completely masked. In other words, the molecules jiggle about so much that they do not settle.

Since most molecules cannot be sedimented under the influence of gravity, inspection of Equation (13-3) suggests that if g were made larger by several orders of magnitude, the ratio $\bar{E}_t/E_p$ would be reduced so that gravitational effects become predominant. This is exactly what a centrifuge does; it provides a large "g" artifically. The means by which a centrifuge provides large "g-forces" can be seen from fundamental physical principles.

Suppose a particle of mass m is located a distance r from the axis of rotation of a solid framework rotating about the axis at an angular velocity ω (radians per second), as depicted in Figure (13-1A). In order that the particle remain at the same distance r from the axis, i.e., so that the inertia of the particle does not cause it to move away trangentially from its circular path, a centripetal force, directed along the radius toward the axis of rotation, is required. The centripetal force F_c is given by

$$F_c = m\omega^2 r \tag{13-7}$$

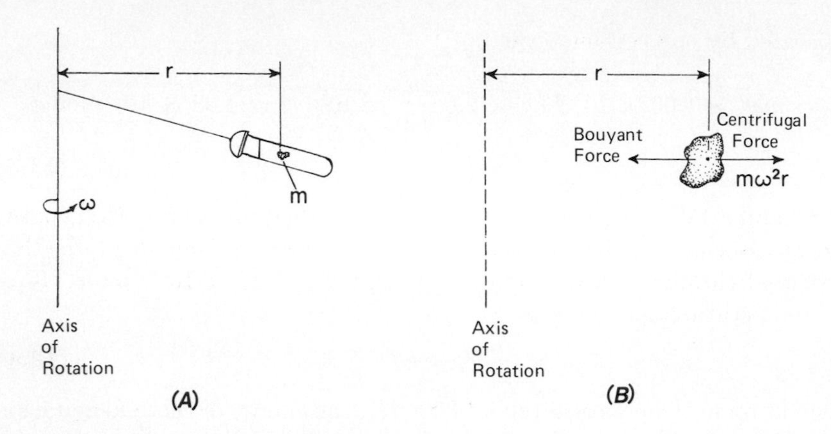

FIGURE 13-1. (A) Schematic representation of a particle of mass m suspended in a fluid that is being centrifuged. (B) The radial forces acting on the mass during centrifugation.

The tendency of the particle to move away from the axis, due to its inertia, is often expressed in terms of a *centrifugal force*, equal in magnitude to the centripetal force, but directed outward; i.e., $\mathbf{F}_{\text{centrifugal}} = -\mathbf{F}_{\text{centripetal}}$. According to Newton's second law, $F = ma$, the acceleration is

$$a_c = \omega^2 r \tag{13-8}$$

From a point of reference at the particle, the acceleration of the particle is equivalent to its being in a gravitational field where the acceleration due to this new "gravity" is $\omega^2 r$. Thus, particles at a distance r from the axis of rotation of a centrifuge spinning with angular speed ω are said to be subjected to "g-fields," and the magnitudes of these fields are usually expressed as multiples of $g\,(= 9.8 \text{ m/sec}^2)$. Accelerations ($\omega^2 r$) of 100,000 g are common in centrifuges.

Consider now a particle (Figure 13-1B) suspended in a liquid medium and subjected to centrifugation. In addition to the centrifugal force $m\omega^2 r$, directed outwardly, a buoyant force opposes this artifical gravitational force according to Archimedes' principle: an object immersed in a fluid is buoyed opposite to the force of gravity by a force equal to the weight of the fluid displaced by the object. If we assume the density of the fluid medium to be ρ_m and the density of the immersed particle to be ρ, the buoyant force F_B (the weight of the fluid displaced) is given by

$$F_B = m_F g' = m_F \omega^2 r \tag{13-9}$$

where m_F is the mass of the fluid displaced by the particle of mass m, and g' is the acceleration due to the artificial gravity of the centrifuge, $\omega^2 r$.

The mass of the fluid displaced m_F is expressed in terms of the particle mass m and the densities of the particle (ρ) and the medium (ρ_m) according to

$$m_F = \rho_m \times \text{volume}$$

and

$$\text{volume} = \frac{m}{\rho}$$

so

$$m_F = m\left(\frac{\rho_m}{\rho}\right) \tag{13-10}$$

and the buoyant force directed toward the axis of rotation is therefore

$$F_B = \frac{m\rho_m\omega^2 r}{\rho} \tag{13-11}$$

Then the net force F acting on the particle is the centrifugal force $m\omega^2 r$ less the buoyant force F_B:

$$F = m\omega^2 r - \frac{m\rho_m\omega^2 r}{\rho} = m\left(1 - \frac{\rho_m}{\rho}\right)\omega^2 r \tag{13-12}$$

The particle suspended in the fluid medium of density ρ_m therefore acts as if it had an effective mass m' given by

$$m' = m\left(1 - \frac{\rho_m}{\rho}\right) \tag{13-13}$$

and subjected to a gravitational acceleration $g' = \omega^2 r$. Notice that if the particle and the medium in which it is immersed have the same density, the effective mass of the particle becomes zero, and according to Equation (13-12), the net force on the particle becomes zero. This case will be important in some of our subsequent considerations.

Several techniques of centrifugation and sedimentation have been developed. The simplest of these is explained by the principles that have been reviewed here. When a suspension of particles, whose density is greater than that of the medium, is centrifuged, the particles eventually form a *pellet* of the particles packed at the bottom of the container holding the suspension. The fluid above the pellet is then called the *supernatant*, which is decanted away to secure the particles. The particles collected in this manner are said to have been *pelleted* or *sedimented*. We shall now consider some further techniques of centrifugation.

Method of Sedimentation Equilibrium

After a centrifuge has been run for a long time, perhaps even days, all particles of greater density than the fluid medium tend to collect at the bottom of the containing vessel. However, the Brownian motion caused by the heat of the sample causes the particles to diffuse, and some of the particles actually move toward the axis of rotation in opposition to the centrifugal force. When the effects of centrifugal sedimentation and inward diffusion have come to equilibrium, the particles will be distributed throughout the fluid with the greatest concentration of particles at the farthest distance from the axis of rotation. The equilibrium distribution of the particles is called the *Boltzmann distribution,* which relates the number of particles per unit volume n_1 at a radius r_1 to the number per unit volume n_2 at r_2 according to

$$\frac{n_1}{n_2} = e^{-\Delta E/kT} \tag{13-14}$$

where ΔE is the difference between the energies of the particles at r_1 and r_2, k is the Boltzmann constant and T is the absolute temperature. Since the average kinetic energy of all the particles is zero (this is implicit in the equilibrium condition), the energy difference is the difference in potential energies of particles at r_1 and r_2. But the potential energy difference is just the work that would be required to move a particle from r_2 to r_1 against the net force (centrifugal force less the buoyant force) $m'\omega^2 r$. Then ΔE is given by

$$\Delta E = \int_{r_2}^{r_1} F\,(-dr) = \int_{r_2}^{r_1} m'\omega^2 r\,(-dr) = -\int_{r_2}^{r_1} m'\omega^2 r\,dr$$

$$= -m'\omega^2 \int_{r_2}^{r_1} r\,dr = \tfrac{1}{2}m'\omega^2(r_2^2 - r_1^2) \tag{13-15}$$

in which the element of displacement is negative because the particle is assumed to move opposite to the direction dr, which is radially outward. Substituting ΔE of Equation (13-15) into the Boltzmann distribution of Equation (13-14), we obtain

$$\frac{n_1}{n_2} = \exp\left[-\frac{m'\omega^2}{2kT}(r_2^2 - r_1^2)\right] \tag{13-16}$$

which may be written (by taking the natural logarithm of both sides) as

$$\ln\left(\frac{n_2}{n_1}\right) = \frac{m'\omega^2}{2kT}(r_2^2 - r_1^2) \tag{13-17}$$

It is perhaps instructive to consider the magnitudes of the numbers one may expect from use of this analysis in a typical experimental situation. Suppose a laboratory centrifuge rotor spins with an angular speed ω of 25,000 rpm (about 2618 rad/second) with the sample at an average radius of 10 cm from the rotor axis. Further suppose that the particles are molecules of MW 50,000, whose specific gravity is 1.5, and that these particles are suspended in water. Let us ask the ratio of the number of particles at $r = 10.2$ cm from the axis to the number at 10.0 cm. We may summarize the pertinent parameters as follows:

$$\omega = 2618 \text{ rad/second}$$

$$\omega^2 = 6.85 \times 10^6 \text{ sec}^{-2}$$

$$r = 10 \text{ cm} = 0.1 \text{ m}$$

$$\omega^2 r = 6.85 \times 10^5 \text{ m/sec}^2 \cong 70{,}000\ g$$

$$\rho_m = 1 \text{ gm/cm}^3$$

$$\rho = 1.5 \text{ gm/cm}^3$$

$$1 - \frac{\rho_m}{\rho} = \frac{1}{3}$$

$$\text{MW} = 5 \times 10^4 \text{ gm/mole}$$

$$m = \frac{5 \times 10^4 \text{ gm/mole}}{6.02 \times 10^{23} \text{ molecules/mole}} = 8.3 \times 10^{-23} \text{ kg/molecule}$$

$$m' = m\left(1 - \frac{\rho_m}{\rho}\right) = 8.3 \times 10^{-23} \text{ kg/molecule} \times \tfrac{1}{3} = 2.77 \times 10^{-23} \text{ kg}$$

$$\frac{n_2}{n_1} = \frac{n_{10.2}}{n_{10.0}}$$

$$r_2 = 10.2 \text{ cm} = 0.102 \text{ m}$$

$$r_1 = 10.0 = 0.1 \text{ m}$$

$$k = 1.38 \times 10^{-23} \text{ joule/°K}$$

$$T = 300 \text{ °K}$$

$$kT = 4.14 \times 10^{-21} \text{ joule}$$

$$r^2 = 0.010404 \text{ m}^2$$

$$r_1^2 = 0.010000 \text{ m}^2$$

$$r_2^2 - r_1^2 = 4.04 \times 10^{-4} \text{ m}^2$$

Using these values in Equation (3-17), we find

$$\ln\left(\frac{n_{10.2}}{n_{10.0}}\right) = \frac{2.77 \times 10^{-23}\,\text{kg} \times 6.85 \times 10^{6}\,\text{sec}^{-2}}{2 \times 4.14 \times 10^{-21}\,\text{joule}} (4.04 \times 10^{-4}) m^2 = 9.26$$

or

$$\frac{n_{10.2}}{n_{10.0}} = 10{,}500 \qquad (13\text{-}18)$$

Thus at 70,000 g there are about 10^4 times more molecules per unit volume at 10.2 cm from the rotor axis than at 10.0 cm. In other words, practically all the molecules are at the bottom of the sedimentation cell. The effect of rotor speed is clearly seen if the same calculation is repeated at 10,000 rpm (roughly 1000 rad/second). Then $\omega^2 = 10^6$, and $\ln(n_{10.2}/n_{10.0})$ is reduced by a factor of 6.85 to a value of 1.35, so that $n_{10.2}/n_{10.0}$ becomes 3.9. At this lower rotor speed there are only about four times as many molecules per unit volume at the lower position.

This sedimentation-equilibrium technique is useful, not only because it is quantitative, but because its results do not depend on the shape of the suspended molecules or on the amount of water bound to these molecules. As we shall see in the next section, these factors affect the *rate* of sedimentation, but not the final equilibrium distribution.

Method of Sedimentation Velocity

A particle suspended in a fluid medium will move outward through the medium when spinning in a centrifuge. The rate at which it moves, i.e., its velocity relative to the medium, depends on the angular speed and radius of the centrifuge, the density and viscosity of the medium, and the size of the particle. Although the velocity of an individual particle cannot be observed, it is possible to observe the motion of those particles that start out at the surface of the medium and form a moving boundary (see Figure 13-2A) as the centrifuge spins. This boundary moves outward from the centrifuge axis at a velocity that can be deduced from photographs of the suspension taken (through a transparent holder) at different times after spinning is begun.

The individual particles forming the boundary accelerate outward under the influence of the centrifugal force as spinning continues. The forces on a typical particle are seen in Figure 13-2B. As the particles move through the fluid, each one experiences an inward frictional force F_f that for low particle velocities is proportional to velocity. The particles of the boundary accelerate quickly to a speed v at which the frictional force is then great enough that the three forces shown in Figure 13-2B sum (vectorially) to zero. The boundary layer is seen moving outward at the velocity v. Since F_f is proportional to this velocity, we may write

$$F_f = fv \tag{13-19}$$

where f is a constant of proportionality. The net force due to the centrifugal force and the buoyant force was given in Equation (13-12), and since this is equal in magnitude to F_f, we find

$$fv = m\left(1 - \frac{\rho_m}{\rho}\right)\omega^2 r \tag{13-20}$$

Then the average velocity of particles comprising the boundary is

$$v = \frac{m\omega^2 r}{f}\left(1 - \frac{\rho_m}{\rho}\right) \tag{13-21}$$

Experimentally, both v and $\omega^2 r$ are measurable, and their ratio is known as the *sedimentation constant s*,

$$s = \frac{v}{\omega^2 r} \tag{13-22}$$

Notice that $\omega^2 r$ is an acceleration with units of m/sec^2, and the average velocity v has units of m/second. The sedimentation constant therefore has units of time. The unit $s = 10^{-13}$ second is one *svedberg* (S), and particles are often identified as 5 S or 12 S particles in describing the results of centrifugation experiments.

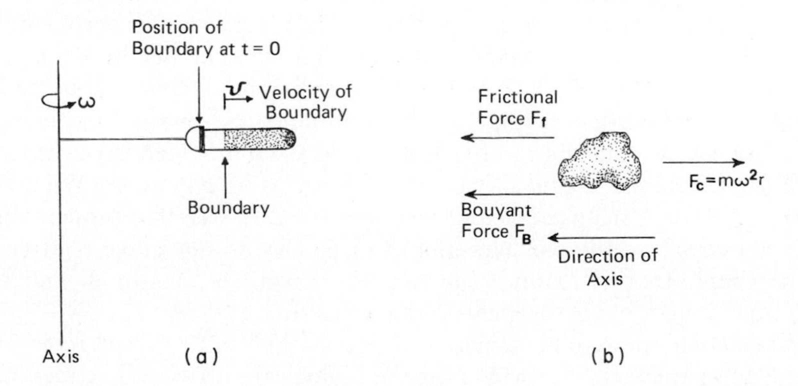

FIGURE 13-2. (A) A moving boundary used in the method of velocity sedimentation. (B) The forces on a moving particle during centrifugation.

If Equation (13-21) and (13-22) are combined, the measurable quantity s is given by

$$s = \frac{m}{f}\left(1 - \frac{\rho_m}{\rho}\right) \tag{13-23}$$

We may discover how the molecular weights of biological particles (biological molecules, viruses, etc.) can be determined from centrifugation experiments by proceeding from Equation (13-23). The constant f can, in effect, be measured. Einstein first showed that f is related to the diffusion constant D (see Chapter IV) by

$$f = \frac{kT}{D} \tag{13-24}$$

The diffusion constant D can be measured by diffusion experiments similar to those described in Chapter IV, by measuring the distance through which particles diffuse by Brownian motion in a measured time interval. By incorporating Equation (13-24) into Equation (13-23), we obtain

$$s = \frac{mD}{kT}\left(1 - \frac{\rho_m}{\rho}\right) \tag{13-25}$$

If this equation is multiplied by Avogadro's number N ($= 6.02 \times 10^{23}$) in both numerator and denominator, it appears as

$$s = \frac{NmD}{NkT}\left(1 - \frac{\rho_m}{\rho}\right) = \frac{MD}{RT}\left(1 - \frac{\rho_m}{\rho}\right) \tag{13-26}$$

where $Nm = M$ is the molecular weight and Nk is the molar gas constant R ($= 8.36$ joule/°K/mole), which we have also encountered in Chapter IV.

Every quantity except M in Equation (13-26) is either an experimentally measurable quantity or a known constant. The measurements of s and D have been described. Both T and the density of the medium ρ_m are easily measured. The particle density ρ can be obtained by suspending the particles in a series of media of known densities, each of which is centrifuged until one medium is found in which the particles do not move. We noted earlier that when the particle and medium densities were the same, no net force is exerted on the particles and the particles do not move relative to the medium. In this manner the particle density is measured, and the molecular weight of the particles is determined.

The method of sedimentation velocity has been useful in molecular weight measurements of a wide range of biological particles. The measurements of molecular weights of relatively small particles like ribonuclease ($M = 1.3 \times 10^4$) and large particles like the bacteriophage T2 ($M =$

3.2×10^7) have been measured by this technique, and the results compare favorably with those obtained from other methods.

Density Gradient Sedimentation

Suppose a solution of a very low molecular weight solute, perhaps a simple salt, dissolved in water, were spun in a centrifuge. In this case, the effect of thermal randomization is comparable in magnitude to the sedimenting effect of centrifugation. Therefore, the salt would hardly be sedimented at all. Instead of the salt being pelleted from solution, we would find that a very gentle gradient of densities is established in the solution with the densest region at the outer end of the container. (We observe a similar density gradient vertically in the atmosphere, in which the density is greatest at the earth's surface.) In fact, by controlling the concentration of the salt in the water, the change in density per unit length along the length of the container can be adjusted so that the gradient is as gentle as desired. The density of the solution can be calibrated in terms of position along the container. If, then, particles, whose density lies within the range of the salt solution, are introduced into the solution and the suspension centrifuged, the particles will come to equilibrium within a narrow band at a position where the solution density is the same as that of the particles. Figure 13-3 illustrates this situation. Then the density of suspended particles can be measured directly by the position of the particle band along the calibrated gradient.

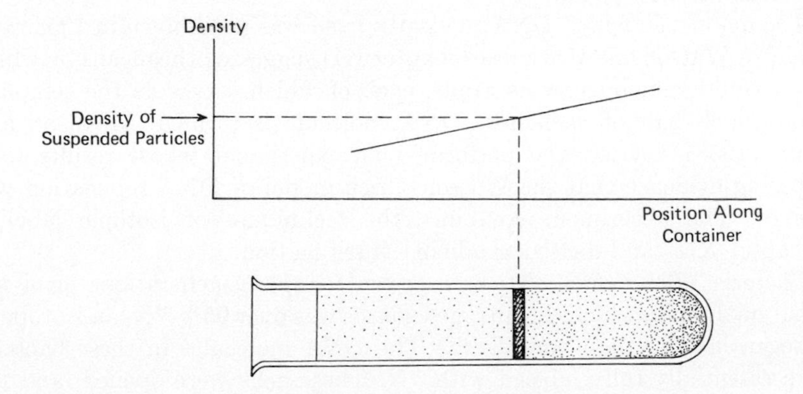

FIGURE 13-3. A density gradient. The solid curve of the graph above the tube represents the density of the liquid medium as a function of position along the tube. The solid band within the tube is composed of particles whose density has caused them to be positioned at the region of the same density in the medium.

The particles collected at a particular location along a density gradient do not form a sharp band. Because of Brownian motion they diffuse to some extent outward and inward from the band center. The smaller these particles are, the farther (and faster) they diffuse in both directions from the center. Therefore, a more diffuse band at a particular position (density) is made up of less massive molecules, since density is mass per unit volume. Indeed, the distribution of the particles within a band can be measured by appropriate photographic techniques, and, by diffusion analysis similar to that we have used in Chapter IV, the molecular weight of the particles can be obtained. Thus, from a single run in a density gradient system, both density and molecular weight of suspended particles may be deduced.

One of the useful aspects of the density gradient technique is not necessarily analytical, but preparative in nature. By introducing a mixture of many kinds of particles into an appropriate gradient solution and centrifuging the conglomeration, a distinct separation of the components of the mixture can be achieved. This method is used, for example, to isolate various components of cells, which are crushed before being introduced into the gradient solution. One way in which separated bands of particles are removed intact is to freeze the sample while spinning, but after it has been spun for hours (most modern preparative centrifuges are equipped with internal refrigeration systems). The frozen sample can then be sliced to isolate each particle group. Notice that although separations in density gradient systems provide no absolute quantitative information, they do quantitatively relate the densities of the components. This characteristic is demonstrated in the following description of one of the classic, crucial experiments of biophysics.

The manner in which DNA replicates itself was not known in 1958. The model of Watson and Crick (see Chapter III) suggested the means by which DNA could split into two strands, each of which serves as the template upon which half of each new DNA molecule is formed. Meselson and Stahl (1958) devised and performed an experiment whose results were imposing evidence that the Watson–Crick model of DNA replication was correct. The experiment combines the techniques of isotopic labeling (Chapter XII) and density gradient centrifugation.

Bacteria (*Echerichia coli*) were grown for many generations on a nutrient medium in which the nitrogen source was over 95% ^{15}N, an isotope of nitrogen heavier than normal ^{14}N. The DNA molecules in these bacteria were essentially fully labeled with ^{15}N. These cells were opened, and the DNA was carefully extracted and centrifuged at 140,000 g in a CsCl density gradient. The position of DNA along the gradient was determined by observing the transmission of ultraviolet light through sections of the gradient and tracing out the light intensity as a function of position along the gradient. The detector and recording mechanism in such an arrangement is called a *photodensitometer*. DNA absorbs strongly in the ultraviolet region, and the photodensitometer can precisely locate a band of DNA along the gradient. The band of DNA containing ^{15}N appeared on a photodensitometer trace at a position as indicated in Figure 13-4a.

The bacteria whose DNA was laden with ^{15}N were then transferred to a medium in which the nitrogen source contained the less dense ^{14}N. After one generation, i.e., after the DNA had undergone one doubling, the DNA was removed from the cells and centrifuged in a CsCl gradient identical to the one used for ^{15}N DNA. The resulting photodensitometer trace appeared as shown in Figure 13-4b, at a position of lower density. Finally, a group of bacteria grown for many generations on normal nutrients (^{14}N) was used to provide DNA for a similar gradient run. The position on the gradient of the DNA with ^{14}N only is shown in Figure 13-4c. The density of the nucleic acid molecules found after one replication of DNA is exactly halfway between the densities of the ^{15}N DNA and of the ^{14}N DNA. Thus we may conclude that half the nitrogen in the DNA of the first generation group was

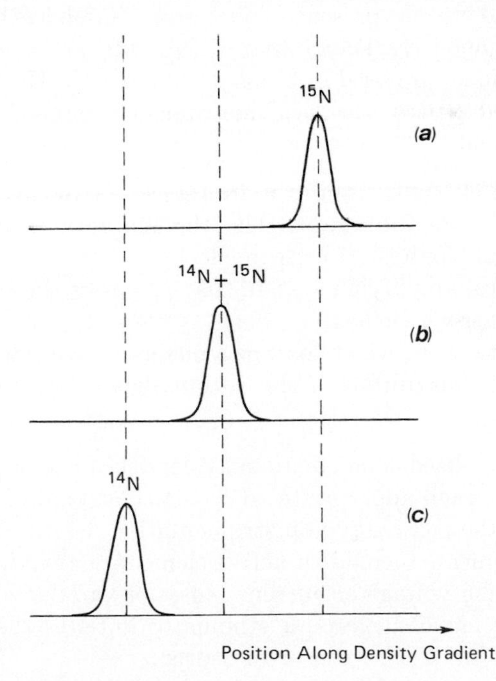

FIGURE 13-4. Photodensitometer traces of DNA bands in the density gradient experiment described in the text. Trace (a) is obtained for the parent generation sample of DNA. Trace (b) is obtained using DNA from the first generation. Trace (c) is obtained using DNA from a control group.

obtained from the parent (^{15}N) group and the other half from the nutrient material containing ^{14}N. This result implies that after each parent DNA molecule splits according to the Watson–Crick model, it does not split further, but incorporates nucleotides from the nutrients to make the new half of each double helix that comprises the two complete daughter DNA molecules. This experiment was among the first to support the Watson–Crick model of DNA that has come to be accepted as the basis of modern molecular genetics.

Biophysical Instrumentation

The early history of biophysical and biomedical research is a record of the development of instrumentation for the observation and measurement of forces, movements, and electrical changes in living systems. From the time when Galvani (1791) observed the twitching of a frog's muscle when touched by metal near an electric generator, through the discoveries of electroencephalography (Caton, 1875) and electrocardiography (Einthoven, 1903), and into the present, bioinstrumentation has been developed and refined continuously. Today many elaborate and sophisticated instrumental techniques are used in biophysical research. Here we shall consider a generalized type of electrical instrument composed of three characteristic components:

1. A sensing element, usually a *transducer* (a device that converts energy from one form to another; the final form is most frequently electrical in biophysical instruments)
2. An electrical amplifying system that processes the signals received from the sensing element
3. A read-out device, which may be a meter or a recording device, that displays the magnitude of the output signal, sometimes as a function of time

Each of these generalized components may consist of one or more of a wide variety of devices, each appropriate to a particular kind of measurement.

In addition to the passive-type instrumentation described above, many biophysical instruments include an active element, a *stimulator*, which is a device for producing voltage or current pulses of variable amplitudes and durations. As the name suggests, a stimulator initiates the physiological responses that are to be observed or measured.

We shall consider some of the common types of devices in each category. It should become clear that the variety of biophysical instrumentation is limitless, and the purpose of this section is to familiarize the students with the names and function of some of the more common research devices.

Transducers

Almost every physical relationship between any two parameters, one of which can be directly observed, can be used as the basis for a transducer. The most useful transducers for biophysical instrumentation convert force or displacement into a change in some electrical parameter. For example, a displacement might simply move a contact along a slidewire, varying the resistance between the contact point and another point on the wire. By incorporating the resistance of the variable length of wire into an electronic circuit, the energy associated with the displacement can be converted into electrical energy.

Every force or pressure (force per unit area) applied to a real system produces some displacement. It is, therefore, impossible to dissociate displacement from force. Most force and pressure transducers use the displacement associated with force or pressure to relate these quantities to some electrical parameter.

THE STRAIN GAUGE

One of the common transducers for the measurement of displacement is the *strain gauge*, a device based on the principle that the elongation of a wire changes its resistance. The resistance R of a length of wire is directly proportional to its length L and inversely proportional to the cross-sectional area A of the wire:

$$R = \frac{\rho L}{A}$$

where the constant of proportionality ρ (whose units are ohm m) is called the resistivity of the material composing the wire. Thus, a strain gauge is an electromechanical device in which small changes in length ΔL result in proportional changes in resistance ΔR. In fact the *strain*, defined as the change in length per unit length, or $\Delta L/L$, can be expressed as a proportionality with $\Delta R/R$:

$$\frac{\Delta L}{L} = F \frac{\Delta R}{R}$$

where the constant of proportionality F is called the *gauge factor*, a characteristic of the material again. By choosing appropriate materials, strain gauges can be constructed over a wide range of sensitivities. For example, nickel has a gauge factor of 12.1, while manganin is characterized by $F = 0.04$.

The strain gauge may be used as a *force transducer* simply by calibrating the change in resistance in terms of the force applied to stretch the wire.

Similarly, the strain gauge may serve as a *pressure transducer* by calibrating it in terms of force per unit area. One of the interesting uses of the strain gauge in biomedical research is as an *accelerometer*, a device for measuring acceleration. This is accomplished by attaching a mass to the strain gauge; when the system of mass plus gauge is accelerated, the force exerted on the gauge, according to Newton's second law, is proportional to the acceleration.

PHOTOELECTRIC TRANSDUCERS

Devices that respond to photons incident upon them by producing electric currents or differences in potential are *photoelectric transducers*, and the active element in such devices is a *photosensitive* material. Among the photosensitive devices are three classes; *photoemissive*, *photoconductive*, and *photovoltaic*. The vacuum *phototube* is a photoemissive device that ejects electrons from a cathode that is irradiated by light of appropriate frequencies. The *photomultiplier tube* is an elaborate version of a phototube in which the number of ejected electrons is multiplied as they strike successive elements called dynodes, metal surfaces that eject many secondary electrons for each electron incident upon them. *Photocells* are semiconductor devices in which the current flow (resulting from release of electrons in the material) occurs within the solid material. Most modern photocells are made of germanium, silicon, or gallium arsenide semiconductor materials, which are said to be *photoconductive*. Photovoltaic devices are more complex semiconductor devices that utilize the potential difference across a junction of two different types of semiconductors. Photons incident on the device cause electrons to be injected into the junction, altering the potential difference across it, and causing electrons to flow in an external circuit. Many variations of materials and configurations of this basic junction device are perhaps recognized when called solar batteries, silicon rectifiers, or phototransistors.

Any of the above photosensitive devices can be used as the active element of a transducer, and many modern biophysical instruments employ a wide variety of phototransducers. For example, a sensitive displacement transducer can be constructed in a simple arrangement similar to that in Figure 13-5. The intensity of light falling upon the photosensitive element is controlled by movement of the shutter attached to the moving object under study. Variation in light intensity on the photodevice produces a variation in electrical output (usually a voltage change) from the transducer. Such a transducer is appropriate, and commonly used, in measurements of muscle contraction similar to those described in Chapter IX.

Pressure transducers using photosensitive elements are common because they are less expensive than strain gauge devices arranged to sense pressure

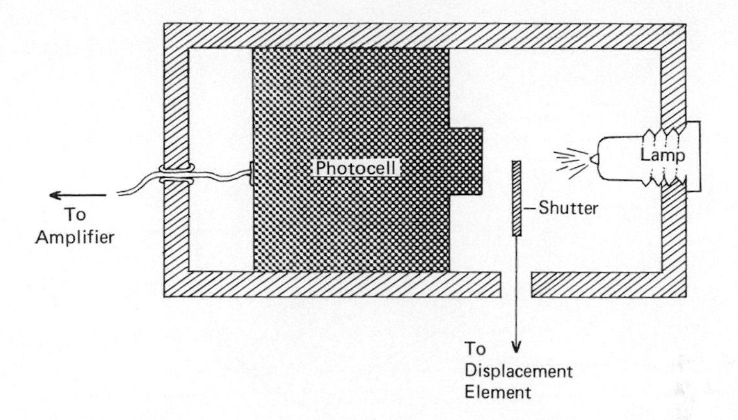

FIGURE 13-5. The shutter arrangement for a typical displacement phototransducer.

changes. One simple pressure transducer utilizing a photosensitive element is the *Bourdon tube*, shown schematically in Figure 13-6. Here again a shutter arrangement controls the incident light flux and therefore the electrical response. The inside of the coiled tube is subjected to pressure variations in a fluid medium, which fills the hollow coil. An increase in pressure causes the tube to uncoil, moving the shutter.

DIFFERENTIAL TRANSDUCERS

Several forms of transducer for measuring displacement utilize the electromagnetic coupling between windings of a transformer, shown schematically in Figure 13-7. The primary winding has a constant-amplitude alternating voltage applied across the input terminals. The secondary winding is center-tapped, and the two sections of the secondary are wired so that the voltages induced across them are 180° out of phase. The output terminals are across the entire secondary winding. The magnetic core is a piece of ferromagnetic material, like iron, that greatly increases the magnetic coupling (beyond that due to free space) between the primary and secondary windings. This core is initially positioned so that one section of the secondary winding nulls the voltage developed across the other section. Thus, if the object under study is rigidly attached to the magnetic core, any linear movement of the object from the initial position causes the core to move from the position in which no secondary output voltage is developed. As the core shifts nearer to one of the sections of the secondary, it causes greater magnetic coupling between that section and the primary winding, and a net voltage appears across the secondary terminals. Such an arrangement is called a *differential transformer* because it provides a dif-

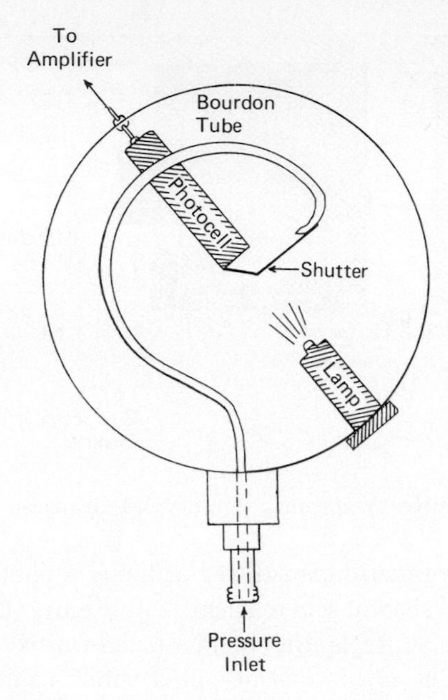

FIGURE 13-6. A Bourdon tube pressure transducer. As the pressure in the Bourdon tube increases, the shutter is moved across the face of the photocell, exposing the photocell to more light.

ferential output voltage, i.e., a voltage that is the difference between those developed across the individual sections of the secondary.

Other forms of differential transducers utilize pressure differentials. For example, the device in Figure 13-8 is called a *pneumotach* (literal translation—air speed), which is usually calibrated to measure the rate of flow of air in liters/min. The chamber of this device is separated by several screens of fine mesh through which the air must travel. The pressure on the entrance side is slightly greater (0.1 mm Hg is typical) than on the exit side. Two pressure transducers, placed on opposite sides of the screens, provide output voltages that are compared electronically so that the difference between their magnitudes comprises the final output signal.

Instrument Components and Calibration

In most biophysical instruments, the primary signal is generated in a transducer. Occasionally the output of the transducer is sufficient to provide an adequate measure of the parameter being studied. More often, however, it is necessary to further process the transducer signal before it can

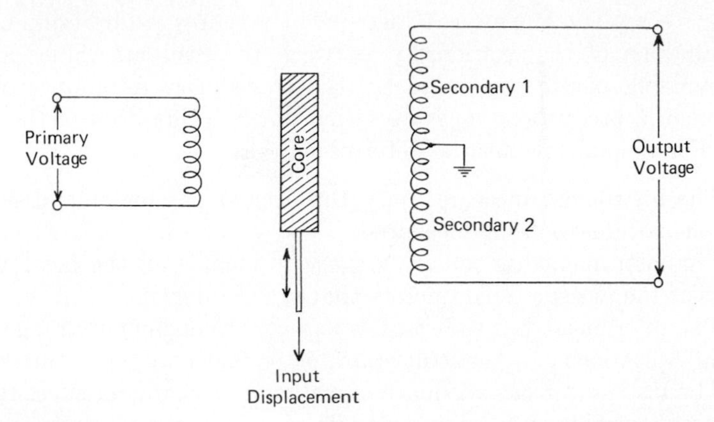

FIGURE 13-7. A differential transducer of the electromagnetic coupling type. Displacement of the core changes the relative amounts of coupling of magnetic flux from the primary to each of the secondary windings; this change results in a voltage output proportional to displacement.

be displayed or recorded in a meaningful way. The signal modifier between the transducer and the final display device is, in general, an electronic *amplifier*, the characteristics of which depend critically on the specific requirements of each experiment. The function of the amplifier, of course, is to increase the magnitude of the transducer signal until it is sufficient to operate a *read-out device*. Perhaps the most frequently used read-out device is the oscilloscope, though meters or manometers are frequently appropriate for many purposes. For permanent recordings of measurements, various types of direct-writing recorders are commonly used; in more sophisticated systems, magnetic tape is used to record data rapidly and continuously.

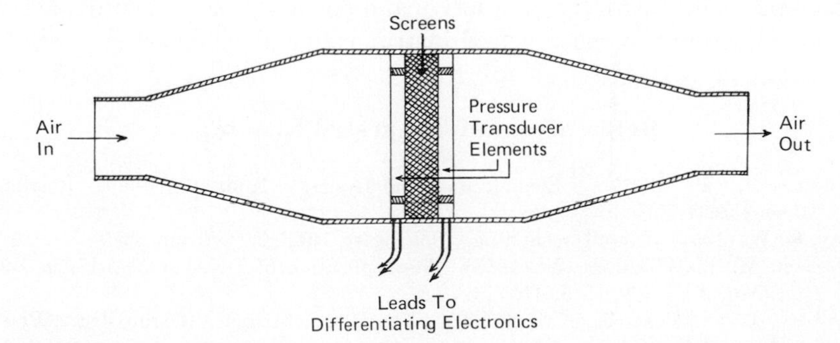

FIGURE 13-8. The pneumotach, a device that measures velocity of a gas by sensing the pressure difference across a series of mesh screens.

Every complete biophysical instrument requires calibration, i.e., the determination of the relationship between the read-out values and the corresponding parameter sensed by the transducer. A number of basic experimental precautions must be taken during calibration of the instrument. For example, it must be determined that:

1. The instrument measures only the desired parameter and does not affect the parameter being measured.
2. The instrument has sufficient *range* to include all the possible magnitudes of the measured parameters that are of interest.
3. The instrument has sufficient *sensitivity* throughout its range to observe all variations of a particular parameter that may be significant.
4. The instrument has adequate *signal-to-noise* characteristics. In other words, it is capable of providing signals of sufficient magnitude that they are easily distinguished from the inherent noise (random variations) of the instrument.
5. The instrument is capable of responding fast enough to changes in the measured parameter so that it faithfully reproduces all variations in the parameter that are of interest. That is to say, the instrument has an adequate *frequency response.*

The satisfactory incorporation of all the above characteristics of an instrument requires that the experimenter acquaint himself with the characteristics of the transducers, amplifiers, and read-out devices. He must also familiarize himself with each parameter to be measured—its probable range of values, speed of variation, how it is affected by incorporation into the experimental environment, etc.

A number of textbooks are available that are devoted entirely to biophysical and biomedical instruments. The student is urged to utilize these general sources. However, it is wise experimental procedure to completely familiarize oneself with the instruction and operating manuals provided by the manufacturers of experimental instruments and components.

References and Suggested Reading

Epstein, H. T. (1963). "Elementary Biophysics." Addison-Wesley, Reading, Massachusetts.

Gray, G. W. (1951). The ultracentrifuge. *Sci. Amer.* **184,** 42. (Offprint 82).

Meselson, M., and Stahl, F. W. (1958). The replication of DNA in *Escherichia coli. Proc. Nat. Acad. Sci. U. S.* **44,** 671.

Svedberg, T., and Pedersen, K. O. (1940). "The Ultracentrifuge." Oxford Univ. Press, London and New York.

Uber, F. M. (1950). "Biophysical Research Methods." Wiley (Interscience), New York.

Yanof, H. M. (1965). "Biomedical Electronics." Davis, Philadelphia, Pennsylvania.

Glossary*

A

abduction. Movement away from the median plane.

absolute muscle strength. The maximum tension that can be developed by a skeletal muscle.

absolute visual threshold. The minimum quantity of light that reaches the eye and produces a visual sensation.

absolutely refractory period. A short period of time during the initial stages of muscle contraction when the muscle cannot be stimulated successfully by any stimulus, however strong.

absorption coefficient. A mathematical function that characterizes a material's ability to absorb electromagnetic waves.

absorption edge. A discontinuity in absorption coefficient (as a function of wavelength) occurring at wavelengths that correspond to energies sufficient to eject electrons from particular atomic shells.

accelerometer. A device for measuring acceleration.

accommodation. The ability of the eye to focus on objects at various distances from the eye.

acetylcholine (ACh). A substance produced at the ends of nerve fibers and motor end plates of muscles. It functions as the transmitter agent in synaptic conduction.

action potential. A series of changes in potential that occurs across the membrane of a neuron following stimulation.

active site. A specific region on an enzyme at which the enzyme interacts with its substrate.

active transport. The movement of substances across cellular membranes in opposition to a concentration gradient and with an expenditure of metabolic energy.

adaptation. The modification of an organism to adjust to new conditions or environment.

adduction. Movement toward the medial plane.

adenosine triphosphate (ATP). A compound present in all living cells that provides energy for a wide variety of life functions. It is produced within organisms from energy derived from sunlight or food.

aerobic. Referring to conditions requiring free oxygen.

afferent neuron. A neuron that conducts signals toward the central nervous system.

* These terms are defined only in the sense in which they are discussed in the text.

afterimage. The persistence of the sensation of light after stimulation has ceased.

afterpotential. Small changes in potential that occur in the latter stages of the action potential on a neuron.

Airy disk pattern. The diffraction pattern resulting from the passage of light from a point source through a circular aperture.

allele. One of a pair of genes at corresponding loci (positions) in a pair of homologous chromosomes.

alpha particle. A helium nucleus, consisting of two protons and two neutrons.

alpha helix. The secondary structure of a protein; the spiral structure of an amino acid chain.

alpha wave. A characteristic pattern in a normal electroencephalogram with a frequency near 10 Hz.

amino acid. The structural unit of proteins. Each amino acid is characterized by the presence of an amino group (—NH_2) and a carboxyl group (—COOH).

amplitude. The maximum departure from its average value of a cyclic disturbance that comprises a wave.

anabolic. Referring to synthetic reactions in metabolism that convert simple substances to more complex ones.

anaerobic. Referring to conditions that do not require free oxygen.

anaphase. A stage in cell division (mitosis or meiosis) in which the halves of each pair of chromosomes move away from the equatorial plane.

anterior. Toward the head or front end. In erect animals, toward the front or ventral side of the body.

anticodon. A triplet group of bases (either adenine, guanine, cytosine, or uracil) on tRNA that forms base pairs with a corresponding group (codon) on mRNA.

antigen. A substance that stimulates an organism to produce antibodies with which the antibodies then reacts.

antinode. A region of maximum excursion or disturbance in a standing wave.

aorta. The large artery through which blood exits the heart on the way to the systemic circulatory system.

aqueous humor. The transparent fluid in the anterior cavity of the eyeball.

Archimedes' principle. The assertion that an object immersed in a fluid is bouyed opposite to the force of gravity by a force equal to the weight of the fluid displaced by the object.

arteriole. A small artery.

atomic mass number. The total number of protons and neutrons in the nucleus of an atom.

augmented lead. Any of the three supplemental leads of an electrocardiograph whose axis bisects a pair of axes of the three basic leads.

auricle. The pinna; the external ear.

autonomic nervous system. The portion of the peripheral nervous system that supplies nerves to structures not under voluntary control.

axon. A long, extended part of a neuron that conveys pulses away from the cell body of the neuron.

axoplasm. The cytoplasmic fluid inside an axon.

B

background. Counts obtained in making measurements of nuclear radiation that do not originate from the source being measured.

bacteriophage. A virus that infects bacteria.

base change. An effect on DNA caused by ionization of a base (purine or pyrimidine) such that the base forms normally forbidden pairs during DNA replication and results in heritable changes in the genetic code.

base deletion. A genetic coding error caused by the complete dissociation of a base (purine or pyrimidine) from the DNA molecule.

basilar membrane. A membrane in the cochlea on which the organ of Corti rests. The membrane separates the cochlea duct from the scala tympani.

beat frequency. The repetition rate of the variation in loudness that results when two sound waves of the same amplitude but slightly different frequencies are heard simultaneously.

Beer's law. The assertion that the absorption coefficient of a substance is proportional to the concentration of that substance at low concentrations.

benign. Of mild character; not malignant.

binaural beat. A sensation of variation in loudness perceived when two sounds of slightly different frequencies are introduced into different ears of the same person.

binomial nomenclature. The system by which living organisms are named with two Latin words, the first of which designates the genus of the organism, and the second of which specifies the particular species within that genus.

biogenesis. The doctrine that all living things come only from preexisting living things.

Biogenetic law. The assertion that the embryo of an organism, during its development, passes through the same stages its species passed during its evolutionary history.

biology. The science or study of living things.

bleaching. The chemical process in which rhodopsin in the retina of the eye is broken down into vitamin A and protein after absorbing light. During the process the purple rhodopsin becomes a pale yellow.

bleeding time. The time required for blood, shed as the result of a pin prick, to clot.

Boltzmann distribution. A mathematical representation that relates the equilibrium spatial distribution of particles to energy differences of the particles at different positions and to the temperature of the particles.

Bourdon tube. A pressure transducer using a photosensitive element.

Bragg's law. A mathematical expression relating the wavelength and angle of reflection of an X-ray beam from a crystal to the spacing of atoms within the crystal.

bremsstrahlung. Literally, braking radiation. The X-radiation resulting from the sudden deceleration of electrons striking a target material.

brilliance. That attribute of a color by which it is classed as equivalent to a value of gray, which ranges from black (zero brilliance) to white.

C

canaliculi (singular, canaliculus). Small canals extending between lacunae in the matrix of bone.

candela. The basic unit of light intensity, specifying a quantity of energy passing through a unit area in a unit time.

capillary. A small, thin-walled blood vessel.

capsid. The protein shell enclosing the nucleic acid in most varieties of viruses.

capsomer. A structural subunit of a capsid.

cardiac vector. The direction of the net polarization wave that progresses through the heart during the cardiac cycle.

catabolic. Referring to metabolic reactions in which complex molecules are broken down into smaller ones.

caudal. Near or in the direction of the tail of an animal.

cell. The fundamental unit of a living system. Cells consist of protoplasm and, in most cases, a nucleus enclosed in a cell membrane.

cell theory. The doctrine that all living organisms are composed of cells or cell products and that the cells comprise the structural and function elements of the organism.

central dogma. The thesis that RNA is "transcripted" only from DNA, and protein is "translated" only from RNA and that the transfer of genetic information proceeds only in the direction from DNA to RNA to protein.

central nervous system. The brain, brain stem, and spinal cord.

central nervous system syndrome. Convulsion, coma, and evidence of central nervous system failure that follow exposure to high dosages of ionizing radiation.

centrifugal force. A force that tends to impel a thing or parts of a thing outward from the center about which it is rotating.

centriole. A small body near the nucleus of a cell which, in the process of cell division, forms the center of the aster rays.

centromere. A constricted region of a chromosome that marks the junction of the arms of the chromosome.

cerebral cortex. The outer portion of the brain.

chiasma (plural, chiasmata). The location of the interchange of genetic material between two chromosomes during meiosis.

chloroplast. A plastid containing chlorophyll. Chloroplasts are the sites of photosynthesis.

chromatid. In meiosis, one of the four elements, i.e., the "arms," of a chromosome whose DNA has doubled.

chromatin. The filamentous form of DNA which, in cell division, forms the substance of chromosomes.

chromoplast. A plastid containing color pigments other than green.

chromosome. Rod-shaped structures in cells containing DNA. Chromosomes are visible during cell division.

coagulation. The clotting of blood.

coagulation time. The time required for drawn blood to form a clot in a test tube.

cochlea. The snail-shaped body in the internal ear containing the receptors of hearing.

codon. A group of three successive nucleotides on a molecule of mRNA which serves as a code for the placement of specific tRNA molecules and amino acids into proper sequence during protein synthesis.

coenzyme. The nonprotein portion of some enzymes that is necessary for activation of the enzyme.

color contrast. The apparent increase in the saturation of complementary colors when close to each other over that perceived when viewed separately.

colorimeter. An instrument for making comparisons of absorption of various wavelengths (colors) of visible light in materials by selecting the wavelengths with colored filters.

compound microscope. An instrument composed of two or more convergent lenses used to produce magnified images of small objects.

cone. A photoreceptor cell in the retina of the eye that functions in color vision.

conjunctiva. A membrane covering the inner surface of the eyelids and the anterior surface of the eyeball.

continuous spectrum. That portion of the X-ray spectrum produced by deceleration of electrons striking the target of an X-ray tube.

contraction time. The time interval between the initiation of muscle contraction and the time the muscle reaches maximum tension.

core enzyme. That portion of the enzyme RNA polymerase that acts as a catalyst with or without another component (the sigma factor) of the enzyme.

corepressor. A molecule that activates the repressor, which in turn prevents the initiation of the synthesis of RNA from DNA.

cornea. The anterior, transparent portion of the sclera (structural framework) of the eyeball.

covalent bond. A strong chemical bond formed by the mutual sharing of electrons between atoms.

cranial nerve. One of twelve pairs of nerves attached to the central nervous system at the base of the brain and at the brain stem.

cross. The result of mating individuals of a species; to interbreed; to mate individuals of a species.

crossing over. A process that takes place during synapsis in which two homologous chromosomes mutually exchange corresponding sections of the chromosomes.

crystalline lens. The lens of the eyeball, located between the aqueous humor and vitreous humor.

curie. A unit of radioactivity, equal to 3.7×10^{10} disintegrations per second.

cuvette. A sample holder used in spectrophotometers.

cyclosis. Streaming movement of protoplasm in leaves of plants, some algae, and protozoa.

cytology. The study of cells.

cytoplasm. The material of a cell outside the nucleus.

D

decibel. A unit of sound pressure level, expressing the relationship between two sound levels in terms of the logarithm of the ratio of the levels.

dendrite. A short branch of a neuron that conducts impulses toward the cell body of the neuron.

deoxyribonucleic acid. See DNA.

desmosome. A passageway out of and between plant cells through which protoplasm can communicate between cells.

diastole. The relaxation phase of the cardiac cycle during which the heart chambers dilate and fill with blood, cf. systole.

differential staining. The application to cells of dyes that stain specific chemicals within the cells.

differential transformer. A transducer that provides an output voltage proportional to the displacement of a movable transformer core between the windings of the transformer.

diffusion. The process by which particles suspended in a fluid medium spontaneously move from a region of high concentration to a region of lower concentration.

dihybrid cross. The mating of an individual having two pure-bred dominant characteristics with another individual having two pure-bred recessive genes in the same characteristic.

diploid. Having double the number of chromosomes present in a gamete.

dispersion. The separation of light into its component wavelengths (colors) by refraction or diffraction.

distal. Away from the point of attachment, or away from the body: cf. proximal.

disulfide bond. A bond between sulfur atoms, especially those in amino acids that secure the three-dimensional structure of proteins.

DNA (deoxyribonucleic acid). The substance in the nucleus of cells containing genetic information in the sequence of its structural units. DNA regulates and specifies the protein synthesis process.

DNA polymerase. The enzyme that catalyzes the replication of DNA.

dominant. Describing a hereditary characteristic or its gene which, in the heterozygous condition (i.e., both a dominant gene and a recessive gene present) is expressed to the exclusion of the recessive characteristic.

dorsal. Pertaining to the back or upper surface of the body.

dose. The quantity and type of radiation directed at or absorbed by a biological system.

dosimetry. The measurement of ionizing radiation.

E

eclipse. That phase of virus reproduction during which the host cell contains no virions.

efferent neuron. A neuron that conducts signals away from the central nervous system.

Einthoven's law. The relationship between the lead potentials for the three standard electrocardiographic leads: $V_{II} = V_I + V_{III}$.

electrocardiogram. The recorded pattern of an electrocardiograph.

electrocardiograph. An instrument for recording the changes in potential on the surface of the body that results from polarization changes on and within the heart.

electromagnetic wave. Traveling waves composed of varying electric and magnetic fields.

electron volt. A unit of energy equal to the work done in moving an electron through a difference in potential of 1 volt.

electrostatic lens. An arrangement of electrostatically changed tubes used to focus electrons in electron microscopes.

endplate. The junction between a motor neuron and the muscle it controls.

endoplasmic reticulum. A complex network of small membranous tubules within the cytoplasm of cells.

enzyme. A protein-containing substance produced by living cells that is capable of catalyzing specific chemical reactions.

erythrocyte. A red blood cell.

extension. The straightening of a limb.

F

fascia. A layer of connective tissue forming a sheath for covering muscles, nerves, and blood vessels.

fatigue. The gradual weakening of tension in a muscle contracted for an extended period.

fiber tract. Bundles of axons within the central nervous system.

fibrin. Threads of insoluble material that form blood clots.

fibrinogen. A soluble component in blood that becomes, when mediated by the protein thrombin, fibrin.

filament. A strand within myofibrils of muscles. Filaments contain two protein chains, actin and myosin.

fixing. The killing and preparation of cells so that the cellular contents are minimally disturbed by subsequent treatment.

fluorescence. Visible radiation emitted by some atoms after the electrons of those atoms have been excited by ultraviolet light.

fluorescent antibody. A group of atoms, coupled to certain locations in specimens, that becomes luminous when excited by ultraviolet light.

focal point. The point along the optical axis of a lens from which rays incident on the lens parallel to the axis diverge, or toward which they converge.

fovea. An indentation in the retina of the eye on the axis of the eyeball. The fovea is rich in cone cells.

free energy. That portion of released energy that is available to do work as a reaction proceeds toward equilibrium.

frequency. The repetition rate of a periodic motion, measured in Hertz (Hz) or cycles per second.

frequency response. A measure of the capability of an instrument to respond fast enough to changes in the parameter being measured so that it faithfully reproduces the variations in that parameter.

frontal plane. A plane that divides the body of an organism into anterior (front) and posterior (back) portions.

funiculus. A bundle of axons and their covering.

G

gamete. A sex cell; an ovum or spermatozoon.

gamma ray. A high energy electromagnetic wave originating in the nucleus of an atom.

ganglion. A mass of nervous tissue composed primarily of the cell bodies of neurons.

gastrointestinal syndrome. A result of exposure of an organism to relatively high radiation levels, consisting of damage to the intestinal lining, bleeding, and diarrhea.

gauge pressure. Pressure over and above atmospheric pressure.

gene. A hereditary factor; a specific region on a DNA molecule capable of determining a specific trait.

genetics. The science of heredity.

genotype. The genetic constitution of an individual, in contrast to the appearance or expressed characteristics of that individual; cf. phenotype.

genus. A taxonomic division that includes related species.

ghost. A bacteriophage that no longer contains its DNA.

glucose. A simple sugar, $C_6H_{12}O_6$, found in plants as a product of photosynthesis and in higher animals as a product of the digestion of carbohydrates.

glycogen. A multiple sugar, $(C_6H_{10}O_5)_n$, often called animal starch, used as an energy storage substance in higher animals.

glycolysis. The breakdown of carbohydrates into pyruvic and lactic acids in the metabolic processes of cells.

glycosidic bond. A covalent bond linking sugars together.

Golgi complex. An organelle in the cytoplasm of cells that is abundant in secretory cells.

grana (singular, granum). Disk-like artifacts within chloroplasts.

group transfer reaction. A coupled reaction, involving no oxidation or reduction, between molecules that exchange functional groups.

H

hair cell. One of the cells in the organ of Corti of the inner ear that have hair-like projections.

half-life. The time required for the activity of a radioactive substance to halve itself.

haploid. Having only one of each of the pair of chromosomes characteristic of the somatic (nongerm) cells of a species. Having half the diploid number of chromosomes.

Haversian system. A structural unit of compact bone, consisting of lamellar layers of bone around a central canal.

helicotrema. The opening between the scalae (canals) at the small (apex) end of the cochlea.

hemopoiesis. The formation of blood cells; the ability of the body to maintain the proper numbers of red blood cells.

heterozygote. An organism in which the gene pairs are unlike (usually referring to a specific trait).

histiocyte. A cell in connective tissue that has the ability to engulf particles.

homologous chromosomes. A pair of chromosomes in which identical genes or their alleles are located at corresponding positions.

homozygote. An individual in which the genes for a specific trait on homologous chromosomes are alike.

host. An organism upon or in which another organism (a parasite) lives and/or obtains nourishment.

hue. The attribute of color, determined by wavelength of the light, by which the color differs characteristically from the gray of the same brilliance.

hyaloplasm. The unstructured portion of the cytoplasm of a cell.

hydrocarbon. A chemical composed of only carbon and hydrogen.

hydrogen bond. A relatively weak chemical bond, usually between H and O or between H and N atoms in biological systems.

hydrolysis. A chemical reaction in which water is one of the reactants.

hydrostatic. Referring to the maintenance of pressure in a fluid.

hypermetropia. Farsightedness.

hypertonic. Pertaining to solutions in which the concentration of a particular solute is greater than in a cell immersed in the fluid.

hypotonic. Pertaining to solutions in which the concentration of a particular solute is less than in a cell immersed in the fluid.

I

icosahedron. A closed regular solid figure with twenty faces, each of which is an equilateral triangle.

***in vitro* (literally, in glass).** In a laboratory situation, outside the normal environment of a living system.

***in vivo*.** In a living system.

incomplete dominance. The failure of one allele a pair on homologous chromosomes to completely subvert the effects of the other.

incus. Anvil; one of the ossicles of the middle ear.

index of refraction. The ratio of the speed of light in a vacuum to the speed of light in a particular medium.

inducer. A substance that inactivates a repressor, thus permitting the initiation of synthesis of RNA from DNA.

induction. The process in which a specific structure develops within an organism as a result of chemicals tramsmitted from another part of the organism.

inferior. In a lower position; below.

insertion. The point of attachment of a muscle to a relatively more movable part, cf. origin.

interphase. The period between mitotic divisions of a cell.

ionic bond. A chemical bond between two atoms formed when one or more electrons are transferred from one atom to the other, resulting in an electrostatic attraction between the atoms.

iris. The pigmented annular diaphragm in front of the lens of the eye containing an opening, the pupil, through which light enters the eye.

irritability. The ability of living matter to respond to stimuli.

isometric. A condition of muscle contraction in which no change in the length of the muscle occurs, cf. isotonic.

isotonic. A condition of muscle contraction in which the muscle shortens while maintaining constant tension, cf. isometric.

isotope. One of two or more nuclides of an element that have the same number of protons but different numbers of neutrons.

K

karyoplasm. The materials within the nucleus of a cell.

karyotype. The shapes of the individual chromosomes of a set of chromosomes within the nucleus of a cell.

kingdom. One of two major categories, animal and plant, into which living things are classified.

L

lacuna (plural, lacunae). A small cavity within bone matrix containing a cell.

Lambert's law. A mathematical statement of the relationship between intensity of electromagnetic radiation and the thickness of an absorbing material through which it passes.

lambert. A unit of luminance characterizing the brightness of a source of light.

latent period. The time period after the viral infection of an organism takes place but before the number of viruses increases.

latent time. The time interval from the initiation of stimulation of a muscle to the time that contraction begins.

lateral. Directed to the side; away from the median plane of the body.

lattice constant. The separation between a pair of atomic planes in a crystal.

Laue powder pattern. Reflections of X-rays from all sets of atomic planes of a crystal resulting from irradiating a powdered form of the crystal with monochromatic X-rays.

law of Bergonie and Tribondeau. The assertion that the more primitive cells are the more likely they are to be susceptible to radiation damage.

law of independent assortment. The principle that each pair of genes is assorted at random and is inherited independently of other pairs.

law of segregation. A genetic principle that in the formation of gametes, the two genes of each pair are segregated from one another with each gamete receiving only one gene of a given pair.

lead. A pair of electrodes used as detectors of potential difference at two points on the surface of the body in electrocardiography.

leukocyte. A white blood cell.

lever arm. The perpendicular distance from the line of action of a force to the axis of rotation about which the force tends to cause rotation.

lipid. Any of a group of widely different organic compounds, all of which are insoluble in water. The group includes fats, oils, waxes, and sterols.

lobe. A spatial division of the brain.

locus. A position or place, especially along the length of a chromosome.

longitudinal wave. A wave in which the vibrations are in the direction of propagation of the wave.

loudness. The subjective perception of intensity of sound energy.

lumen. A unit of light flux, equal to the flux from a standard candle through 1 steradian.

luminance. The brightness of a source of light.

lymph. The clear fluid of the interstitial spaces among cells of the human body that is collected by the lymph vessels.

lymphatic system. The network of vessels that conduct lymph from the body to the bloodstream.

lymph nodes. Bean-shaped bodies in the lymphatic system that serve as filters of the lymph fluid.

lymphocytes. Agranular white blood cells formed in lymph nodes.

lysis. The destruction of cells by enzymes that cause the bursting of the plasma membrane of the cell.

lysozyme. An enzyme that causes lysis.

M

magnetic lens. An arrangement of coils of wire carrying current and thereby producing magnetic fields shaped so as to focus electrons in an electron microscope.

malleus. A hammer; one of the ossicles of the middle ear.

maturation period. That phase of viral replication that begins with the accumulation of viral DNA in the host cell and ends when mature viruses have formed within the host.

meatus. A natural opening in the body.

medial. Located toward the median plane.

median plane. The plane passing through the center of the body dividing it into right and left halves.

meiosis. A form of cellular division in which the chromosome number is reduced from the diploid number to the haploid number, as in the formation of sex cells.

meristematic. Referring to the region of plants where embryonic tissue cells divide and differentiate.

messenger ribonucleic acid. See mRNA.

metabolism. The chemical energy changes that occur in a living system and that are involved in the life processes of the system.

metaphase. The phase of cell division (mitosis) in which the chromosome pairs are arranged with their centromeres along the equatorial plane.

microphonic potential. A graded electrical potential difference that occurs across the organ of Corti in the inner ear and correlates with the frequency of sound entering the ear.

microtome. A special knife for slicing thin sections of tissue for microscopic examination.

minimum resolvable separation. A measure of the resolution of an optical system; the smallest distance between two objects that can be resolved by a microscope.

mitochondrion (plural, mitochondria). An organelle, present in all cells, that plays a major role in the supply of energy for metabolic activities of the cell.

mitosis. Cell division in which the daughter cells have the same number of chromosomes as the parent cell.

Monera. A proposed kingdom of plant forms having no true nuclei.

monochrometer. An instrument that selects and transmits a single wavelength (color) of light from a broad spectrum of wavelengths.

monohybrid cross. A mating of parents having genes that differ in only one characteristic.

morphology. The form and structure of an organism.

mRNA (messenger ribonucleic acid). A linear form of RNA used to transmit information from nuclear DNA to the site of protein synthesis in the cytoplasm.

multiple alleles. Three or more forms of an allele that may occupy the same locus of a chromosome.

muscle fiber. An element of a gross muscle, bounded by a sarcolemma (a sheath enclosing a cell with numerous nuclei). Fibers are up to 4 cm in length and 100 μm in diameter.

mutagenic. Capable of causing mutations.

mutation. An abrupt change in the inheritable characteristics of an organism caused by a change in the genetic material of the organism.

myelin sheath. A white sheath covering axons and composed of layers of myelin laid down by Schwann cells.

myocardium. The heart muscle.

myofibril. The smallest element of muscle structure capable of contraction.

myopia. Nearsightedness.

N

natural selection. The principle that the fittest survive; the means by which Darwin suggested that species originate.

neuroglia. The supporting, non-nervous cells of the nervous system.

neuron. A nerve cell.

node. The region of minimum distrubance in a standing wave.

node of Ranvier. One of a series of constrictions in the myelin sheath that surrounds some axons.

nuclear membrane. A porous double-layered membrane forming the surface of the nucleus of a cell.

nucleocapsid. The capsid of a virus and the nucleic acid it encloses.

nucleolus. One or more stainable bodies inside the nucleus of a cell, composed primarily of RNA.

nucleon. A proton or a neutron.

nucleoside. A purine or pyrimidine base attached to a ribose or deoxyribose sugar.

nucleotide. A structural unit of nucleic acids consisting of a sugar, either ribose or deoxyribose, a phosphate group (PO_4), and one of the purine or pyrimidine bases.

nucleus (atomic). The massive central portion of an atom containing protons and neutrons.

nucleus (cellular). A prominent mass within most cells that contains the genetic material (DNA) of the cell.

nuclide. A particular species of atomic nucleus.

numerical aperature. A measure of the light-gathering power of a lens.

nyctalopia. Night blindness.

O

objective lens. The lens of a microscope closest to the object being viewed.

ocular lens. The lens of a microscope nearest the eye of the user.

operator. An operator gene; a gene that is situated close to some number of structural genes and which controls the activity of the structural genes.

operon. An operator gene and the structural genes under its control.

optic disk. The blind spot on the retina of the eye; the passageway through which nerves exit the retina.

optic nerve. The nerve from the eye to the brain, originating at the optic disk.

optical path. The product of the physical path length of a ray of light and the index of refraction of the medium through which the light is passing.

organ. A part of an organism having a definite form and function.

organ of Corti. The organ that converts sound signals into nerve impulses, located on the basilar membrane inside the cochlea.

organelle. Any of the specialized structures within a cell.

organic evolution. The evolution of living things; the production of new species through natural selection.

organism. An individual plant or animal.

origin. The relatively fixed end of a muscle, cf. insertion.

osmosis. The flow of water through a semipermeable membrane resulting from differences in concentration of solutions on either side of the membrane.

osmotic pressure. The pressure required to prevent water from passing into a solution from which it is separated by a membrane permeable only to water.

ossicle. One of the small bones, the malleus, the incus, or the stapes, in the middle ear.

oval window. A membrane between the middle and inner ear against which the stapes transmits pressure signals which are carried into the cochlea through a fluid.

ovum (plural, ova). An egg cell.

oxygen debt. The state of muscles which have performed work at a rate such that insufficient oxygen has been supplied to the muscles and lactic acid has accumulated.

P

parasite. Organisms that live in or on other organisms from which they receive some advantage without compensation.

peplomer. A subunit of a peplos.

peplos. A membranous envelope covering some viruses, such as those of the herpes group.

peptide bond. The chemical bond that joins two amino acids together.

pericardium. The membranous sac that encloses the heart.

period. A characteristic of a wave or of oscillatory motion; the time required to execute one cycle of a wave or oscillation.

pH. A scale ranging from 0 to 14 indicating the degree of acidity or alkalinity of a solution. Water has a pH of 7; acid solutions have pH less than 7; and bases have pH greater than 7.

phagocyte. A cell having the ability to engulf particles.

phase contrast. A technique of light microscopy that provides for visual contrast in materials that are not differentially absorptive.

phase plate. A plate having a circular groove inscribed in it that is inserted into a microscope at the focal plane of the objective lens to introduce a quarter wavelength shift of phase into the light passing through the groove. Light phase-shifted by the phase plate interacts with light diffracted by the specimen to form a phase contrast image.

phenotype. The expressed characteristics of an individual, in contrast to its genetic constitution, cf. genotype.

phon. The loudness level of a sound equal to the intensity in decibels above the threshold level of a 1000-cycle tone of equal loudness.

phosphorylation. The addition of a phosphate group (PO_4) to a compound.

photocell. A semiconductor device in which charged carriers are released to flow as current when released by incident light.

photodensitometer. An instrument that records a measurement of light intensity as a function of position.

photometer. An instrument capable of comparing absorption or emission characteristics of two substances.

photometry. The science of the measurement of light intensity.

photon. A discrete packet of electromagnetic energy.

photopic. Referring to the light-adapted eye, cf. scotopic.

photosynthesis. The process by which plants convert energy from the sun into molecules useful to living systems as a food (energy) source.

physiology. The functioning of an organism or any of its parts.

pinocytosis. The process in which a cell takes material into itself by engulfing it and opening a temporary passageway through the cell's plasma membrane.

pitch. The quality of sound directly related to frequency.

place theory. A theory of hearing based on any model that postulates that frequency discrimination depends on frequency sensitivity at different locations along the basilar membrane of the ear.

plasma. The straw-colored liquid of the blood in which the blood cells are immersed.

plasma membrane. The cell membrane that covers and contains the contents of the cell.

plastid. A body in the cytoplasm of cells, especially plant cells, having various structures, pigmentations, and functions.

platelet. A small body in the blood that functions in the clotting of blood.

pneumotach. A differential transducer that measures the rate of air flow.

point-heat theory. The notion that the energy of ionizing radiation is absorbed as heat in a small localized region of a cell.

polypeptide. A chain of amino acids linked together by peptide bonds.

polyribosome. A group of ribosomes attached to the same molecule of mRNA during the protein synthesis process.

polysaccharide. Long, often branching, chains of simple sugars linked together by glycosidic bonds.

porphyrin. A complex organic compound containing four small 4-carbon rings and usually a metal; a component of hemoglobin and chlorophyll.

posterior. At or toward the back of the body.

power. The rate at which work is done.

precordial lead. One of the leads of an electrocardiograph that are positioned on the chest.

projection lens. The lens of an electron microscope that projects the final image onto a fluorescent screen or film.

prophase. The initial stage of mitosis, characterized by the formation of chromosomes from the chromatin in the nucleus of a cell.

protein. A characteristic substance of all living systems, composed of amino acid chains.

Protista. A proposed kingdom comprised of most algae and characterized by possession of a true nucleus and chromosomes.

protoplasm. A generic term for living substance, used primarily to refer to the living substance of a cell.

protozoan. A one-celled animal.

proximal. Nearest or nearer to the body or to the point of attachment, cf. distal.

Punnett square. A checkerboard-like diagram for analyzing the distribution of chromosomes from gametes in a genetic cross.

purine. A purine base; either adenine or guanine, components of nucleic acids.

pyrimidine. A pyrimidine base; either cytosine, thymine, or uracil, components of nucleic acids.

Q

quantum. A photon; a discrete packet of electromagnetic energy.

R

radioactive decay. The decrease in activity of radioactive material with time.

radioautography. The process of recording on film the location of radioactive material within tissues of an organism.

radiocarbon dating. The process of determining the age of once-living material by measurement of the radioactivity of the carbon-14 in the material.

real image. An image formed by one or more lenses in which the light rays actually pass through the image.

receptor. The ending of a peripheral nerve that acts as a sensor.

recessive. Referring to a genetic characteristic, which, in the presence of a dominant gene for the same characteristic, fails to be expressed.

reflex arc. A nervous pathway from the point of stimulation, through the spinal cord, and to an effector (responding) organ.

refraction. The bending of light in traversing an interface between media of different indices of refraction.

refractory period. The time after the stimulation of a nerve or muscle during which it is no longer capable of responding.

regulator gene. The section of DNA that directs production of a protein called a repressor, which in turn determines the activity of a group of structural genes.

Reisner's membrane. One of the major membranes in the cochlea that separates two of the cochlear chambers.

relative biological effectiveness (RBE). A numerical factor indicating the effectiveness of any form of radiation on a specific biological material.

relaxation time. The time interval between the time of maximum tension of a contracted muscle and the time at which the resting state is restored.

release factor. A protein that recognizes terminator codons on mRNA and causes the amino acid chain being synthesized to terminate.

replica. A thin layer of plastic material, formed over a surface, then stripped off and used as the specimen in electron microscopy.

resolution. A measure of the ability of a microscope to resolve, or visually separate, two closely spaced points.

resonance. The production of large amplitudes in an oscillatory system by injecting energy into the system at its natural frequency.

respiration. The series of chemical reactions by which a living cell obtains energy from nutrients.

responsiveness. The capacity to react to stimuli.

retina. The layer of nervous tissue on the inner posterior surface of the eyeball upon which images are focused.

rhodopsin. The light-sensitive pigment in the rods of the retina.

ribonuclease (RNase). The enzyme that breaks up RNA chains into individual nucleotides.

ribonucleic acid. See RNA.

ribosomal ribonucleic acid. See rRNA.

ribosome. Small bodies, usually attached to the endoplasmic reticulum within a cell, that are rich in RNA and function in the synthesis of protein.

RNA (ribonucleic acid). One of the nucleic acids, containing ribose sugar, which plays a prominent role in protein synthesis.

rod. The long rod-shaped part of a rod cell in the retina of the eye.

roentgen. A unit of radiation quantity; that quantity of X-radiation or γ-radiation that produces one electrostatic unit of charge in air at standard conditions.

rotation. The turning of a body part about the long axis of the body.

rRNA (ribosomal RNA). A form of RNA found in the ribosomes, the sites of protein synthesis.

S

saccharide. A simple sugar.

sarcomere. A contractile unit of striated muscle.

scala tympani. A major canal of the cochlea filled with a fluid, perilymph.

scala vestibuli. A major canal of the cochlea.

Schwann cell. One of the cells that form the myelin sheath around axons of nerve cells.

sclera. The white fibrous outer coating of the posterior portion of the eyeball; the structural framework of the eyeball.

scotopic. Pertaining to the dark-adapted eye, cf. photopic.

self-absorption. The effect of finite thickness of a radioactive material in which the material itself absorbs part of the emitted radiation.

septum. A wall dividing a cavity into two or more parts, as the septum of the heart.

serum. The plasma of blood from which fibrin has been removed.

sex chromosome. One of a pair of chromosomes that determines the sex of its carrier.

sex linkage. The relationship between the characteristics whose genes lie on sex chromosomes and the chromosomes themselves, sometimes resulting in the characteristic being expressed more frequently in one sex than the other.

shock. The generalized condition in which blood volume in the circulatory system is greatly reduced.

simple microscope. A single magnifying lens.

skeletal muscle. A muscle that moves the bones of the body.

sodium pump. The mechanism that restores the ion concentration characteristics of the resting state to neurons after a depolarization pulse has passed along the neuron.

species. A subgroup of a genus; consisting of a group of individuals which resemble each other structurally and functionally and which, in a natural environment, interbreed to produce fertile offspring.

spectrometer. An instrument for measurement of the wavelengths of emitted or absorbed electromagnetic radiation.

spectroscope. An instrument that displays various wavelengths of the visible spectrum so that they may be observed with the eye.

spermatozoon (plural, spermatozoa). The male reproductive cell of an animal.

sphygmomanometer. An instrument for measuring blood pressure with a cuff-like device around the arm.

spike potential. The largest of the series of peaks within the action potential on an axon.

spleen. A large organ near the stomach that functions in the destruction of old red blood cells, storage of blood, and as a blood filter.

standing wave. A wave reflected at both ends along its direction of propagation, characterized by points along the wave (nodes) where minimum disturbance of the medium occurs and by points (antinodes) midway between the nodes where maximum disturbance occurs.

stapes. A stirrup; one of the ossicles of the middle ear.

starch. A multiple sugar, $(C_5H_{10}O_5)_n$, composed of numerous glucose units and used as a food storage substance in plants.

steroid. One of a group of compounds that includes sterols, sex hormones, and the D vitamins.

stimulator. A device for producing voltage or current pulses of various or variable amplitudes and durations.

strain. Change in length per unit length.

strain gauge. A transducer used to measure displacement.

stroma (plural, stromata). A structural framework, especially of red blood cell or in chloroplasts, where it supports the grana.

structural gene. A section of a DNA molecule which codes for a particular enzyme.

substrate. The chemical substance upon which an enzyme acts.

summation. The increase in tension that occurs in a muscle in response to a stimulus that occurs after a muscle is already in a state of contraction due to an earlier stimulus.

superior. Above or in a more elevated position, cf. inferior.

supernatant. The fluid above particles that have been precipitated or centrifuged from a suspension.

synapse. The connection between the axon of one neuron and the dendrite or cell body of another neuron.

synapsis. The joining in pairs of homologous chromosomes during meiosis.

systole. The contraction part of the cardiac cycle, cf. diastole.

T

target theory. The supposition that "key molecules" exist in cells that, upon direct action of ionizing radiation, results in biological changes in the cells.

taxonomy. The science of classification.

telophase. The final stage of mitosis, characterized by the appearance of two new nuclei.

tendon. An elastic strand of connective tissue by which a muscle is attached to bone or other structure.

test cross. The mating of an individual of unknown genotype for a particular trait with a homozygous recessive for the same trait. The outcome of the cross determines the unknown genotype.

tetanus. A continuously contracted state of a muscle.

timbre. A subjective quality of sound that depends on the relative contributions of various harmonics to a complex (nonsinusoidal) sound wave.

tissue. An aggregation of cells having roughly the same structure and which function together, as muscle tissue, nervous tissue, etc.

tonus. The state of slight tension in muscles.

torque. That which produces or tends to produce rotation.

tracer. A specific, identifiable nuclide that can be incorporated into a living system and later located.

transcription. The process by which mRNA is produced from DNA bearing encoded genetic information.

transducer. A device that converts energy from one form into another.

transfer ribonucleic acid. See tRNA.

translation. The process in which an amino acid chain is formed at the direction of the coded information on mRNA.

transverse wave. A wave in which the vibrations are in a direction perpendicular to the direction of propagation of the wave.

trichromatic theory. Any of the theories of color vision that assumes the presence in the eye of three sets of sensory mechanisms, corresponding to three primary colors.

triplet code. A group of three successive nucleotides on a molecule of DNA which serves to code the transcription of mRNA.

triton. A nucleus of ^3H, composed of a proton and two neutrons.

tRNA (transfer ribonucleic acid). A three-dimensional form of RNA that attaches to an appropriate amino acid, transfers it to the site of protein synthesis, and by base-pairing of the tRNA with mRNA, positions the amino acid on the ribosome so that the amino acid assumes its predetermined position in the protein chain.

tunic. A coating or covering, especially referring to layers of the eyeball.

turgor pressure. The pressure on the walls of plant cells exerted by water inside the cell. The wilting of plants is due to lack of turgor pressure.

twitch. A single contraction of a muscle in response to a stimulus of short duration.

tympanum. The eardrum; a membrane between the middle ear and outer ear.

U

ultracentrifuge. A high speed centrifuge; a device for rapidly spinning suspensions to separate the materials of different densities.

V

vacuole. A cavity within a cell, usually filled with a fluid (cell sap).

valve. One or more flaps that open to permit flow of blood in one direction, but not in the opposite direction.

van der Waals forces. Nonspecific attractive forces between atoms in close proximity.

vasoconstriction. The reduction in diameter of blood vessels.

vasodilation. The relaxation of blood vessels with an accompanying increase in diameter.

vein. A blood vessel that transports blood from the tissues of the body toward the heart.

vena cava. One of the great veins that return blood to the heart.

ventral. Toward the chest or underside, cf. dorsal.

venule. A small vein.

virion. A complete virus, capable of reproducing itself.

visual acuity. The ability to distinguish details of an object.

vital staining. The application of stains or dyes to living organisms to visualize specific parts.

vitamin. An organic substance that is essential in small quantities for the normal growth and development of an organism.

vitreous humor. A transparent jelly-like substance that fills the large cavity within the eyeball.

W

wave. A cyclic disturbance in a medium that propagates through the medium without bodily moving the particles of the medium along with the wave itself.

wavelength. The distance between successive disturbance maxima in a wave.

X

X-ray. High energy electromagnetic waves that originate in the extranuclear electrons of atoms.

Z

zygote. A fertilized ovum.

Index

A

B

D

E

N

S

T

U

V

W

X

Y

Z

B C 7 D 8 E 9 F 0 G 1 H 2 I 3 J 4